Equine nutrition and feeding

Equine nutrition and feeding

David Frape

Longman
Scientific &
Technical

Longman Scientific & Technical,
Longman Group UK Limited,
Longman House, Burnt Mill, Harlow,
Essex CM20 2JE, England
and Associated Companies throughout the world.

Published in the United States of America
by Churchill Livingstone Inc., New York

First published 1986
Reprinted 1990, 1993

British Library Cataloguing in Publication Data
Frape, David L.
   Equine nutrition and feeding.
   1. Horses – Feeding and feeds
   I. Title
   636.1'0852   SF285.5
ISBN 0-582-45014-4

Library of Congress Cataloging-in-Publication Data
Frape, David L. (David Lawrence)
   Equine nutrition and feeding.

   Bibliography: p.
   Includes index.
   1. Horses – Feeding and feeds. I. Title.
SF285.5.F73   1986   636.1'085   85-13231
ISBN 0-582-45014-4

Set in Linotron 202 10/11pt Erhardt
Produced by Longman Singapore Publishers (Pte) Ltd.
Printed in Singapore

Contents

CHAPTER 7
Feeding the breeding mare   134

CHAPTER 8
Growth of the horse   156

CHAPTER 9
Feeding for performance and the metabolism of nutrients during exercise   169

Foreword

David Frape has extensive experience in applying scientific principles to the feeding and nutrition of horses. This book therefore represents a rare opportunity for those concerned with the management and welfare of horses to obtain an authoritative view of the intricacies of feeding and of diet composition with its inevitable influence on the growth and performance of horses of all ages and purposes. The text provides a comprehensive coverage of the subject and is laid out in a concise, logical manner.

Although the subject of feeding horses remains an art, the scientific principles are, nonetheless, very important and must be understood by those responsible. In particular, the book provides the resources for stud managers, trainers, veterinary graduates and university students to bring themselves up to date with current knowledge of the subject. It will also help to improve their understanding and competence for advice in an area where there has been, unfortunately, too little information available in the past. In particular, they will be able to follow the references for further reading, which is an essential ingredient of continuing education.

The contents follow a logical path, from a comprehensive view of the digestive system to the utilization and role of nutrients for maintenance, work and growth. On the other side of the coin, the ingredients of animal feeds are balanced by a description of the nutrient requirements under varying conditions of use.

The appendices contain a wealth of analytical material, including descriptions of dietary composition and energy requirements. These provide a useful reference source for those wishing to bring the art and science of feeding into a balanced perspective.

I congratulate David Frape on his industrious contribution to the subject of nutrition in the horse and I am sure that the book will become a valuable addition to the literature.

P. D. Rossdale, MA PhD FRCVS
Newmarket

January 1985

Introduction

Computing the life of a horse to be half the age of man that must last from 10 to 20 years.

W. Gibson, 1726

(William Gibson lived and worked as a farrier off Oxford Street, London W1)

The book will give the reader in simple terms a basic and straightforward exposition of the scientifically established principles of the feeding and nutrition of horses and ponies (the pony is a horse not in excess of $14\frac{1}{2}$ hands). Horses belong to the family Equidae, which includes asses and zebras – all monodactyls, that is they have one functional digit on each leg. The Equidae belong to the order Perissodactyla, or odd-toed ungulates, the grouping of which depends to a great extent on the skeletal structure of their legs. The various families within the order are also in part defined by their dental formulae. Both of these characteristics – the structure of the legs and the teeth – affect their feeding habits and behaviour and all are affected by, or are related to nutrition.

Horses evolved as grazing and browsing non-ruminant herbivores. By domestication during the past few thousand years man has considerably changed and adapted the mode of life of horses and ponies to his needs. A profound knowledge of husbandry was assiduously gathered by trial and error by past generations. During this century people without access to this fund of knowledge have acquired and maintained horses and ponies in a wide variety of natural and unnatural environments. Many diet-related problems have arisen, particularly where horses are fed domestic diets in surroundings alien to their natural instincts. Therefore, the first objective of this text will be to explain causes of some of these problems, as far as we understand them, and to show, as far as possible, how they may be avoided.

Many horses are raised in Western Europe for the purpose of producing meat for human consumption. However, the principal role of the adult horse and the major ultimate purpose for raising horses is as animals ideally suited to physical work. This is either as domestic riding horses or for drawing implements and wheeled vehicles. In considering the dietary needs and nutritional problems of horses in this book, the role of the horse as a working animal, either in the guise of an athlete or in that of the pleasure horse, will be kept firmly in mind. Of course, the role of diet in a number of subsidiary objectives must be discussed, where evidence permits. These include reproduction, milk production, and growth of the foal and adolescent horse in all its forms.

The horse and mule population of the world is believed to have reached a peak of about 75 million in 1918, of which a third were accounted for in the United States, where they were predominantly heavy working animals. Through mechanization in agriculture the population in western countries declined progressively until the 1960s, but the past two decades have seen a progressive increase in the numbers of pleasure horses, reflecting some of the recreational pursuits of an affluent society.

In today's competitive world horses and man compete for food. The horse as a pleasure animal competes with grazing ruminants raised for their meat and milk potential, and competes with non-ruminants such as pigs or ourselves for cereals and pulses. The traditional diet of the horse has been derived not only by trial and error and by local availability, but also by myth. In order to ensure the continued existence of the horse in large numbers for our pleasure, a more objective assessment of the animal's dietary needs becomes essential, not only to establish its actual needs but also to take full advantage of the range of raw feed materials now available throughout the world. It is hoped that some of the following discussions will contribute to that assessment.

The amount of scientifically obtained evidence on the feeding and nutrition of horses has increased appreciably during the past ten years or so but there remain yawning gaps in our scientific knowledge. Little information is available from other members of the Equidae, or indeed from the Perissodactyla as a whole. Where necessary, therefore, inferences about diet and nutrition will be drawn from experimental work on other domesticated species.

The text will be of value to practising veterinary surgeons, to students, to those responsible for stables, and indeed to all individuals with their own horse. Some basic biological and nutritional knowledge will be assumed and it is not the objective of this book to retrace steps through the basic biological sciences that are well covered in many standard texts.

Acknowledgements

In writing this book I am indebted to a number of people. First, I should like to thank Professor Alastair N. Worden MB, FRCPath, FRCVS for encouragement and for helpful criticism of the glossary and Dr P. D. Rossdale FRCVS and his partners at Beaufort Cottage Stables, Newmarket for their readiness for discussion and advice. I am grateful to Mr Norman Comben MRCVS for his guidance with antiquarian books and, together with the staff of the Royal College of Veterinary Surgeons Library, for gaining access to many. I should also like to thank Mr R. A. Jones BVM&S, MRCVS and R. N. W. Ellis, MRCVS for data. Many friends in the horse world, including Mr Harold Knight, Mr Luca Cumani, Mr Bill Johnson, Mr Michael Dickinson and Mr Chris Bartle, kindly provided help, advice and permission to take photographs of their horses and the staff of many studs and training stables gave much appreciated assistance. Finally, I should like to thank Mrs Noreen Clark for the dedicated typing of the manuscript.

I thank Dr. P. D. Rossdale, FRCVS, for the reproduction of photographs used in Plates 7.2 (b) and (c) and acknowledge the following sources for use of information in constructing some of the figures and tables:

Agriservices Foundation, Clovis, California (Figs 6.2 and 6.3; Tables 8.1, 8.2, 8.3 and 8.4); American Institute of Nutrition (Table 7.2); American Society of Animal Science (Figs 7.1, 7.2, 7.3 and 7.4; Tables 1.6, 3.1, 3.4, 6.1, 6.2, 6.3 and 7.2); American Veterinary Medical Association (Table 4.3); Animal Health Trust, Newmarket, England (Table 3.4); Baillière Tindall Ltd (Figs 6.2, 6.3 and 8.3); Beaufort Cottage Laboratories, Newmarket, England (Tables 3.1 and 3.5); British Veterinary Association (Figs 8.1 and 11.2; Tables 8.5, 9.2 and 11.5); Cambridge University Press (Fig. 1.4; Tables 1.3, 10.1 and 10.2); Cornell Veterinarian Inc. (Figs 1.5 and 3.2; Table 3.2); Miller Publishing Co. (Fig. 3.1); Longman Group Ltd (Fig. 2.1; Table 10.1); Ministère de l'Agriculture, Service des haras et de l'equitation, Le Lion d'Angèrs (Figs 6.9, 6.10, 11.2, 11.3 and 11.4); Ministry of Agriculture, Fisheries and Food, London (Table 10.7); National Academy of Sciences, Washington (Tables 6.3, 6.4, 6.6 and 6.7); Paul Parey, Hamburg (Tables 1.4 and 7.2); Pergamon Press Ltd (Table

11.4); Dr D. H. Snow MRCVS, Newmarket, England (Fig. 9.13); South African Veterinary Medical Association (Table 9.1); Terra-Verlag, Konstanz, FRG (Table 9.3); Veterinary Publications Inc, Princeton, NJ (Table 9.3).

CHAPTER 1

The digestive system

A horse which is kept to dry meat will often slaver at the mouth. If he champs his hay and corn, and puts it out again, it arises from some fault in the grinders ... there will sometimes be great holes cut with his grinders in the weaks of his mouth. First file his grinders quite smooth with a file made for the purpose.

Francis Clater, 1786

The domesticated horse consumes a variety of feeds ranging in physical form from forage with a high content of moisture to cereals with large amounts of starch, and from hay in the form of physically long fibrous stems to salt licks and water. By contrast, the wild horse has evolved and adapted to a grazing and browsing existence, in which it selects succulent forages containing relatively large amounts of water, soluble proteins, lipids, sugars and structural carbohydrates but little starch. Short periods of feeding occur throughout most of the day and night, although generally these are of greater intensity in daylight. In domesticating the horse, man has generally restricted its feeding time and introduced unfamiliar materials, particularly starchy cereals, protein concentrates and dried forages. The art of feeding gained by long experience is to ensure that these materials meet the varied requirements of horses without causing digestive and metabolic upsets. Thus, an understanding of the form and function of the alimentary canal is fundamental to a discussion of feeding and nutrition of the horse.

The mouth

The lips, tongue and teeth of the horse are ideally suited for the prehension, ingestion and alteration of the physical form of feed to that suitable for propulsion through the gastrointestinal tract in a state that facilitates admixture with digestive juices.

The upper lip is strong, mobile and sensitive and is used during grazing to place forage between the teeth in contrast to the process in the cow where the tongue is used for this purpose. By contrast, the horse's tongue moves ingested material to the cheek teeth for grinding. The lips are also used as a funnel through which water is sucked.

As distinct from cattle, the horse has both upper and lower incisors enabling it to graze closely by shearing off forage. Horses tend to masticate their feed for extended periods and the lateral and vertical movements of the jaw, accompanied by profuse salivation, enable the cheek teeth to pulverize feed into small particles coated with mucus suitable for swallowing.

The horse has two sets of teeth. The first to appear, the deciduous, or temporary milk, teeth erupt during early life and are replaced during growth by the permanent teeth. The permanent incisors and permanent cheek teeth erupt continuously to compensate for wear and their changing form provides a basis for assessing the age of a horse. In the gap along the jaw between the incisors and the cheek teeth the male horse normally has a set of small canine teeth. The gap, by happy chance, securely locates the bit. The dental formula and configuration of both milk and permanent teeth are given in Fig. 1.1. The lower cheek teeth are implanted in the mandible in two straight rows that diverge towards the back. The space between the rows of teeth in the lower jaw is less than that separating the upper teeth (Fig. 1.1). This accommodates a sideways, or circular, movement of the jaw that effectively shears feed. The action leads to a distinctive pattern of wear of the biting surface of the exposed crown. This pattern results from the differences in hardness which characterize the three materials (cement, enamel and dentine) of which teeth are composed. The enamel, being the hardest, stands out in the form of sharp prominent ridges. It is estimated that the enamel ridges of an upper cheek tooth in a young adult horse, if straightened out, would form a line more than a foot (30 cm) long. This irregular surface provides a very efficient grinding organ.

Horses and ponies rely more on their teeth than we do. People might be labelled concentrate eaters; concentrates require much less chewing than does roughage. Even among herbivores, horses and ponies depend to a far greater extent on their teeth than do the domesticated ruminants – cattle, sheep and goats. Ruminants swallow grass and hay with minimal chewing and then depend on the activity of bacteria in the rumen to disrupt the fibre. This is then much more readily fragmented during chewing the cud. Horses have no rumen and therefore when grass and hay are consumed the material must be ground (assuming the teeth are sound) to particles of less than 1.6 mm in length before swallowing. In order to achieve this the number of chewing movements is considerably greater than that required for chewing concentrates. Horses make between 800 and 1200 chewing movements per kg of concentrates, whereas 1 kg of long hay requires between 3000 and 3500 movements. In ponies, chewing is even more protracted – they require 5000 to 8000 chewing movements per kg of concentrate alone, and very many more for hay (Meyer, Ahlswede & Reinhardt 1975).

Vertical section

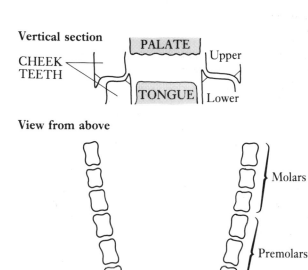

View from above

Molars

Premolars

Wolf tooth

Canine (48 months in stallion or gelding)

Corner incisor (48–72 months)

Lateral incisor (36–72 months)

Central incisor (30 months)

Dental formula

Deciduous :$2(Di \frac{3}{3} Dc \frac{0}{0} Dp \frac{3}{3}) = 24$

Permanent :$2(I \frac{3}{3} C \frac{1}{1} P \frac{3 \text{ or } 4}{3} M \frac{3}{3}) = 40$ or 42

Fig. 1.1 Configuration of permanent teeth in the upper or lower jaw (the molars and premolars in the lower jaw are slightly closer to the midline). The deciduous teeth on each side of each jaw are: 3 incisors, 1 canine, 3 molars. The deciduous canines are vestigial and do not erupt. The wolf teeth (present in the upper jaw of about 30% of fillies and about 65% of colts) are often extracted as their sharp tips can injure cheeks when a snaffle bit is used. Months (in parentheses) are approximate ages at which permanent incisors and canines erupt, replacing the deciduous teeth.

The physical presence of feed material in the mouth stimulates the secretion of a copious amount of saliva. Some 10–12 litres are secreted daily in an animal fed normally. This fluid seems to have no digestive enzyme activity, but its mucus content enables it to function as an efficient lubricant preventing 'choke'. Its bicarbonate content, amounting to some

50 mequiv/litre, provides it with a buffering capacity. The concentration of bicarbonate and sodium chloride in the saliva is, however, directly proportional to the rate of secretion and so increases during feeding. The continuous secretion of saliva during eating seems to buffer the digesta in the proximal region of the stomach permitting some microbial fermentation with the production of lactate. This has important implications for the well-being of the horse (see Ch. 11).

The stomach and small intestine

The stomach of the adult horse is a small organ, its volume comprising about 10 per cent of the gastrointestinal tract (Fig. 1.2, Pl. 1.1). In the suckling foal, however, the stomach capacity represents a larger proportion of the total alimentary tract. Most digesta are held in the stomach for a comparatively short time, but this organ is rarely completely empty and a significant portion of the digesta may remain in it for 2–3 hours. Some digesta pass into the duodenum shortly after eating starts, when fresh ingesta enter the stomach. Expulsion into the duodenum is apparently arrested as soon as feeding stops. When a horse drinks, a high proportion of the water passes along the curvature of the stomach wall so that mixing with digesta and dilution of the digestive juices it contains are avoided. This process is particularly noticeable when digesta largely fill the stomach.

The entrance to the stomach is guarded by a powerful muscular valve called the cardiac sphincter. Although a horse may feel nauseous, it rarely vomits, partly because of the way this valve functions. This too has important consequences. Despite extreme abdominal pressure the cardiac sphincter is reluctant to relax in order to permit the regurgitation of feed or gas. On the rare occasions when vomiting does occur, ingesta usually rush out through the nostrils, owing to the existence of a long soft palate. Such an act may indicate a ruptured stomach.

The daily secretion and release of gastric juice into the stomach amounts to some 10–30 litres, and seems to be stimulated by the physical presence of feed in the organ but not by the sight of food. Secretion of gastric juice continues even during fasting, although the rate seems to vary from hour to hour. The act of eating stimulates the flow of saliva – a source of sodium, potassium, bicarbonate and chloride ions. Saliva's buffering power retards the rate at which the pH of the stomach contents decreases. This action, combined with a stratification of the ingesta, brings about marked differences in the pH of different regions (about 5.4 in the fundic region and 2.6 in the pyloric region). Fermentation occurs in the oesophageal and fundic regions of the stomach but particularly in that part known as the saccus caecus. Fermentation yields primarily lactic acid. As digesta approach the pylorus at the distal end of the stomach, the gastric

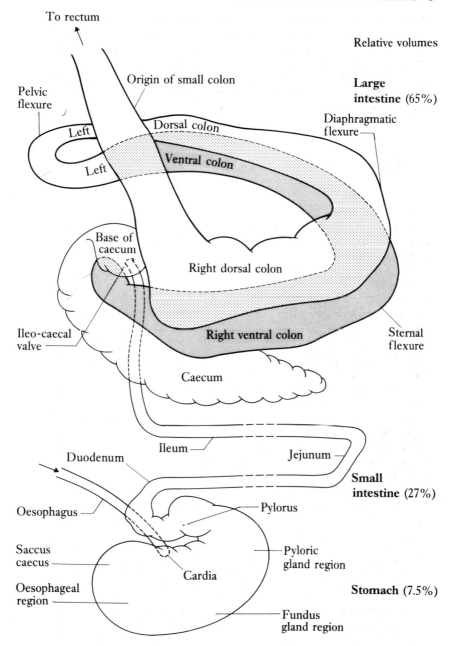

Fig. 1.2 Diagram of gastrointestinal tract.

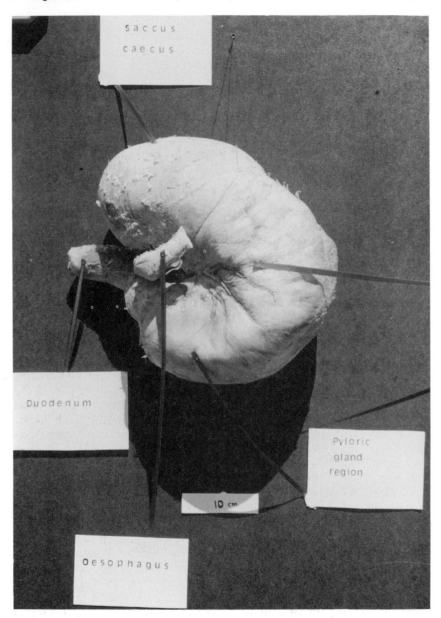

Plate 1.1 The stomach of a 550-kg Thoroughbred mare, capacity 8.4 litres, measuring about 20 × 30 × 15 cm. Acid fermentation of the stomach contents takes place in the saccus caecus (top of the picture).

pH falls owing to the secretion of hydrochloric acid (HCl), which potentiates the proteolytic activity of pepsin. The activity of pepsin is some 15–20 times greater than in the fundic region. Because of the stomach's small size and consequentially the relatively short dwell time, the degree of protein digestion is slight.

Large quantities of pancreatic juice are secreted as a result of the presence of food in the stomach in response to stimuli mediated by vagal nerve fibres and by endocrine (hormone) activity. In fact, although secretion is continuous, the rate increases by some four to five times when feed is first given. This secretion, which enters the duodenum, has a low order of enzymatic activity, but provides large quantities of fluid and sodium, potassium, chloride and bicarbonate ions. Some active trypsin is, however, present. There is conflicting evidence for the presence of lipase in pancreatic secretion and bile, secreted by the liver, probably exerts a greater, but different, influence over fat digestion. The horse lacks a gall bladder, but stimulation of bile secretion and pancreatic juice is caused by the presence of gastric HCl in the duodenum. Secretion of both fluids ceases after a fast of 48 hours. Bile is both an excretion and a digestive secretion. As a reservoir of alkali it helps preserve an optimal reaction in the intestine for the functioning of digestive enzymes secreted there. In the horse, the pH of the digesta leaving the stomach rapidly rises to slightly over 7.0.

A high proportion of the energy sources consumed by the working horse contains cereal starches. These consist of relatively long, branched chains, the links of which are α-D-glucose molecules joined as shown in Fig. 1.3. Absorption into the bloodstream depends on the disruption of the bonds linking the glucose molecules. This is contingent entirely upon enzymes secreted in the small intestine. These are held on the brush border of the villi in the form of α-amylase (secreted by the pancreas) and as α-glucosidases (secreted by the intestinal mucosa) (see Table 1.1). The concentration of α-amylase in the pancreatic juice of the horse is only 5–6 per cent of that in the pig, whereas the concentration of α-glucosidases is comparable with that in many other domestic mammals. The α-glucosidases include sucrase, the disaccharidase present in concentrations five times that of glucoamylase and capable of digesting sucrose. Another important disaccharidase in the intestinal juice is the β-glucosidase, neutral β-galactosidase (neutral or brush-border lactase), which is necessary for the digestion of milk sugar in the foal. This enzyme has a pH optimum around 6.0 and its activity decreases as the horse matures, being absent from the brush border of horses more than 4 years old. In horses of this age, induction of neutral lactase by lactose feeding also seems impossible so that the sugar is fermented; large quantities of lactose may thus cause digestive upsets. If a suckling foal, or one given cow's milk, lacks an active form of this enzyme, it suffers from diarrhoea.

The amount of protein hydrolysed in the small intestine is about three times that in the stomach. Proteins are in the form of long folded chains,

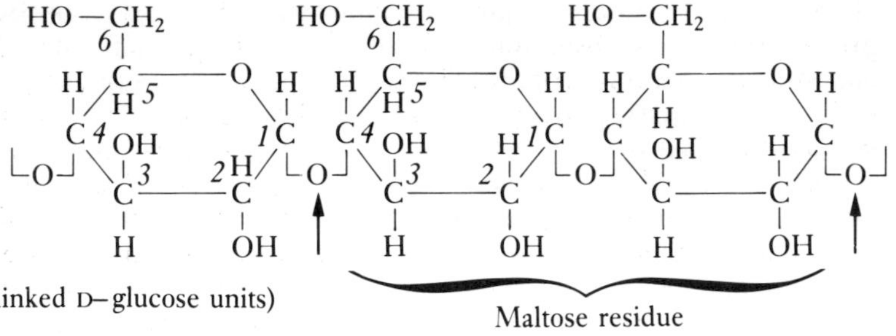

Starch

(α 1–4 linked D–glucose units)

Maltose residue

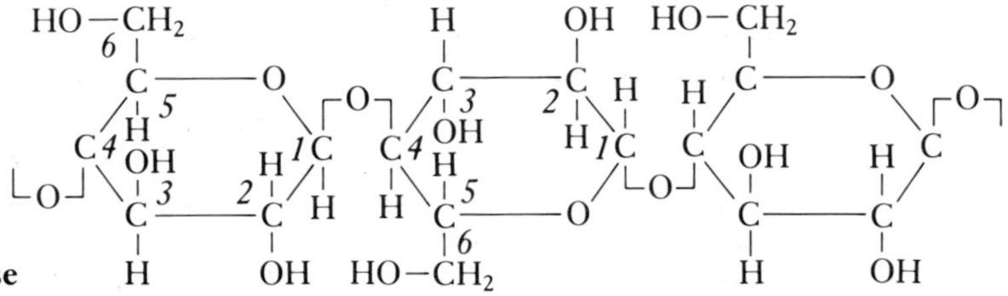

Cellulose

(β1–4 linked D–glucose units)

Fig. 1.3   Diagrammatic representation of three glucose units in two carbohydrate chains (the starch granule also contains amylopectin, which has both 1–4 linkages and 1–6 linkages). The arrows indicate the site of intermediate digestion.

Table 1.1   Carbohydrate digestion in the small intestine of the horse

Substrate	Enzyme	Product
Starch	α-Amylase	Limiting dextrins (about 34 glucose units)
Limit dextrins	α-Glucosidases (glucoamylase, maltase and isomaltase)	Glucose
Sucrose	Sucrase	Fructose and glucose
Lactose	Neutral-β-galactosidase	Glucose and galactose

the links of which are represented by amino acid residues. For proteins to be digested and utilized by the horse, these amino acids must usually be freed, although the gut mucosal cells can absorb dipeptides. Details of this process are little known, but the enzymes responsible are amino-peptidases and carboxy-peptidases secreted by the wall of the small intestine.

The horse differs from the ruminant in that the composition of its body fat is influenced by the composition of dietary fat. This suggests that fats are digested and absorbed from the small intestine before they can be altered by the bacteria of the large intestine. The small intestine is the primary site for the absorption of dietary fat and long-chain fatty acids. Bile continuously draining from the liver facilitates this by promoting emulsification of fat, chiefly through the agency of bile salts. The emulsification increases the fat–water interface so that the enzyme lipase may more readily hydrolyse neutral fats to fatty acids and glycerol. These are readily absorbed, although it is possible that a considerable proportion of dietary fat, as finely emulsified particles of neutral fat (triglycerides), is absorbed into the lymphatic system. American evidence (Kane, Baker & Bull 1979) suggests that horses do in fact digest fat quite efficiently and that the addition of edible fat to their diet has some merit, particularly in so far as endurance work is concerned (see Ch. 9).

The small intestine is about 21 m (70 feet) long. Even so, digesta are propelled through it quite rapidly in the adult horse, some appearing in the caecum within 45 minutes after a meal. Much of the digesta moves through the small intestine at the fast rate of 30 cm (1 foot) nearly per minute. It is therefore surprising how much digestion and absorption apparently occur in that organ and, although differences in the composition of digesta entering the large intestine can be detected with a change in diet, it is a considerably more uniform material than that entering the rumen of the cow. This fact has notable practical and physiological significance in the nutrition and well-being of the horse. The nature of the material leaving the small intestine is described as fibrous feed residues, undigested feed starch and protein, microorganisms, intestinal secretions and cell debris.

The large intestine

Grazing herbivores have a wide variety of mechanisms and anatomical arrangements for making use of the chemical energy locked up in the structural carbohydrates of plants. A characteristic of all grazing and browsing animals is the enlargement of some part of the gastrointestinal tract to accommodate fermentation of digesta by microorganisms. More than half the dry weight of faeces is bacteria and the bacterial cells in the digestive tract of the horse number more than ten times all the tissue cells in the body. No domestic mammal secretes enzymes capable of breaking down the complex molecules of cellulose, hemicellulose, pectin and lignin into their component parts suitable for absorption, but, with the exception of lignin, intestinal bacteria achieve this. The process is relatively slow in comparison to the digestion of starch and protein. This means that the flow of digesta has to be arrested for sufficient time to enable the process to reach a satisfactory conclusion from the point of view of the energy economy of the host animal. During the weaning and post-weaning periods of the foal and yearling, the large intestine grows faster than the remainder of the alimentary canal to accommodate a more fibrous and bulky diet. At the distal end of the ileum there is a large blind sack known as the caecum, which is about 1 m long in the adult horse and has a capacity of 25–35 litres. At one end there are two muscular valves in relatively close proximity to each other, one through which digesta enter from the ileum and the other through which passage from the caecum to the right ventral colon is facilitated. The right and left segments of the ventral colon and the left and right segments of the dorsal colon constitute the great colon, which is some 3–4 m long in the adult horse and has a capacity of more than double that of the caecum. The four parts of the great colon are connected by bends known as flexures. In sequence, these are the sternal, the pelvic and the diaphragmatic flexures. Their significance probably lies in changes in function and microbial population from region to region and in acting as foci of intestinal impactions. Digestion in the caecum and ventral colon must depend almost entirely on the activity of their constituent bacteria and ciliate protozoa. High levels of alkaline phosphatase activity, known to be associated with high digestive and absorptive action, are found in the large intestine of the horse unlike those of the cat, dog and man.

The diameter of the great colon varies considerably from region to region but reaches a maximum in the right dorsal colon where it forms a large sacculation with a diameter of up to 500 mm. This structure is succeeded by a funnel-shaped part below the left kidney where the bore narrows to 70–100 mm as the digesta enter the small colon. The latter continues dorsally in the abdominal cavity for 3–4 m before the rectum, which is some 300 mm long, terminates in the anus (Fig. 1.2).

Intestinal contractions

The walls of the large and small intestines contain longitudinal and circular muscle fibres essential (1) for the contractions necessary in moving the digesta by the process of peristalsis in the direction of the anus, (2) for allowing thorough admixture with digestive juices and (3) for bathing the absorptive surfaces of the wall with the products of digestion. During abdominal pain these movements may stop so that the gases of fermentation accumulate. In contrast to the small intestine the walls of the large intestine contain only mucus-secreting glands, that is they provide no digestive enzymes.

The extent of intestinal contractions increases during feeding – large contractions of the caecum expel digesta into the ventral colon but separate contractions expel gas, which is hurried through much of the colon. The reflux of digesta back into the caecum is largely prevented by the sigmoid configuration of the junction.

Passage of digesta through the large intestine is mainly a function of movement from one of the compartments to the next through a separating barrier. Considerable mixing occurs within each compartment, but there seems to be no retrograde flow between them. The barriers are: (1) the ileocaecal valve already referred to; (2) the caecal-ventral colonic valve; (3) the ventral–dorsal colonic flexure (pelvic flexure), which separates the ventral from the dorsal colon; and (4) the dorsal small colonic junction at which the digesta enter the small colon. Resistance to flow increases in the same order, that is the last of these barriers provides the greatest resistance, with implications concerning impactions (Ch. 11). This resistance is much greater for large food particles than for small particles. The time taken for waste material to be voided after a meal is such that in ponies receiving a grain diet, 10 per cent is voided after 24 hours, 50 per cent after 36 hours and 95 per cent after 65 hours. Most digesta reach the caecum and ventral colon within 3 hours of a meal, so that it is in the large intestine where unabsorbed material spends the greatest proportion of time. The rate of passage in ruminants is somewhat slower, and this partly explains their greater efficiency in digesting fibre. In the horse, passage time is influenced by the physical form of the diet; for example, pelleted diets have a faster rate of passage than chopped or long hay, and fresh grass moves more rapidly than hay.

Microbial digestion

There are three main distinctions between microbial fermentation of feed and digestion brought about by the horse's own secretions: (1) the β-1,4 linked polymers of cellulose* (Fig. 1.3) are degraded by the intestinal microflora but not by the horse's own secretions; (2) during their growth the microorganisms synthesize dietary indispensable (essential) amino acids; and (3) they are net producers of water-soluble vitamins of the B group and of vitamin K_2.

In the relatively small fundic region of the stomach, where the pH is about 5.4, there are normally some 1000 million bacteria per gram of ingesta. The species present are those that can withstand moderate acidity, common types being lactobacilli, streptococci and *Veillonella gazogenes*. In the large intestine the bacterial populations are highest in the caecum and ventral colon. Here, the concentration of cellulose-digesting bacteria is six

---

* The cell walls of plants contain several carbohydrates (including hemicellulose) that form up to half the fibre of the cell walls of grasses and a quarter of those of clover. These carbohydrates are also digested by microorganisms, but the extent depends on the structure and degree of encrustation with lignin, which is indigestible to both gut bacteria and horse secretions.

to seven times higher than in the terminal colon. About 20 per cent of the bacteria in the large intestine can degrade protein. Ciliate protozoa in the large intestine number about one millionth of the bacterial population. Although ciliates are individually very much larger than bacteria and they thus contribute a similar total mass to the large intestinal contents, their contribution to metabolism is less as this is roughly proportional to their surface area.

Numbers of specific microorganisms may change by more than a hundredfold during 24 hours in domesticated horses being given, say, two discrete meals per day. These fluctuations reflect changes in the availability of nutrients (in particular, starch and protein) and consequentially changes in the pH of the medium. Thus a change in the dietary ratio of cereal to hay not only will have large effects on the numbers of microorganisms but will considerably influence the species distribution in the hind gut.

Although frequency of feeding may have little impact on digestibility per se, it can have a large impact on the incidence of digestive disorders and metabolic upsets, which of course, affect the well-being of the horse. These consequences result directly from the effect of diet and digesta upon the microbial populations (bacteria and protozoa).

Despite the proteolytic activity of microorganisms in the hind gut, protein breakdown per litre is about forty-fold greater in the ileum than in the caecum or colon through the activity of the horse's own digestive secretions in the small intestine.

The microbial degradation of dietary fibre, starch and protein yields large quantities of short-chain volatile fatty acids (VFA) as byproducts, principally acetic, propionic and butyric acids (Table 1.2, Fig. 1.4). These acids would soon pollute the medium, rapidly producing an environment unsuitable for continued microbial growth; but an equable medium is maintained by the absorption of VFA into the bloodstream and by the secretion of bicarbonate and phosphate buffers, which enter the large intestine from its wall and from the ileum. In addition to the absorption of products of microbial digestion, there is the vital absorption of large amounts of water and electrolytes (sodium, potassium, chloride and phosphate). The quantities of water absorbed daily from the large intestine decrease progressively from the caecum to the small colon; this decline in water absorption is accompanied by a parallel decrease in sodium absorption. In the pony, 96 per cent of the sodium and chloride and 75 per cent of the soluble potassium and phosphate entering the large bowel from the ileum are absorbed into the bloodstream. Although phosphate is efficiently absorbed from both the small and large gut, calcium and magnesium are not, these being absorbed mainly from the small intestine (Fig. 1.5). This phenomenon has been proffered as a reason why excess dietary calcium does not depress phosphate absorption, but excess phosphate can depress calcium absorption in the horse.

The water content of the small intestinal digesta amounts to some 87–93 per cent, but the faeces of healthy horses contain only 58–62 per

Table 1.2 Effect of diet on the pH, production of volatile fatty acids (VFA) and lactate and on microbial growth in the caecum and ventral colon of the horse 7 hours after the meal.

| Diet | pH | FA (mmol/litre) | | | | Total bacteria per (ml × 10^-7) |
		Acetate	Propionate	Butyrate	Lactate	
Hay	6.90	43	10	3	1	500
Concentrate + minimal hay	6.25	54	15	5	21	800
Fasted	7.15	10	1	0.5	0.1	5

Note: Values given are typical but all except the pH show large variations.

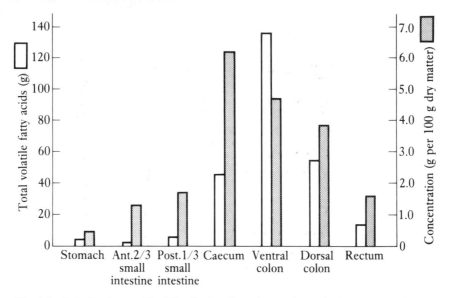

Fig. 1.4   Volatile fatty acids (□) calculated as the total weight (in grams) of acid in organ (as acetic acid) or as the concentration per 100 g of dry matter (▒) in lumen. (After Elsden, Hitchcock et al. 1946).

cent water. The type of diet has a smaller effect on this than might be imagined. For instance, oats produce fairly dry faeces, but bran produces moist faeces, although in fact they contain only some 2 or 3 percentage units more of moisture.

Caecal bacteria from horses adapted to a grain diet are less efficient at digesting hay than are the microbes from hay-adapted horses. An analogous situation exists for hay-adapted caecal microbes when subjected to grain substrate. If such a dietary change is made abruptly in the horse, impactions may occur in the first of these situations and colic, laminitis or puffy swollen legs can result in the second (see Ch. 11.) The caecal microorganisms in a pony or horse tend to be less efficient at digesting hay than are the ruminal microbes in cows. The digestibilities of organic matter and crude fibre in horses given a diet containing more than 15 per cent crude fibre (a normal diet of concentrates and hay), are about 85 and 70–75 per cent respectively, of the ruminant values. This has been attributed to the combined effects of a more rapid rate of passage of residues and differences in cellulolytic microbial species.

Microbial degradation seems to occur at a far faster rate in the caecum and ventral colon than in the dorsal colon (Fig. 1.4) and the rate is also faster when starches are degraded rather than structural carbohydrates. A change in the ratio of starch to fibre in the diet leads to a change in the proportions of the various acids yielded (Table 1.2); and these proportions

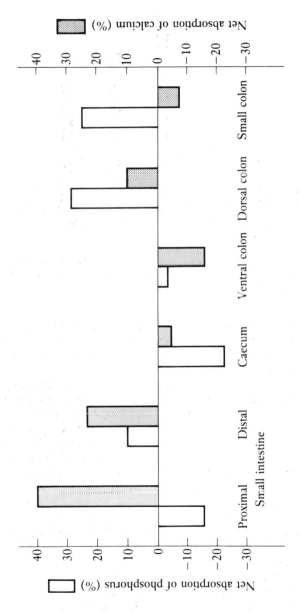

Fig. 1.5 Net fractional absorption of phosphorus (□) and calcium (▨) from various regions of the horse intestine (After Schryver, Hintz & Lowe 1974).

also differ in the organs of the large intestine. Thus, proportionately more propionate is produced as a consequence of the consumption of a starch diet and the caecum and ventral colon yield more than the dorsal colon does. Many bacteria have the capacity to degrade dietary protein, so yielding another blend of VFA.

Whereas most ruminal butyrate is metabolized in the mucosa before entering the bloodstream, in horses all VFA pass readily to the blood. Lactic acid produced in the stomach is apparently not well absorbed from the small intestine. On reaching the large intestine some is absorbed, along with that produced locally, but much is metabolized by bacteria to propionate.

Microbial activity inevitably produces gases – principally carbon dioxide, methane and small amounts of hydrogen – which are normally absorbed, ejected from the anus or they participate in further metabolism. The gases can, however, be a severe burden with critical consequences when production rate exceeds that of disposal.

Although dietary fibre is not degraded as readily by the horse as by ruminants (Table 1.3), the horse utilizes the energy of soluble carbo-hydrates more efficiently by absorbing a greater proportion as sugars. Horses also differ from ruminants in absorbing a higher proportion of dietary nitrogen in the form of the amino acids present in dietary proteins, proportionately less being converted to microbial protein. Only a small proportion of the amino acids present in microbial protein is made available for direct utilization by the horse. For these reasons, young growing horses in particular respond to supplementation of poor-quality dietary protein with lysine, the principal limiting indispensable amino acid (Fig. 1.6).

Table 1.3 Proportion of volatile fatty acids (VFA) in digesta to bodyweight in four herbivores

	Grams VFA per kg bodyweight
Ox	1.5
Sheep	1.5
Horse	1.0
Rabbit	0.5

(From Elsden, Hitchcock et al. 1946)

Urea production

Urea is a principal endproduct of protein catabolism in mammals and much of it is excreted through the kidneys. It is a highly soluble, relatively innocuous compound and a reasonably high proportion of the urea

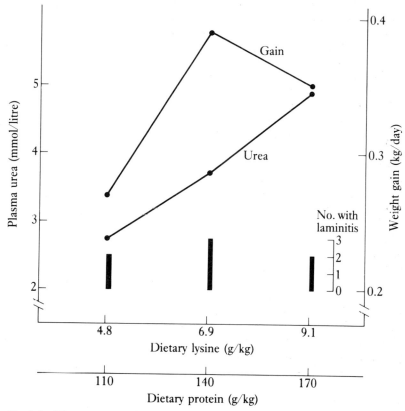

Fig. 1.6 The response of forty-one 6-month-old ponies over a 3-month period to diets containing different amounts of protein and lysine (initial weight 127 kg). In this experiment note that increased protein intake led to elevated protein catabolism and urea production without an increase in incidence of laminitis. (From Yoakam, Kirkham & Beeson 1978).

produced in the liver is secreted into the ileum and conveyed to the large intestine (Table 1.4) where most may be degraded to ammonia by bacteria. The possession by microorganisms of the enzyme urease, which does not occur in mammalian cells, makes this reaction possible. Most of the ammonia produced is re-utilized by the intestinal bacteria in protein synthesis. Some, however, diffuses into the blood, where levels are normally maintained very low by a healthy liver. If ammonia production greatly exceeds the capacity of the bacteria and of the liver to utilize it, ammonia toxicity can arise. The fate of any urea added to the diet is similar.

In summary, several studies have led to the conclusion that digestion and fermentation in and absorption from the large intestine account, in net terms,

Table 1.4 Effect of diet on the flow of nitrogen from the ileum to the caecum in horses

Diet	Nitrogen flow daily (mg N per kg bodyweight)
Concentrate, 3.75 kg daily (1%)*	62
Concentrate, 7.5 kg daily (2%)*	113
Hay	68
Straw	37

(From Schmidt, Lindemann & Meyer 1982)

* Weight of concentrate given as percentage of bodyweight.

for 30 per cent of dietary protein, 15–30 per cent of dietary soluble carbohydrate and 75–85 per cent of dietary structural carbohydrate, although one study found that 80 per cent of the total net disappearance of nitrogen compounds occurs in the large intestine (Reitnour & Salsbury 1972; Robinson & Slade 1974; Wootten & Argenzio 1975; Glinsky, Smith et al. 1976; Glade & Bell 1981; Glade 1983a). The salient causes of variation in values for each of the principal components of the horse's diet are not only the degree of adaptation of the animal, the dietary balance and the chemical constitution of the dietary components but also differences in methodology applied by various investigators in their quantifications.

Further reading

Sissons S & Grossman J D (1961) *The anatomy of the domestic animals.* W B Saunders: Philadelphia & London.

CHAPTER 2

The utilization of absorbed nutrients for maintenance work and growth

A horse whose work consists of travelling a stage of twenty miles three times a week, or twelve every day should have one peck of good oats, and never more than eight pounds of good hay in twenty four hours. The hay, as well as the corn, should, if possible, be divided into four portions.

James White, 1823

Glucose and volatile fatty acids (VFA)

Horse diets rarely contain more than about 4 per cent fat and 7–12 per cent protein so that these represent relatively minor sources of energy in comparison to carbohydrate, which may constitute by weight two-thirds of the diet. Furthermore, protein is required primarily in the building and replacement of tissues and is an expensive source of energy. However, both dietary protein and fat can also contribute to those substrates used by the horse to meet its energy demands for work. Protein does so by the conversion of the carbon chain of amino acids to intermediary acids and of some of the carbon chains to glucose. Neutral fat does so following its hydrolysis to glycerol and fatty acids. The glycerol can be converted to glucose and the fatty acid chain can be broken down by a stepwise process called β oxidation in the mitochondria, yielding ATP and acetate, or more strictly acetyl coenzyme A (acetyl CoA), and requiring tissue oxygen (see Fig. 9.2).

Carbohydrate digestion and fermentation yield predominantly glucose and acetic, propionic and butyric volatile fatty acids (VFA). These nutrients are collected by the portal venous system draining the intestine and a proportion of them is removed from the blood as they pass through the liver. Both glucose and propionate contribute to liver starch (glycogen) reserves, and acetate and butyrate bolster the fat pool (Fig. 2.1) and also constitute primary energy sources for many tissues.

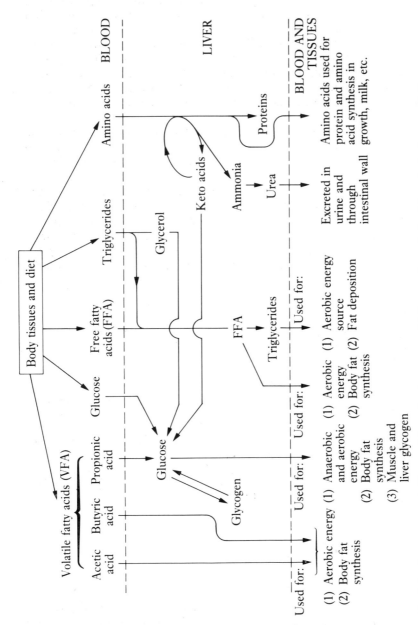

Fig. 2.1   Sources, metabolism and fate of major energy-yielding nutrients derived from body tissues and diet. (After McDonald, Edwards & Greenhalgh 1981).

Healthy horses and ponies maintain a blood-glucose concentration within certain defined limits. This is necessary as glucose represents the preferred source of energy for most tissues. After a meal the concentration of blood glucose increases above its resting level. In ponies, normal healthy resting levels may range between 50 and 60 mg/dl (2.8–3.3 mmol/litre), but horse breeds may generally have higher resting levels with Thoroughbreds in the region of 80 mg/dl (4.4 mmol/litre). Excess glucose not required to meet immediate energy demands for muscular activity may be converted to depot fat, or to liver or muscle glycogen. This process is stimulated by the hormone insulin, which responds to a rise in blood glucose.

Blood insulin reaches a peak somewhat after that of blood glucose. The glucose peak occurs 4–8 hours after a feed and there may be a return to resting levels, say, 2 hours later in ponies, or somewhat quicker in Thoroughbreds. This time interval after glucose loading is called the tolerance time (see Figs 2.2 and 2.3), but a more accurate measure of the differences between species and breeds is the half time – when blood glucose is halfway between the peak and resting levels. Horses and ponies have a lower glucose tolerance than does man or the pig, but a slightly greater one than do ruminants. Thoroughbreds and other hotblooded horses generally have a higher tolerance than do ponies. In other words,

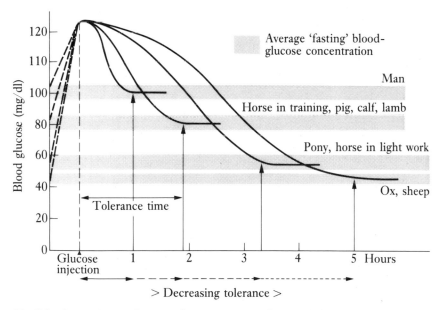

Fig. 2.2 Approximate glucose tolerance times and normal 'fasting' blood-glucose concentrations. (Glucose is injected intravenously to allow comparisons between species with different digestive anatomy and mechanisms. By providing the glucose in the form of a starch meal, the peak is delayed 2–4 hours in the horse.)

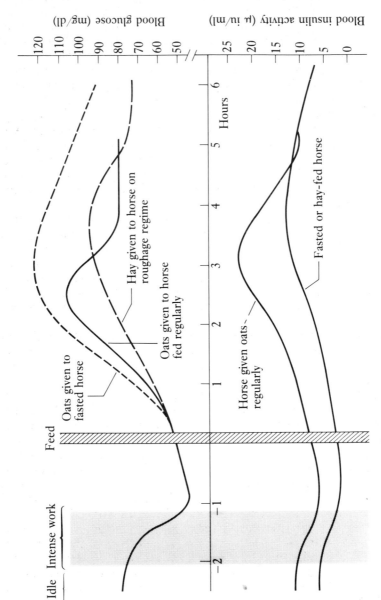

Fig. 2.3 Responses of blood glucose and insulin to feed.

pony tissue tends to be less sensitive to insulin and the insulin activity of the blood of ponies tends to be lower, although there is considerable adaptation to diet. Ponies can therefore withstand fasting better than Thoroughbreds, and Thoroughbreds become more excitable after feeding.

Insulin prevents excess blood glucose from being spilled out in the urine by increasing the uptake of tissues and so lowering blood levels. However, in order to avoid hypoglycaemia its effects must be counter-balanced by that of other hormones (for example glucagon, the glucocorticoids, epinephrine and norepinephrine). The system is thereby maintained in a state of dynamic equilibrium.

Hard muscular work may require that energy is available for muscular contraction at a rate some forty times that needed for normal resting activity. Thus, rapid changes in the supply of blood glucose could result unless the animal's system responds quickly. There are many changes to accommodate the altered circumstances but our discussion at this point will relate to the supply of nutrients to the tissues.

During a gallop, pulmonary ventilation increases rapidly so that more oxygen is available for transport by the blood to the skeletal and cardiac muscles for the oxidative release of energy. However, this process cannot keep pace with the demand for energy and glucose is therefore broken down to lactic acid releasing energy in the absence of oxygen. The fall in blood glucose stimulates the glucocorticoids and the other hormones which enhances glycogen breakdown so that blood glucose can rise during moderate exercise.

Repeated hard work (training) brings about several useful physiological adaptations to meet the energy demands of muscular work. First, the pulmonary volume and therefore the tidal volume of oxygen increase and the diffusion capacity for gases increases, so that carbon dioxide is disposed of more efficiently from the blood and oxygen is absorbed at a faster rate. This process is greatly assisted by changes in both numbers of red cells (erythrocytes) and the amount of haemoglobin in the blood. There is, therefore, a greater capacity for the oxidation of lactic acid and fatty acids to carbon dioxide. Nevertheless, training is associated with a decrease in insulin secretion, possibly a higher glucocorticoid secretion, larger amounts of muscle glycogen and blood glucose and, because of the greater work capacity, higher concentrations of blood lactate. The glucocorticoids, and possibly epinephrine in the trained animal, stimulate a more efficient break-down (lipolysis) and oxidation of body fat as a source of energy so conserving glycogen and yielding higher concentrations of free fatty acids (FFA) in the blood. The glycerol released during fat breakdown tends to accumulate during hard exercise, possibly because of the raised concentration of blood lactate, and only on completion of hard muscular work is it utilized for the regeneration of glucose (Fig. 2.1).

The energy requirements of extended work can be accommodated entirely within the aerobic breakdown of glucose and by the oxidation of body fat. Thus no continued accumulation of lactate was observed in two

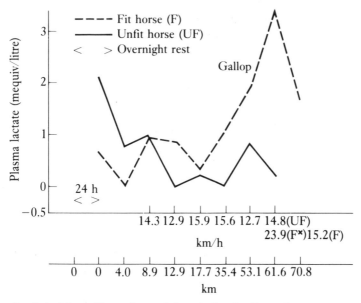

Fig. 2.4 The effect of speed but lack of effect of distance achieved on the concentration of plasma lactate. Horses separated at 53.1 km[x]. Note only fit horse galloped between 53.1 and 61.6 km from start. Unfit horse retired after 61.6 km. (From Frape, Peace & Ellis 1979).

horses subjected to an endurance ride (Fig. 2.4), and although body fat represents the primary source of energy, its relatively slower breakdown means that there is a gradual exhaustion of muscle and liver glycogen associated with a continuous decline in blood glucose (Fig. 2.5), despite elevated concentrations of FFA, in the blood. Exhaustion occurs when blood glucose reaches a lower tolerable limit. In a more general sense hypoglycaemia (low blood glucose) contributes to a decrease in exercise tolerance. Therefore, horses and ponies conditioned to gluconeogenesis – that is the production of glucose from non-carbohydrate sources through adaptation and training (Fig. 2.1), may more readily withstand extended work. Hypoglycaemia may occur when extra hard exercise coincides with a peak in insulin secretion suggesting that the time of feeding relative to racing (or other work) is important (see Ch. 9) and, furthermore, that horses and ponies conditioned to gluconeogenesis through high-roughage diets may more readily cope with sustained anorexia (persistent scarcity of food).

Glucose represents a much larger energy substrate in individuals given a high-grain diet, whereas VFA will do so in those subsisting on roughage. Horses and ponies accustomed to a diet rich in cereals will have, in a rhythmical fashion, greater peaks and lower troughs of blood glucose, than those individuals maintained on a roughage diet owing to differences in insulin secretion and tissue sensitivity to insulin and the differences in rates

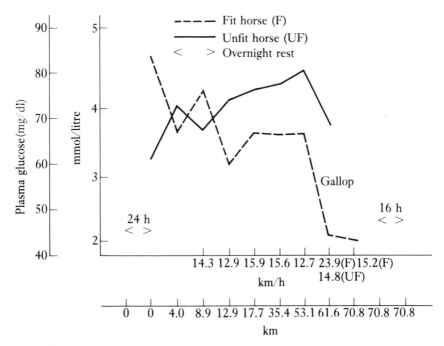

Fig. 2.5 The effect of speed and distance achieved on the concentration of plasma glucose in two horses. (From Frape, Peace & Ellis 1979).

of consumption of the two types of diet. The grain-fed horse at peak blood glucose is more energetic, but may be less so in the trough. The practical corollary of this is that individual horses and ponies accustomed to a diet rich in concentrates should be fed regularly and frequently in relatively small quantities, not only to prevent the occurrence of colic, but also to smooth out the cyclic changes in blood glucose. In Fig. 2.6. the energy transfers of the young working horse are summarized.

There is conflicting evidence about the factors that control appetite and hunger in horses and ponies. It is clear that amounts of FFA in the blood are not significantly different between satiated individual animals and those with a normal hunger (Ralston, Van den Broek & Baile 1979). It also seems that satiety is not directly associated with an elevation in blood glucose (Ralston & Baile 1982a), although individuals with low concentrations of blood glucose tend to eat more and faster (Ralston, Van den Broek & Baile 1979). Ralston & Baile (1982a) suggested that blood-glucose concentration in ponies does not influence the amount of food consumed in a meal but it may influence the interval between voluntary feeds, without affecting the amount consumed when the pony goes to the trough. Supplementary corn oil seems to extend the interval before the next meal and reduce total feed intake 3 to 18 hours after administration (Ralston & Baile 1983). A trigger mechanism controlling the feeling of satiety, or hunger,

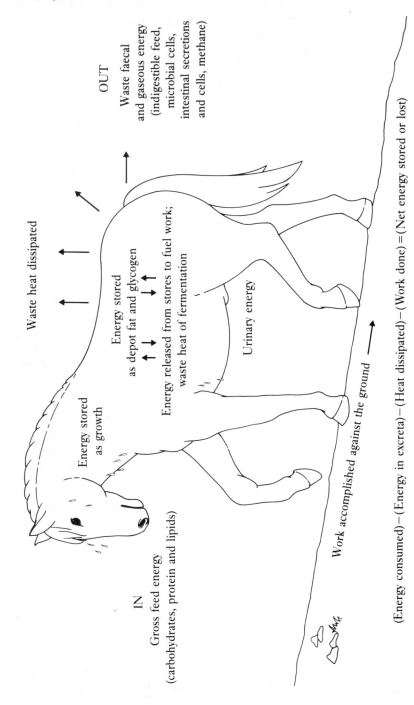

OUT

Waste faecal
and gaseous energy
(indigestible feed,
microbial cells,
intestinal secretions
and cells, methane)

Waste heat dissipated

Energy stored
as depot fat and glycogen

Energy released from stores to fuel work;
waste heat of fermentation

Energy stored
as growth

Urinary energy

IN

Gross feed energy
(carbohydrates, protein and lipids)

Work accomplished against the ground

(Energy consumed) − (Energy in excreta) − (Heat dissipated) − (Work done) = (Net energy stored or lost)

Fig. 2.6   Energy transfer in the young adult working horse.

in the horse or pony may be the amount of digestion products (especially glucose) in the intestine (McLaughlin 1982; Ralston & Baile 1982b; Ralston 1984) and VFA production in the caecum (Ralston, Freeman & Baile 1983). That is, when these products in the intestinal lumen and mucosa attain certain concentrations, eating stops, and this may be mediated by afferent vagal nerve fibres. With access to feed, eating recommences when these concentrations have fallen below a certain threshold. The degree of fill in the stomach and the blood-glucose concentration, according to this evidence, have no influence on eating (Ralston & Baile 1982b; Ralston 1984); but taste (Randall, Schurg & Church 1978; Hawkes, Hedges et al. 1985), visual contact between horses (Sweeting, Houpt & Houpt 1985) and energy density of feed, rate of eating and time of day (Doreau 1978) seem to influence feed intake. The practical interpretation of this for feeding management is considerable and will be discussed in Chapter 6.

Amino acids and non-protein nitrogen

Proteins consist of long chains of amino acids, each link constituting one amino acid residue. In all the natural proteins that have been examined, the links, or α-amino acids, are of about twenty different kinds. Animals do not have the metabolic capacity to synthesize the amino group contained in half the different kinds of amino acid. The horse and other animals can produce certain of them from others by transferring the amino group from one to another carbon skeleton in a process known as transamination. Ten or eleven of the different types cannot be synthesized at all or cannot be synthesized sufficiently fast by the horse to meet its requirements for protein in tissue growth, milk secretion, maintenance, etc. Plants and many microorganisms can synthesize all twenty-five of the amino acids. Thus the horse and other animals must have plant material in their diet, or animal products originally derived from plant food, in order to meet all their needs for amino acids (i.e. they are unable to survive on an energy source and inorganic N). Whether or not microorganisms, chiefly in the horse's large intestine, synthesize proteins, the amino acids of which can be utilized directly by the horse in significant amounts, is a contentious issue. The consensus of opinion is that although this source makes some contribution, probably in the small intestine, only small amounts of amino acid can be absorbed from the large intestine and by far the major part is voided as intact bacterial protein in the faeces (Reitnour, Baker et al. 1970; Slade, Bishop et al. 1971; Reitnour & Salsbury 1972, 1975; Frape 1975; Glade 1983b).

During the digestion of dietary protein, the constituent amino acids are released and absorbed into the portal blood system. The amount of protein consumed by the horse may be in excess of immediate requirements and although there is some capacity for storing a little above those needs in the form of blood albumin, most excess amino acids, or those provided in excess of the energy available to utilize them in protein synthesis, are deaminated in the liver with the formation of urea. The concentration of this product rises in the horse's blood (see Fig. 1.6), although some of the amino-nitrogen may be utilized in the liver for the synthesis of dispensable amino acids (Fig. 2.7). An increase in the blood concentration of urea in endurance horses may simply reflect rapid tissue protein catabolism for gluconeogenesis in glycogen depletion (Fig. 2.8) and, of course, the carbon skeleton of deaminated dietary amino acids is used as an energy source.

The extent to which dietary protein meets the present requirements of the horse depends on its quality as well as its quantity. The more closely the proportions of each of the different indispensable amino acids in the dietary protein conform with the proportions in the mixture required by the tissues, the higher is said to be the quality of the protein. If a protein, such as maize gluten, containing a low proportion of lysine, is consumed and then digested, the amount of it which can be utilized in protein synthesis will be in proportion to its lysine content. As the lysine is 'limiting', little

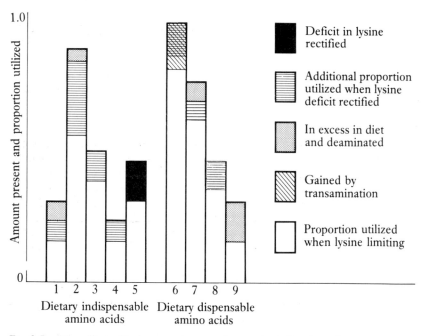

Fig. 2.7   The effect of lysine supplementation on the utilization of a dietary protein that is grossly limiting in its lysine (5) content. Only nine of a possible twenty-five amino acids are shown. Amino acids 3 and 4 limit further utilization of a lysine-supplemented diet. Addition of supplementary 3 and 4 now would decrease deamination of 1, 2, 7 and 9.

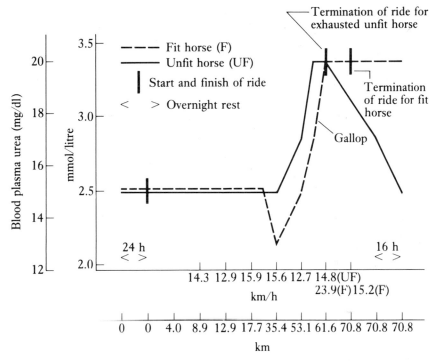

Fig. 2.8 The effect of speed and distance on the concentration of blood urea. The rise in plasma urea of unfit horse resulted from a stress other than galloping. (Data from Frape, Peace & Ellis 1979).

of it will be wasted, but, conversely, the other amino acids, both dispensable and indispensable, will be present in excessive quantities and so will be deaminated to an alarming extent. If the relative deficiency of lysine in the gluten is made good by supplementing the diet with either a good-quality protein such as fishmeal or with synthetic lysine, then the amounts of each of the amino acids available in the blood plasma will more closely conform with the requirement so that proportionately more of those amino acids can be used for protein synthesis (see Figs 1.6 and 2.7). It has been shown recently that the proportions of amino acids in the common sources of feed proteins given to horses and ponies are such that lysine is the indispensable amino acid most likely to limit the tissue utilization of the protein (Wirth, Potter & Broderick 1976; Comerford, Edwards et al. 1979; Ott, Asquith et al. 1979; Ott, Asquith & Feaster 1981).

Tissue proteins are broken down to amino acids and resynthesized during normal maintenance of adult or growing animals. This process is not fully efficient and, together with losses of protein in the sloughing off of epithelial tissues and in various secretions, there is a continual need of dietary protein to make good the loss. However, these losses are relatively small in comparison with the protein synthesis of normal growth or milk production and proportionately less lysine is required. It follows that less protein or protein of poorer quality is needed for maintenance than is necessary for growth or milk secretion. Nevertheless, it has been shown that the protein needs of the adult horse for maintenance are less when good-quality protein is provided than when poor-quality protein is given (Slade, Robinson & Casey 1970). For example, adult Thoroughbred mares were shown to remain in nitrogen balance when given 97 g of fish protein per day but they required 112 g for balance when the protein source was maize gluten (Slade, Robinson & Casey 1970). In summary, the protein value of a diet is the product of the amount and the quality of its protein.

Another attribute of dietary protein that should not be neglected when alternative feeds are available, is digestibility. For example, leather is a rich source of protein, but valueless to the horse because of its low digestibility. In practice, the problem arises only when byproducts are being considered. Most dietary proteins have an apparent digestibility coefficient of about 0.8. A related feature to that of digestibility is availability of amino acids and of lysine in particular. In practice, a reduction in lysine availability is encountered when skimmed milk, fishmeal and meat meals are overheated during processing. The excessive heating in milk leads to a reaction between some of the lysine and unsaturated fats. These reactions lead to products from which the animal's digestive system cannot recover the lysine.

Non-protein nitrogen

Urea is synthesized in the liver from amino acids present in excess of need so that a rise in dietary protein above requirements is associated with a rise in plasma urea (see Figs 1.6 and 2.7). In ponies on diets containing from 6 to 18 per cent of protein between 200 and 574 mg of urea N per kg metabolic bodyweight ($W^{0.75}$) daily are recycled and degraded in the intestinal tract. In a pony weighing about 150 kg, this range is equivalent to 54–154 g of crude protein daily. While urea is within the tissues of a horse it cannot be degraded or otherwise utilized. However, when provided with an adequate source of dietary energy, microorganisms – chiefly in the large gut – utilize it in protein synthesis (Reitnour 1982). In the absence of an adequate supply of energy, which is normally present as fibre, starch and protein, a proportion of the ammonia (or, under acid conditions, ammonium ion) diffuses back into the blood and may not be utilized either by the horse or by its captive microorganisms. A fine balance is required, for in the absence of sufficient nitrogen, microbial growth cannot occur at a maximum rate and therefore a maximal rate of fibre breakdown and utilization will not prevail. Whereas circulating urea is non-toxic to the horse, except when very high concentrations affect osmolality, the absorbed ammonia is highly toxic. A healthy liver copes adequately with low concentrations by the formation of dispensable amino acids, through transamination reactions and by urea synthesis. However, if liver failure occurs, and this is more frequent in older horses, ammonia intoxication can occur without any increase in blood urea.

Limited evidence (see Fig. 1.6) does not support the widely held view that excessive protein consumption, *per se*, predisposes horses to laminitis. The flow of urea and other nitrogen compounds into the large intestine from the ileum varies with the quantity of diet and its type (see Table 1.4). These digesta are relatively impoverished of nitrogen in horses receiving a diet of straw. The provision of non-protein nitrogen or, for that matter, of protein as a supplement to this diet, results in an increased flow of nitrogen and a stimulation of microbial growth in the large intestine. Urea, or more effectively biuret, added to low-protein diets in concentrations of 1.5 to 3 per cent, has increased nitrogen retention in both adult and growing horses with functioning large intestines, and pregnant mares subsisting on poor pasture apparently benefit from the consumption of supplementary urea. The large intestine of the horse can absorb small quantities of amino acids, including lysine, and the addition of urea or biuret to low-protein, poor-quality hay diets may increase dry matter and fibre digestion as well as nitrogen retention by stimulating microbial growth.

In summary, it would seem that horses and ponies with functioning large intestines and given diets containing less than 7–8 per cent crude protein can make some use of supplementary non-protein nitrogen as an adjunct to that secreted back into the small intestine in digestive secretions and more directly from the blood. In ruminants, large amounts of soluble nitrogen entering the rumen lead to a rapid production of ammonia and therefore to ammonia toxicity. However, in horses, ammonia toxicity caused by excessive dietary non-protein nitrogen, or protein, is less likely chiefly because much of the nitrogen is absorbed into the bloodstream before it reaches the regions of major microbial activity – the large intestine.

Protein for maintenance and growth

Maintenance

The protein needed by the horse for body maintenance can be defined as the amount of protein required by an individual making no net gains or losses in body nitrogen and excluding any protein that may be secreted in milk. In these circumstances the animal must replace shed epithelial cells and hair, it must provide for various secretions and keep all cellular tissues in a state of dynamic equilibrium. The losses are a function of the lean mass of body tissues, depicted as a direct proportion of metabolic body size. For most purposes, the latter is considered to be the bodyweight (W) raised to the power 0.75 and evidence suggests that horses daily require about 2.7 g of digestible dietary protein per kg $W^{0.75}$. A horse weighing 400 kg would therefore need daily about 240 g of digestible protein, or 370 g of dietary crude protein. This assumes that the protein has a reasonable balance of amino acids, although, as already pointed out, the lysine content of the protein for maintenance need not be as high as in that required for growth.

Weight gain

A young horse with a mature weight of 450 kg normally gains 100 kg between 3 and 6 months of age at the rate of 1 kg/day. Growth rate in kg per day declines during the succeeding months and it therefore gains the next 100 kg when aged between about 6 and 12 months and 75 kg between 12 and 18 months (Hintz 1980a). From a very young age the rate of gain per unit of bodyweight decreases continuously, while the daily maintenance requirement increases. As the weaned foal grows, an increasing proportion of that daily gain is composed of fat and a decreasing proportion is lean. It is thus apparent that the dietary requirement for protein and the limiting amino acid lysine decline with increasing age in the growing horse. For colts aged 3 months, a maximum rate of gain has been achieved with diets containing 140–150 g protein and 7.5 g lysine per kg. Diets may differ in the amount of digestible energy they provide per kg. For obvious reasons it is more accurate to state the protein requirements as a proportion of the digestible energy (DE) provided. Current evidence suggests that Thoroughbred and Quarter Horse yearlings require 0.45 g lysine per megajoule (MJ). A compounded stud nut for young growing horses may contain about 12–13 MJ of DE per kg and oats about 11 MJ of DE per kg. However, hard hay, containing 50–60 g protein per kg may provide 7.5–8 MJ of DE per kg. If the yearling consumes a mixture (approximately 50:50) of concentrates and hay, the diet provides on average 10 MJ of DE per kg and the minimum lysine requirement is 4.5 g/kg of total diet. Hay of 50–60 g/kg protein may contain only 2 g/kg of digestible lysine and therefore the concentrate should contain at least 7 g lysine per kg in order to meet the minimum requirement. A yearling consuming 9 kg daily of total feed of this type would receive about 40 g of lysine.

Much of the growth of horses may take place on pasture. Leafy grass protein of several species has been shown to contain 55–59 g of lysine per kg. During the growing season the protein content in the dry matter of grass varies considerably from 110 g to 260 g/kg in the leaf, whereas the flowering stem contains only 35–45 g/kg. Thus the lysine content of the grazed material as a fraction of air-dry weight can vary from 5 to about 13 g/kg, and if a leafy grass diet is supplemented with a concentrate mixture, the lysine and protein requirements may be met by cereals as a source of that protein. Because the quality of pastures varies so much, the use of cereals alone may mean that the protein and lysine requirements are not always met and, of course, the mixture may be inadequate as far as several other nutrients are concerned (Ott, Asquith et al. 1979; Ott, Asquith & Feaster 1981). Table 10.1 (p. 208) gives some analytical data for pasture in several months of the grazing season.

Studies of the protein requirements of growing horses have failed to show any relationship between amounts of dietary protein and the incidence of laminitis (founder) (Yoakam, Kirkham & Beeson 1978), but have shown that there is a greater incidence of the problem when starch intake is excessive (Garner, Coffman et al. 1975; Garner, Hutcheson et al. 1977; Moore, Garner et al. 1979). Restricted feeding of concentrates is therefore advocated and horses should be adjusted gradually to lush pasture by gradually increasing their time of access daily. (A method for *ad libitum* feeding is outlined in Ch. 8.)

Further reading

McDonald P, Edwards R A & Greenhalgh J F D. (1981) *Animal nutrition.* Longman: London & New York.

CHAPTER 3

The roles of major minerals and trace elements

Grass is the first nourishment of all colts after they are weaned ... Whereas when they are fed with corn and hay, but especially with the first, ... it exposes them to unspeakable injuries.

William Gibson, 1726

Calcium and phosphorus

The functions of calcium (Ca) and phosphorus (P) are considered together because of their interdependent role as the main elements of the crystal apatite, which provides the strength and rigidity of the skeleton. Bone has a Ca:P ratio of 2:1, whereas in the whole body of the horse the ratio is approximately 1.7:1, because of the P distribution in soft tissue. Bone acts as a reservoir of both elements, which may be tapped when diet does not meet requirements. The elements of bone are in a continual state of flux with Ca and P being removed and redeposited by a process that facilitates the reservoir roll and enables growth and remodelling of the skeleton to proceed during growth and development. The acute role of Ca relates to its involvement in a soluble ionic form for nerve and muscle function. Consequently $[Ca]^{2+}$ concentration in the blood plasma must be maintained within closely defined limits.

The flux and distribution of Ca and P in the body are regulated by two hormones in particular, functioning antagonistically at the blood–bone interface, the intestinal mucosa and the renal tubules (see also under vitamin D, Ch. 4). The horse kidney seems to play a greater part in controlling concentrations of Ca in the blood than does the intestinal tract and this may have practical significance for diet and renal disease. The mean values and ranges for serum Ca and P are listed in Table 3.1.

Serum phosphate values vary without untoward physiological effects. For example, strenuous exercise can depress blood phosphate to half the 'resting value' for about 3 hours. Nutritional secondary hyperparathy-

Table 3.1 Mean values and ranges for serum electrolytes (mmol/litre) in horses of different ages

		Birth–36 h	3 weeks	Yearlings	Horses in training	Mares at stud
Ca	Mean	3.2	3.2	3.3	3.2	3.2
	Range	2.7–3.6	2.5–4.0	2.7–4.0	2.6–3.9	2.9–3.9
PO₄	Mean	2.5	2.5	1.8	1.3	1.1
	Range	1.2–3.8	1.6–3.4	1.4–2.3	1.1–1.5	0.5–1.6
Na	Mean	136	137.5	138.5	138.5	138.5
	Range	126–146	130–144	134–143	134–143	134–143
K	Mean	4.8	4.5	4.3	4.3	4.3
	Range	3.7–5.4	3.6–5.4	3.3–5.3	3.3–5.3	3.3–5.3
Mg	Mean	0.83	0.81	0.78	0.78	0.78
	Range	0.57–1.10	0.66–1.10	0.62–1.10	0.62–1.10	0.62–1.10
Cl	Mean					
	Range	Normal range for all ages 89–109				

(Modified from published tables of the Beaufort Cottage Laboratories, Newmarket, with permission of S. W. Ricketts)

roidism is a diet-related clinical disorder of horses in which serum phosphate is slightly raised and serum $[Ca]^{2+}$ values are slightly depressed (Figs 3.1 and 3.2).

Failure of the osteoid, or young bone, to mineralize is called rickets

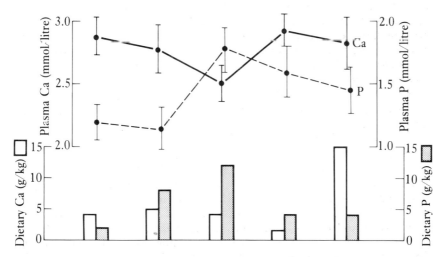

Fig. 3.1 The effects of dietary calcium (□) and phosphorus (□) on mean concentrations in blood plasma, ± standard error. (After Hintz & Schryver 1972).

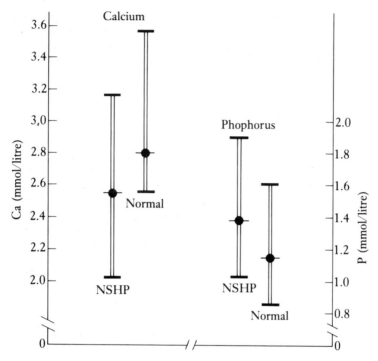

Fig. 3.2   Mean and range of serum calcium and phosphorus concentrations in spontaneous nutritional secondary hyperparathyroidism (NSHP). (From Krook 1968). Notice how in cases of NSHP Ca is depressed and P elevated.

in young and osteomalacia in adult horses. In extreme cases when mineral is being resorbed from bone, the outcome may be generalized osteodystrophia fibrosa in which fibrous tissue is substituted for hard bone and characteristic enlargement of the facial bones (big-head) may occur. In the presence of vitamin D in each of these conditions the body, through the agency of parathormone, is endeavouring to maintain homeostasis of blood Ca by accelerating the removal of Ca from the bones. Diets based on wheat-bran and cereals are rich in organic P and low in Ca, predisposing horses to these conditions. The tendency to lower blood $[Ca]^{2+}$ leads to increased bone resorption, increased renal excretion of phosphate, to an increased rate of bone-mineral exchange and to a greater susceptibility of bones to fracture. Deposition of Ca salts in soft tissue, including the kidney (nephrocalcinosis), may also be apparent. Horses have no 'horse sense' when it comes to selecting a diet containing a balanced mixture of Ca and P – they prefer the palatability of a P-rich diet, but this is not available for selection in the natural grazing environment.

Hypocalcaemia may occur in adult horses as a post-exertional stress when the horse develops tetanic spasms, incoordination or even inability to stand. Extended work and overheating lead to a rise in blood pH and possibly depress the concentration of $[Ca]^{2+}$ in the blood. Elevated body temperature *per se* can bring about a loss of 350 to 500 mg of Ca per hour in sweat and this rate of loss may exceed the capacity for blood replenishment by bone mobilization.

Dietary Ca and P levels

Excessive amounts of dietary Ca do not seem to initiate the kinds of dietary problem encountered in other domestic species. However, one experiment in which Shetland foals aged 4 months were given a diet containing 25 g/kg of Ca, and a Ca:P ratio of 6:1, for 4 years (Jordan, Myers et al. 1975), resulted in a slight enlargement of the marrow region of the long bones and thinning of the cortical area, together with less bone mineral per unit of cortex. Lameness is frequently a characteristic of resorption of cortical bone through loss of support for tendons and ligaments. Excessive amounts of dietary Ca may make bones brittle, increasing bone storage of Ca. But investigations in which diets containing from 7 g to 27 g/kg of Ca have been compared for up to 2 years, show that differences are small; bone density is increased by high-Ca diets and the cortex of the long bones is slightly thinner (Schryver, Craig & Hintz 1970; Schryver, Hintz & Lowe 1971, 1974; Schryver, Hintz et al. 1974).

Ca deficiency ailments are noticeable in horses grazing certain pastures, or given hay made from the same species of tropical and subtropical grass; for example, *Setaria* spp., which are rich in oxalates (see Ch. 10).

The equine kidney plays a vital part in Ca homeostasis and daily urinary Ca excretion shows a direct relationship with intake. Diets rich in Ca yield urine containing a precipitate of Ca salts; the urinary loss of Ca in a 300 kg yearling given a diet containing 20 g of Ca per kg was 20–30 g in 6–8 litres of urine daily (that is, 0.36% Ca). The absence of calculus formation in the kidney demonstrates the horse's ability to deal with large amounts of Ca despite the low solubility of the element; conversely, a dietary deficiency of Ca yields urine almost devoid of the element. In contrast, endogenous loss of Ca in faeces, representing the minimal obligatory loss that must be replenished from dietary sources, is largely unaffected by the dietary amounts. Urinary losses of Ca decrease by 50–75 per cent in extended work (Schryver, Hintz & Lowe 1975, 1978), whereas sweat losses increase – during 20 minutes of hard work, ranges of 80–145 mg of Ca and 11–17 mg of P have been found in sweat (Schryver, Van Wie et al. 1978). On the other hand, horses and ponies idle for long periods retain less Ca than those worked when the dietary P concentration is excessive. Horses must absorb about 2.5 g of Ca per 100 kg bodyweight daily to balance the obligatory loss. Those given two to three times their Ca requirement absorb nearly a half of it, whereas at the requirement level the absorption rate may lie between one-half and two-thirds of intake.

Absorption of half the total ingested necessitates a daily intake of 5 g per 100 kg bodyweight for maintenance of the adult, or about 2.5 g/kg diet. Similarly, assuming average digestibility values the horse must be provided daily with about 2 g of P per 100 kg bodyweight or 1 g/kg diet for maintenance.

The lack of impact of dietary Ca on the efficiency of P absorption in the horse may be related to the fact that Ca and P are absorbed from different regions of the intestine (see Fig. 1.5). However, dietary Ca can affect absorption of other elements. For example, excessive Ca can depress the absorption of magnesium, manganese and iron owing to competition at common absorption sites, or possibly to the formation of insoluble salts. Meyer and colleagues in Germany have reported that between 50 and 80 per cent of dietary Ca and between 45 and 60 per cent of magnesium are absorbed in the small intestine (Meyer, Schmidt et al. 1982), whereas there is a net secretion of these elements into the large intestine. They have further demonstrated that the site of P absorption varies with composition of diet. No P is absorbed in the upper small intestine of horses fed solely on roughage, whereas some is in those given only concentrates. Large amounts of phosphate secreted into the caecum and ventral colon probably act as a buffer to volatile fatty acids produced there, and the colon is the major site of absorption and reabsorption of phosphate.

The net availability of Ca in a variety of feeds has been estimated to lie between 45 and 70 per cent except where significant amounts of oxalates are present. The dietary level of phosphate influences Ca absorption. When dietary P, as inorganic phosphate, was raised from 2.0 g to 12 g/kg, Ca absorption was decreased by more than 50 per cent in young ponies receiving a diet otherwise adequate in Ca (4 g/kg diet) (Schryver, Hintz & Craig 1971).

Dicalcium phosphate or bone flour-P is digested to the extent of 45–50 per cent, but P in salts of phytic acid, the predominant source in cereal and legume seeds, is only 30 per cent available despite the presence of large numbers of intestinal bacteria secreting phytase. Phytate-P constitutes at least 75 per cent of the total P in wheat grain and 54–82 per cent of the P in beans. Large amounts of vitamin D in the diet can increase the utilization of phytin-P, but as these are almost toxic they cannot be recommended.

In summary, the maintenance requirement for the major minerals Ca and P is that necessary to balance losses in the faeces and urine, as well as unspecified 'dermal losses'. There is an additional need for growth and, in the breeding mare, for mineralization of the foetal skeleton and for lactation. Each kilogram of lean body tissue in the horse contains about 20 g of Ca and 10 g of P; the amounts required in the diet to allow for maintenance and growth are shown in Table 3.2. Mare's milk contains on average about 900 mg of Ca and 350 mg of P per kg (see Fig. 7.3). A 500-kg mare may produce a total of 2000 kg of milk in a lactation extending over

Table 3.2 Minimum daily requirements for calcium and phosphorus of growing horses

Age (months)	Bodyweight (kg)	Weight gain (kg/day)	Ca (g/day)	P (g/day)
3	100	1	37	31
6	200	0.72	33	27
12	300	0.50	31	25
18	375	0.30	28	23
Mature	450	0	23	18

(From Schryver, Hintz & Lowe 1974)

5–6 months – a total lactation deficit of 1.8 kg of Ca and 0.7 kg of P derived from skeletal reserves and feed. Daily dietary Ca and P requirements with average availabilities of 50 and 35 per cent, respectively, are 10 g of Ca and 5.5 g of P to balance such losses.

Limestone flour, dicalcium phosphate and bonemeal are reliable sources of Ca and the second two of P also, although bonemeal should be used only if it has been steamed and marketed in paper or plastic sacks. The requirement of 10 g of Ca is met by 28 g of limestone or 40 g of dicalcium phosphate, which also meets the P needs entirely.

Magnesium

Magnesium (Mg) is a vitally important ion in the blood; it forms an essential element of intercellular and intracellular fluids, it participates in muscular contraction and it is also a cofactor in several enzyme systems.

Bone ash contains 0.8 per cent of Mg in addition to 36 per cent of calcium and 17 per cent of phosphorus. Mg is absorbed from the lower half of the small intestine but there is an 'obligatory loss' of Mg secreted into the intestinal tract amounting to about 1.8 mg/kg of bodyweight daily; a further obligatory loss of about 2.8 mg/kg occurs in the urine, and the maintenance requirement to offset these losses is about 13 mg/kg of bodyweight daily, or about 2 g/kg of diet.

A dietary deficiency of Mg leads to hypomagnesaemia associated with nervousness, sweating, muscular tremors, rapid breathing (hyperpnoea), convulsions, heart and skeletal muscle degeneration and, in chronic cases, mineralization of the pulmonary artery caused by deposition of calcium and phosphate salts. Normal blood serum values are given in Table 3.1.

Mg, naturally present in feed, is available in proportions ranging from 45 to 60 per cent, the more 'digestible' sources being milk and possibly lucerne. Large amounts of dietary phosphorus seem to depress absorption of Mg slightly, but not as effectively as dietary oxalates do. The inorganic sources of Mg – magnesium oxide, magnesium sulphate and magnesium carbonate – are all about 70 per cent absorbed in the horse. An increase in the dietary level of Mg from 1.6 to 8.6 g/kg was shown to increase calcium absorption without an effect on phosphorus. Table 3.3 gives the daily requirements for Mg in horses weighing 400 and 500 kg.

Table 3.3 Daily requirements for magnesium (g)

Mature weight (kg):	400	500
Adult		
Rest	5.6	7.0
Medium	6.5	8.0
Last 90 days of gestation	6.5	8.0
Peak lactation	6.6	8.1
Growing: age (months)		
3	5.5	6.8
6	5.2	6.3
12	5.3	6.5
18	5.6	7.0

Potassium and sodium

A dietary deficiency of potassium (K) may reduce appetite and depress growth rate; a reduction in plasma K (hypokalaemia) occurs and in extreme deficiency there may be clinical muscular dystrophy and stiffness of the joints. Spontaneous changes in serum $[K]^+$ can be precipitated as a result of strenuous exercise; this will be discussed in Chapter 9.

Foals require about 10 g of K per kg diet, but adults need less – about 5–6 g of K per kg. Cereals are relatively poor sources but hay contains between 15 and 25 g of K per kg; thus most diets should contain ample if at least one-third is in the form of good-quality roughage. Animals in heavy work generally consume more cereals thus lowering dietary K when losses in sweat would normally be increasing.

Lush pastures can contain large amounts of K in the dry matter and so theoretically may 'interfere' with magnesium metabolism. Although pasture grass may contain as much as eighteen times more K than sodium (Na), supplementary Na in the form of common salt for grazing stock is normally unnecessary. Forages are a richer source of Na than are cereals and within normal ranges the one element tends to inhibit the loss of the other in the urine, conserving the body's resources of Na in the grazing stock. Young foals may become deficient in K as a result of persistent diarrhoea and this in turn tends to precipitate acidosis. The means of assessing potassium status, the major causes of K depletion and their therapy are discussed in Chapter 11. Diets providing 2–4 g of Na per kg should adequately meet the requirement for Na, except during periods of excessive sweating in very hot weather. Diets containing 5–10 g of common salt per kg will amply meet the normal Na requirement.

Chloride

Where the requirements for common salt (sodium chloride) are met, it is unlikely that a deficiency of chloride will occur. The major source of loss, particularly in hot weather, is sweat where even at moderate rates of work horses may lose 100 g of salt per day (60 g chloride).

Trace elements

Most stabled horses in the United Kingdom now receive supplements containing variable quantities of trace elements, and the horse seems able to cope with some measure of abnormal intake without showing clinical signs of toxicity or deficiency. Those trace elements of prime importance in the diet of horses are discussed here. Cobalt is considered on page 57 under vitamin B_{12}.

The variety of geological strata underlying UK soils yields grazing areas that cause clinically recognizable signs of specific deficiencies in cattle and sheep (Suttle, Gunn et al. 1983). There is biochemical evidence to show that horses and ponies in these areas may similarly reflect their nutritional environment. Tables 3.4 and 3.5 give average serum values for some trace elements.

Abnormalities in leg growth and development of foals and yearlings have been reported to be associated with dietary deficiencies of copper, manganese (H. F. Hintz, personal communication 1983) and selenium, (Caple, Edwards et al. 1978; Sandersleben & Schlotke 1977), and toxicities of iodine (Drew, Barber & Williams 1975) and lead (Osweiler, Van Gelder & Buck 1978).

Copper

Evidence has accumulated in England and Wales that hypocupraemia occurs widely in grazing cattle with the greatest prevalence in parts of Somerset, Derbyshire, Lincolnshire and Cumbria. The cause in some areas has been attributed to an excess of molybdenum (Mo) derived from the

Table 3.4   Values for five trace elements (μmol/litre) in normal serum and milk (4 months' lactation)

		Blood serum	Milk*		
			Partum	1–8 days	9–120 days
Cu	Thoroughbred (USA)	24–35	16	10	4
	Thoroughbred (UK)	8–18			
	Quarter Horse	5–31			
Fe	Thoroughbred	28±7			
	Quarter Horse	28			
	Arabian	23	24	17	12
	Standardbred	30±7			
	Shetland	19			
Zn	Thoroughbred (UK) pastured	17±7	98	52	36
	Thoroughbred (UK) stabled	26±8			
Pb		7–8			
	Thoroughbred suckling	1.0			
	Weanling–adult	1.9			
	Standardbred suckling	0.8			
	Weanling–adult	1.6			

(After Ullrey, Struthers et al. 1966; Blackmore & Brobst 1981)
*   Quarter Horse and Arabian mares.

underlying Mo-rich strata, particularly Lias clays and marine black shales of Jurassic and Carboniferous ages. The high levels of Mo associated with relatively low copper (Cu) lead to Cu:Mo ratios in the herbage narrower than 6:1, causing Cu deficiency on the so-called 'teart' pastures. Hypocupraemia may also occur outside these regions. In these cases deficiency is attributed to low-Cu levels *per se* in the soils and herbage.

The horse is not as susceptible to Cu deficiency as are ruminants, but symptoms have been described, such as erosion of the articular cartilage of joints, and anaemia and haemorrhage in parturient mares (Stowe 1968). The copper-containing enzyme lysyl oxidase controls cross linkages in cartilage protein. It has been suggested that interference from Mo will cause anaemia and failure of osteoblastic function, which results in thinning of the cortex (shank) of long bones in the horse. Mo interference may be implicated in a widening of the growth plate and fractures of long bones

Table 3.5 Normal mean blood characteristics of horses*

Sample	Characteristic	Units (per litre)		Yearlings	Horses in training	Mares	Foals
Plasma	Albumin	g		27	34	27	NA
Serum	AST	iu		140	160	140	NA
Serum	CK	iu		43	68	43	NA
Serum	SAP	iu		118	86	71	NA
Serum	Creatinine	mmol		141	125	141	157
Serum	Cu	μmol	Stable	NA	12.4	NA	NA
			Grass	NA	15.9	NA	NA
Serum	Zn	μmol	Stable	NA	26.0	NA	NA
			Grass	NA	17.0	NA	NA
Serum	Mo	μmol		NA	0.31	NA	NA
Serum	Se	μmol		NA	1.5†	NA	NA
Whole blood	Se	μmol		NA	1.6†	NA	NA
Plasma	α-Tocopherol	mg			4		
Plasma	Retinol	μg			180		
Whole blood	Cyanocobalamin	μg	Stable		3.7 to 6.6		
			Grass		2.8 to 20.0		
Serum	Folate	μg	Stable		4.5 to 12.0		
			Grass		5.3 to 13.5		
Whole blood	Thiamin	μg		28	30	33	24

* Data are averages drawn mainly from Thoroughbreds. There are few data on cyanocobalamin and folate so the ranges quoted are those found by the author. The first lines of data are provided by Beaufort Cottage Laboratories, Newmarket. (See Table 3.1 for serum Ca, PO_4, Mg, Na, K and Cl values.)

† Minimum adequate.

Abbreviations: AST, aspartate amino-transferase; CK, creatine kinase; SAP, serum alkaline phosphatase; iu, international unit; NA, not available.

in foals in southern Ireland. Yet thiomolybdates, which bind copper, have not been detected in horses when dietary Mo concentrations of 10 mg per kg are found. Their formation may depend on the presence of a rumen. In England, hypocupraemia has been detected in grazing horses, although some grazing Thoroughbreds have higher serum Cu levels than do stabled horses (Stubley, Campbell et al 1983).

Cu is transferred to the foetal liver, which, like the neonatal liver, contains more Cu than that of older foals or of their dams. The obligatory losses of Cu in the faeces of ponies amount to about 3.5 mg per 100 kg of bodyweight daily in the presence of low levels of dietary Mo (Cymbaluk, Schryver & Hintz 1981; Cymbaluk, Schryver et al. 1981). However, in order to allow for adverse interactions with other trace elements and to maximize iron retention, a dietary intake of 15–20 mg/kg of feed is recommended. The need of foals is for 25–30 mg/kg of feed. Excessive dietary levels of Cu (up to 800 mg/kg) for 6 months caused no liver damage in ponies and no adverse effect on fertility despite a rise in liver Cu to over 4000 mg/kg of dry matter (Smith, Jordan & Nelson 1975).

The extent to which pasture plants extract trace metals from the soil depends on the soil's pH and moisture content and the plant species. Effects may also be attributable to the root systems of plants as legumes and many herbs have deeper roots than grasses do. The levels of trace elements in herbage are clearly of importance, but the horse will also consume soil while grazing. Soil intake will depend on the density of the herbage. In certain conditions cattle and sheep may consume more than 10 per cent of their daily intake of dry matter as soil.

Zinc

A deficiency of dietary zinc (Zn) in many domesticated animals, including the horse, depresses appetite and growth rate in the young, causes skin lesions and is associated with a depression of Zn concentrations in the blood. Zn participates with many cellular enzymes, either as part of the molecule or as an activating cofactor, and a deficiency thus has fairly widespread physiological effects. In most domestic species Zn deficiency affects bone formation and shape, but there is little direct evidence of this in the horse.

An increase in dietary Zn from 26 to 100 mg/kg progressively increases serum-Zn concentrations. However, the dietary requirement of the horse is probably less than 50 mg/kg and supplements normally used include zinc carbonate or sulphate; these inorganic salts possess a higher availability than do phytate salts of zinc in cereal grains and oilseed meals. The efficiency of absorption of Zn in all forms is probably affected more by diets rich in phytate than is the absorption of other trace elements but, even so, high phytate concentrations are unlikely to depress the utilization of Zn by more than 25 per cent.

Zn is one of the less toxic of the essential trace metals, but where there is industrial pollution of pastures, grazing animals may show signs of toxicity. Toxic dietary concentrations probably exceed 1000 mg/kg; a dietary level of 5.4 g of Zn per kg causes anaemia, epiphyseal swelling, stiffness and lameness, including breaks in the skin around the hooves (Willoughby, MacDonald & McSherry 1972; Willoughby, MacDonald et al. 1972). In other domestic animals moderately high dietary concentrations of Zn can precipitate a copper deficiency and as little as 100 mg of Zn per kg diet have been shown to increase the faecal loss of copper and to lower blood copper by about 10 per cent in the horse.

Manganese

A deficiency of manganese (Mn) is thought to be a cause of enlarged hocks, and, by affecting the growth plate, to shorten legs with characteristic knuckling-over of joints. Excessive use of limestone was reported in the United States to cause a high concentration of calcium in alfalfa and to precipitate a flexural deformity of the legs of growing horses that was rectified by manganese supplements (H. F. Hintz, personal communication 1983). The young also seem to suffer lameness and incoordination of movement if they lack sufficient Mn, and Mn deficiency is a possible explanation of tiptoeing in situations where suckled foals are on pasture containing less than 20 mg of Mn per kg dry matter. A severe deficiency can give rise to resorption *in utero*, or death at birth, and lesser deficiencies may provoke irregular oestrous cycles. The effects on bone growth are probably related to the involvement of Mn in an enzyme system required for the synthesis of chondroitin sulphate in the developing cartilage.

Iron

Most natural feeds, apart from milk, are fairly rich sources of iron (Fe) and deficiences are unlikely unless the horse is anaemic through heavy parasitization. The foal is born with an adequate store of liver Fe and the foal's grazing activity is normally an adequate supplement to the mare's milk, which contains meagre amounts of most trace elements with the possible exception of selenium in mares supplemented with this element. The levels of Fe, zinc and copper in mare's milk are shown in Table 3.4.

A deficiency of Fe causes anaemia. A dietary concentration of 50 mg/kg should be adequate for growing foals. One of 40 mg/kg is said to meet the maintenance requirements of adults (National Research Council 1978).

Dietary Fe toxicity in horses has not been studied; toxic amounts probably exceed 1000 mg/kg of diet. Dietary concentrations of several hundred mg/kg are quite common and seem to have no adverse effect, and no clear-cut effect on zinc, manganese or copper utilization has been established.

Fluorine

Fluorine (F), like calcium, phosphorus and magnesium, is part of the crystalline structure of bones and teeth and so is an essential nutrient. In the horse, the risk of excess (fluorosis) is far greater than that of deficiency. Industrial contamination of pastures, especially from brickworks, causes a softening, thickening and weakening of bones in grazing animals, although a decrease in industrial effluent has ensured that very few cases are recognized today. The horse seems to excrete more fluorine in its faeces than do cattle, but dietary concentrations should not exceed 50 mg/kg.

A world shortage of sources of digestible phosphorus has led to an increase in the use of rock phosphates; as some of these are rich in fluorine a careful scrutiny of their composition is essential before purchase and use.

Iodine

Like fluorine, iodine (I) is the subject of both deficiencies and excesses in practice and the signs presented by a horse are rather similar in both states. Iodine is specifically required in the synthesis of thyroxine in the thyroid gland; pregnant mares may show no external signs of iodine deficiency, but they may exhibit abnormal oestrous cycles and produce weak foals subject to high mortality and with enlarged thyroid glands. A deficiency with these signs is shown by grazing animals in certain inland areas deficient in iodine, or where there has been the persistent consumption of plant material containing goitrogens (see Ch. 5).

Diets supplemented with between 0.1 and 0.2 mg of I per kg should adequately meet the requirements of horses. Amounts of 4–5 mg/kg of diet, providing 48 mg or more per day for pregnant mares, have caused toxicity. The excessive feeding of seaweed is often implicated, manifested in the foal by enlarged thyroid glands, leg weakness and high mortality within the first 24 hours after birth.

Selenium

Selenium (Se) has gained prominence during recent years since the realization that it forms an integral part of the glutathione peroxidase (GSH-P_x) molecule. This enzyme catalyses peroxide detoxification in bodily tissues during which reduced glutathione (GSH) is oxidized; it is closely involved with the activity of α-tocopherol (vitamin E), which protects polyunsaturated fatty acids from peroxidation. The requirement for α-tocopherol and Se is increased in the presence of high levels of dietary polyunsaturated fatty acids – cod liver oil, linseed and corn oils, and, in fact, pasture grass.

Se deficiency produces pale, weak muscles in foals and a yellowing of the depot fat; it is known that this form of muscular dystrophy in foals is related to a subnormal level of blood Se and a depressed activity of the enzyme GSH-P_x. Serum Se values may fall to less than 0.3 μmol/litre in foals and reduced amounts among Thoroughbreds in the United Kingdom have been associated with poor racing performance (Blackmore, Willett & Agness 1979; Blackmore, Campbell et al. 1982). Only in extreme deficiencies, not normally seen in adult animals, is there sufficient muscle damage for extensive membrane leakage of enzymes such as aspartate amino transferase (AAT) and creatine kinase (CK or CPK) to be detected. The peroxidation occurring in muscles after exercise has, however, not been shown to be reduced by supplementary Se.

Se deficiency *in vitro* changes neutrophil function and so may affect resistance to infection. In the western United States concentrations of Se in the blood show a negative correlation with the incidence of reproductive diseases in mares (Basler & Holtan 1981). In this particular study, blood concentrations ranged from 1.2 to 3.1 μmol/g and dietary concentrations ranged from 0.045 to 0.451 mg/kg. Amounts of Se in the serum are closely correlated with whole-blood Se, so serum values in the horse seem on present evidence to be a good and reproducible measure of status. An increase in dietary manganese from 38 to 50 mg/kg increases both Se retention and blood Se (Spais, Papasteriadis et al. 1977).

Feed fats contain variable levels of polyunsaturated fatty acids; vegetable oils are generally much richer sources than are hard animal fats. Although linoleic acid is perhaps the most abundant polyunsaturated fatty acid in vegetable oils, some oils, including those present in grasses, are richer in α-linolenic acid, which is particularly sensitive to peroxidation in tissues and which may therefore be more likely to cause problems than linoleic acid.

The alkali treatment of roughage to improve its digestibility has attracted interest recently. It must however be appreciated that such treatment destroys α-tocopherol and β-carotene, and unless appropriate supplementation is given symptoms of myopathy may occur in animals consuming significant quantities of such roughage.

Horses require about 0.15 mg of available Se per kg of feed to meet their dietary requirement for this element; plant Se is more available than that provided as sodium selenite, although barium selenate apparently possesses a higher availability. Se is highly toxic to animals and also to persons handling the salt in highly concentrated forms. The minimum toxic dose of Se through continuous intake is 3–5 mg/kg of feed and acute toxicity is caused in sheep given amounts equal to, or greater than, 0.4 mg of Se per kg bodyweight in a single dose.

Outside the United Kingdom there are areas where soils can contain in excess of 0.5 mg Se per kg and amounts of between 5 and 40 mg/kg of dry matter are found in certain accumulator plants. Concentrations of up to several thousand mg per kg have been detected in species of milk vetch (*Astragalus*). Various species of woody aster (*Xylorhiza*) and goldenweed (*Oonopsis*), which grow in low-rainfall areas, are also indicator plants, containing relatively high levels of Se. Toxicity is thus more common in dry regions, but horses select grasses rather than these toxic weeds where there is adequate grazing as the indicator plants are unpalatable. Where grass is sparse, animals suffer from 'alkali disease' in which excessive Se causes a loss of hair on the mane and tail, lameness, bone lesions, including twisted legs in foals, and sloughing of hooves. There is no simple remedy apart from removal of the animals from the region if destruction of the seleniferous plants is impractical.

Further reading

Suttle N F, Gunn R G, Allen W M, Linklater K A & Wiener G (eds) (1983) *Trace elements in animal production and veterinary practice* (Proc. Symp. British Society of Animal Production and British Veterinary Association). BSAP (Occasional Publ. no. 7): Edinburgh.

Underwood E J (1977) *Trace elements in human and animal nutrition.* Academic Press: New York & London.

CHAPTER 4

Vitamin and water needs of the horse

The drink of all brute creatures being nothing but water, it is therefore the most simple . . . as it is the proper vehicle of all their food, and what dilates the blood and other juices, which without sufficient quantity of liquid, would soon grow thick and viscid.

William Gibson, 1726

Vitamins are nutrients that horses require in very small quantities, although the actual needs for each differ considerably. For example, the dietary requirement for niacin or for α-tocopherol (vitamin E) may be at least 1000 times that for either vitamin D or vitamin B_{12}. However, measurements of vitamin requirements lack precision; there is little direct evidence on the requirements for any of the vitamins in the horse and assertions are largely based on measurements in other domestic animals.

Like other mammals, horses require vitamins for normal bodily functions. These requirements will be met by vitamins naturally present in feed, supplementary sources, and, in the case of vitamin K and the water-soluble B vitamins, additional amounts are supplied from microbial synthesis in the intestinal tract. The tissue requirements are complicated by the synthesis of ascorbic acid from simple sugars in the horse's tissues, the production of vitamin D in the skin as a reaction to ultraviolet light, the tissue synthesis of niacin from the amino acid tryptophan, and the partial substitution of a need for choline by methionine and other sources of methyl groups.

Dietary requirements for specific vitamins are therefore affected by circumstance; for example, where horses are kept indoors or are maintained in very high northern latitudes, or indeed have highly pigmented skins and thick coats of hair, their dietary requirements for vitamin D will be greater. Young foals possess a poorly developed large intestine so that little dependence on it for B vitamin or vitamin K synthesis may be assumed. Foals grow fast and in common with other domestic animals one must assume that as their tissue requirements exceed those of adults, so their dietary needs are far greater. Tissue demands will also be larger for lactating mares than for barren mares, but as the former are likely to be eating more, this tends to lessen the difference per unit of feed. Adult horses are able to draw on much larger reserves of some vitamins to see them through periods of deprivation. For example, a good grazing season on high-quality grass can satisfy the mare's vitamin A requirement through the ensuing winter. In some instances, however, old mares or other horses have a diminished ability to assimilate nutrients, particularly the fat-soluble vitamins, through a decline in digestive efficiency with age and possibly through the debilitating damage of intestinal parasites, and there is some evidence that the fertility of old barren mares benefits from larger than normal doses of vitamin A.

Inferences drawn from other domestic animals cannot be used in estimating the effect of strenuous work on vitamin needs. It has been asserted, with some justification, that the dietary requirements for certain B vitamins involved in energy metabolism are increased for animals in heavy work, both in total and per unit of feed. This conclusion is reinforced by a frequent decline in appetite during extra hard work. The nutritional requirements of work should, however, not be confused with pharmacological responses. For example, thiamin given parenterally in single doses of 1000–2000 mg is said to have a marked sedative effect on nervous racehorses.

Recommended dietary allowances for vitamins are given in Tables 4.1 and 6.7 and a summary of deficiency signs is given in Table 4.2.

Fat-soluble vitamins

Vitamin A (retinol)

Grazing horses derive their vitamin A from the carotenoid pigments present in herbage. The principal one of these is β-carotene and fresh leafy herbage contains the equivalent of 100 000–200 000 international units (iu) vitamin A per kg of dry matter for most domestic animals (1 iu equals 0.3 μg of retinol, vitamin A alcohol). The horse, however, seems to be relatively inefficient in the conversion of β-carotene to vitamin A, and the carotene in good-quality grass or alfalfa hay is estimated to possess only a fortieth of the value, weight for weight, of retinol (vitamin A). Although fresh pasture herbage would normally provide well in excess of the requirement, hay used for feeding horses in the United Kingdom provides meagre amounts of carotene and particularly where it is more than 6 months old should be considered to contribute none unless it is visibly green.

Various signs of vitamin-A deficiency have been recorded as it has several important functions, among them the integrity of epithelial tissue, normal bone development and night vision. One of the earliest signs of deficiency includes excessive lacrimation (tear production); a protracted

Table 4.1   Adequate concentrations of available vitamins* per kg of total diet (assuming 88 per cent dry matter)

	Mature horse		Mares: last 90 days gestation Stallions	Lactating mare	Weaned foal	Yearling
	Maintenance	Intense work				
Vitamin A (iu)	1600	1600	3500	3000	3000	2500
Vitamin D (iu)	500	500	700	600	800	700
Vitamin E (mg)	50	80	60	60	70	60
Thiamin (mg)	3	4	3	4	4	3
Riboflavin (mg)	2.5	3.5	3	3.5	3.5	3
Pyridoxine (mg)	4	6	5	6	6	5
Pantothenic acid (mg)	5	10	5	8	10	5
Biotin (μg)	200	200	200	200	200	200
Folic acid (mg)	0.5	1.5	1.0	1.0	1.5	0.5
Vitamin B_{12} (μg)	0	5	0	0	15	0

* There is no evidence of a dietary requirement for vitamin K, niacin or ascorbic acid in healthy horses.

Table 4.2   Signs of advanced vitamin deficiency in the horse and pony. The status should always be kept well above that leading to these signs to provide positive benefits

Vitamin A	Anorexia, poor growth, night blindness, keratinization of skin and cornea, increased susceptibility to respiratory infections, infertility especially in older mares, lameness
Vitamin D	Reduced bone calcification, stiffness and abnormal gait, back pain, swollen joints, reduction in serum calcium and phosphate
Vitamin E	Pale areas of skeletal muscles and myocardium. Red cell fragility, reduced phagocytic activity
Vitamin K	Extended blood-clotting time (prothrombin time), but this is rarely seen unless it is induced by drugs
Thiamin	Anorexia, incoordination, dilated and hypertrophied heart, low blood thiamin and elevated blood pyruvate
Folic acid	Poor growth, lowered blood folate
Biotin	Deterioration in the quality of the hoof horn expressed as dish-shaped walls that crumble at the lower edges so that shoe nails fail to hold

deficiency may cause impaired endometrial function in the mare. Figure 4.1 indicates that these clinical signs and symptoms of deficiency occur under fairly extreme conditions of deprivation. As many horses are stabled for most of their time, when they consume little or no fresh herbage, the possibility of this deprived state exists. However, few cases of overt vitamin-A deficiency are recognized amongst stabled horses in western countries as most routinely receive supplementary synthetic sources. There is evidence of responses in several animal species to rates of intake above the minimum requirement level (Fig. 4.1) under the stress of certain chronic transmissible diseases. Some forms of infertility, particularly in elderly mares, may respond to vitamin-A therapy and responses amongst Thoroughbreds in training suffering tendon strain and lameness have been noted (Abrams 1979).

Plasma retinol concentration is normally 15 to 20 μg per dl and a marginal dietary deficiency causes little change in this. A daily intake of 50 000 iu vitamin A in a 500-kg horse should be sufficient in all circumstances. When this amount is increased fifteen or more times, toxic reactions should be expected.

Preformed vitamin A and β-carotene, in common with all other fat-soluble vitamins, are unstable, being subject to oxidation, so that natural feed gradually loses its potency. Synthetic forms of vitamin A are stabilized and, when undiluted and stored in reasonable conditions, they retain more than three-quarters of their potency for several years. The grain/protein

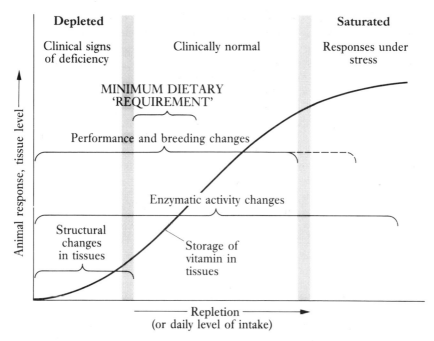

Fig. 4.1 General relationship of animal response to level of vitamin intake.

concentrate portion of the diet supplemented to the extent of 10 000 iu of vitamin A per kg (3.0 mg of retinol per kg) should allow all foreseeable vitamin A demands to be met. In practice, deficiencies may arise from failure to supplement feed or from the provision of badly stored old feed.

Vitamin-A deficiency can also be induced in livestock by other dietary abnormalities. Evidence from several grazing species indicates that signs of deficiency can arise in stock subsisting on poor forage marginal in carotene, which, like much horse hay, contains less than about 7 per cent of crude protein and is deficient in zinc. Authenticated evidence for such interactions in equids is unavailable, but they probably explain the observations of Jeremiah (14:6). Thus, stock under range conditions should also be given adequate supplementary protein and trace elements.

Vitamin D (calciferol)

Vitamin D is intimately involved in the metabolism of calcium and phosphorus, some discussion of which has already been given. However, the widespread occurrence of bony abnormalities in horses and the misunderstandings concerning the interpretation of calcium and phosphate values in blood justifies a short summary of the functioning of vitamin D, which has been elegantly unravelled in recent years.

The vitamin itself is relatively inactive but under the influence of parathormone secreted by the parathyroid gland in the neck it is converted to its active form, a steroid hormone, in the kidney. This hormone has two targets – the small intestine and bone. In the small intestinal tissue it stimulates the absorption of dietary calcium and phosphate partly through the agency of calcium-binding protein. In bone tissue, together with parathormone, it serves to mobilize bone minerals. In the kidney tubules, parathormone stimulates the reabsorption of calcium ions but blocks the reabsorption of phosphate. The vital objective of these two hormones, together with thyrocalcitonin, is to sustain a constant level of blood calcium. It is a fascinating fact that they modulate both calcium and phosphorus nutrition but with different signals. When the diet is deficient in calcium but adequate in phosphorus, then a fall in plasma calcium ions triggers the release of parathormone from the parathyroid gland. This stimulates renal calcium reabsorption and the production of the vitamin D hormone. Intestinal absorption and bone mobilization of both calcium and phosphate are facilitated so that blood calcium and blood pH are returned to normal.

Blood phosphate does not rise because of parathormone's blocking effect on the renal reabsorption of phosphate. Thyrocalcitonin counterbalances and modulates the effect of parathormone by increasing the net deposition of calcium in bone stimulated by a raised serum calcium. By contrast, deficiency of dietary phosphorus depresses blood phosphate, which in turn directly raises ionized calcium in the blood, but also stimulates production of the vitamin D hormone. The combined effect of this is to suppress parathormone production, increasing phosphate retention by the kidney (negating phosphate diuresis), stimulating calcium and phosphate absorption from the small intestine. Blood calcium, however, does not rise excessively as the lack of parathormone increases the urinary loss of calcium (see also Ch. 11).

It is evident that when vitamin D nutrition is adequate for a given age of horse, but not so calcium or phosphorus, blood calcium is held within fairly well-defined limits and phosphate will be more variable. In the absence of vitamin D the efficiency of calcium absorption from the intestinal tract and the mobilization of bone calcium are depressed so that blood-calcium levels will fall. Some mobilization of bone calcium will continue, however, so that osteomalacia, or gradual bone decalcification, occurs in the adult horse and rickets, or reduced calcification of bones, is displayed by the young. There is a loss of appetite, discomfort on standing and lameness, increased risk of bony fractures and a thinning of the cortex of long bones. In young horses the growth-plate (epiphyseal plate) of long bones is irregular, widened and poorly defined and the epiphyses are late in closing.

If the cereal/protein concentrate component of the diet is supplemented with 1000 iu of vitamin D per kg (25 μg of cholecalciferol or ergocalciferol as 1 iu is equivalent to 0.025 μg of cholecalciferol or ergocalciferol), then the daily requirement for vitamin D should be met. Moderately large doses of vitamin D can, to some extent, compensate for low dietary calcium by promoting further calcium absorption, particularly where dietary phosphorus is in excess. However, large doses of vitamin D (in excess of 2000–3000 iu per kg bodyweight daily or in excess of 60 000–100 000 iu per kg diet) will cause similar signs to those of the deficiency and eventual death owing to the effect of vitamin D hormone on bone mineral mobilization. Several plant species not found in the United Kingdom actually synthesize this highly active hormone so that horses grazing areas where they exist will develop rickets and soft-tissue calcification (for example, *Cestrum diurnum*, a member of the potato family, sometimes incorrectly called wild jasmine found in Florida and other subtropical states, including Texas and California, causes this condition).

Vitamin E (α-tocopherol)

Several tocopherols possess vitamin E activity, although by far the most potent is α-tocopherol. As tocopherols have an antioxidant property that protects other substances in food, they are themselves destroyed by oxidation. This is accelerated by poor storage, mould damage and by ensilage of forage or the preservation of cereals in moist conditions. After the crushing of oats or grinding of cereals, the fats are more rapidly oxidized and vitamin E is gradually destroyed unless the material is pelleted. Fresh, green forage and the germ of cereal grains are rich sources of vitamin E, but feeds are frequently supplemented today with the relatively stable acetate ester of α-tocopherol.

The horse stores vitamin E less well than it does vitamin A and the onset of deficiency is accelerated when the diet contains insufficient selenium and is rich in unsaturated fats. Vitamin E functions by protecting unsaturated lipids in tissue from oxidation. In conditions where the intake of selenium and vitamin E is low, which can occur on pasture, mares give birth to foals suffering from myodegeneration. Pale areas in the myocardium and skeletal muscles are apparent on postmortem examination and muscle-cell damage is seen histologically. If the foals survive, damage is said to be irreversible. Other symptoms include steatitis, or yellowing of the body fat, and general fat necrosis with multiple small haemorrhages in fatty tissues. As with other causes of muscle damage, the activity of blood creatine kinase and aspartate amino-transferase rises and probably the fragility of red blood cells increases.

Some benefit is said to accrue from the treatment with selenium–vitamin E injections of 'tying-up', or myositis (see Ch. 11, p. 266) which is sometimes seen after 1 or 2 days' rest in hard-worked horses. The author has not been able to verify this and more substantial evidence is required.

Limited experimental evidence suggests that vitamin E supplements increase amounts of blood glucose and lactate in exercised horses and may help maintain the normal packed cell volume of the blood. A vitamin E deficiency is known to reduce endurance in rats and the vitamin may be particularly important for extended work. Claims have also been made for improvements in the fertility of mares and stallions by vitamin E supplementation of diets containing marginal amounts (Darlington & Chassels 1960). Vitamin E plays a part in the immune response and an increase in the phagocytic activity of foal neutrophils has been induced by supplementation.

Although more vitamin E may be needed when selenium is deficient, both are required nutrients and the amount of vitamin E that should be present in the diet rises in proportion to the level of dietary unsaturated fats (Agricultural Research Council 1981). Recent evidence suggests that the requirement may be as high as 1.5–4.4 mg/kg of bodyweight daily (50–150 mg/kg diet) (Ronéus, Hakkarainen et al. 1985). Typical rations for horses should contain 75 iu of vitamin E per kg (1 iu equals 1 mg DL-α-tocopheryl acetate), although the requirement of very young foals may be slightly greater and that of idle adult horses somewhat less than this.

Vitamin K

Vitamin K_2, along with the B vitamins, is synthesized by functioning gut microorganisms in amounts that should normally meet the horse's requirements. However, this source may be inadequate during the first couple of postnatal weeks, or during extended treatment with sulphonamides.

Vitamin K is essential for blood clotting. Airway haemorrhages in bleeders are an expression of blood vessel fragility and not one of a failure in this mechanism and so may not be controlled by vitamin-K therapy. It is common for racehorses to present evidence of a mild form of haemorrhaging after races.

There is some body storage of vitamin K and natural feeds, particularly leafy material, are fairly rich sources so that no supplementary source is normally necessary.

Water-soluble vitamins

B Vitamins

Normal intestinal synthesis plus the quantities present in typical horse feeds seem to meet the maintenance requirements for riboflavin, nicotinic acid, pantothenic acid, and pyridoxine. Should there be a basic change in diet towards root vegetables and certain byproducts in the future, then an increase in supplementary needs might be needed. The needs of lactating mares and weanling foals should be met if good-quality pasture grass is provided. Additional nutrient demands of exercise are discussed in Chapter 9.

Thiamin

A deficiency of this vitamin (Carroll, Goss & Howell 1949) causes loss of appetite and weight, incoordination of the hind legs, a dilated and hyper-trophied heart, and a decline in blood concentration of thiamin and in the activity of enzymes requiring thiamin as a cofactor. Table 4.3 gives normal values for blood-thiamin levels. This vitamin has been used in the treatment of 'tying-up' but there is no corroboration that a deficiency of it is a cause of the anomaly. Grazing animals on heath land infested with bracken fern (*Pteridium aquilinum*) can exhibit signs of thiamin deficiency if they take to eating bracken. Treatment with thiamin is usually effective.

Table 4.3 Normal blood thiamin concentrations (μg/litre) in standardbred horses

Stallions	2.25 ± 0.16
Geldings	3.03 ± 0.13
Mares and fillies	3.36 ± 0.11
Less than 1 year	2.42 ± 0.26
1–4 years	2.81 ± 0.10
5–10 years	3.53 ± 0.14
10–20 years	3.35 ± 0.17

(From: Loew & Bettany 1973)

About 25 per cent of the free thiamin synthesized in the caecum is absorbed into the blood and a total dietary level of 3 mg/kg seems to meet the requirement. Whether or not the requirement per kg of feed rises during periods of hard work has yet to be demonstrated.

Vitamin B$_{12}$ (cyanocobalamin)

The cyanocobalamin molecule contains the element cobalt. Cattle and sheep grazing areas deficient in this element develop vitamin-B$_{12}$ deficiency as the rumen microorganisms are then unable to manufacture the vitamin. Cobalt therapy rectifies the situation. Horses seem to be more resistant to cobalt deficiency but undoubtedly they require cobalt at a minimum level of about 0.1 mg/kg of diet for adequate intestinal synthesis. Synthesis in foals may be inadequate and, in fact, supplemental vitamin B$_{12}$ has been shown to increase the blood concentration of the vitamin. As it is required for cell replication a deficiency may cause anaemia and a reduction in the number of red blood cells. Although an overt deficiency has not been produced in adult horses, it has been suggested that a response, including a stimulation to appetite, can be obtained in some anaemic animals. Adult horses in training on high-grain rations may need dietary supplementation because a decline in appetite shown by such horses may reflect a build up of blood propionate. This volatile fatty acid is produced proportionately and absolutely in much greater quantities when diets of this type are consumed, and its metabolism to succinate requires an enzyme that depends on cyanocobalamin as a cofactor (Agricultural Research Council 1980). The author has observed by radioimmunoassay that horses in training and those at stud have different blood levels of this vitamin and of folic acid.

Folic acid

This is closely associated with vitamin B_{12} in metabolism and in some domesticated animals a deficiency causes a form of anaemia. Folic acid supplementation of stabled Thoroughbreds receiving an inadequate diet produces an increase in serum folate concentration from about 4 to 9 μg/litre. Recent Australian work on folic acid in the horse confirms the author's observations suggesting that there is an increased utilization of folic acid by horses in hard work. Green forage legumes are rich sources of the vitamin, but its availability in some sources is low.

Biotin

The biotin contained in wheat, barley and milo (sorghum) grains and in rice bran is almost completely unavailable for utilization. That contained in oats is only slightly more digestible. However, all the biotin in maize, yeast and soyabean is accessible, together with most of that in grass and clover foliage.

Horses with hoof horn that tends to crumble at the lower edges of the wall respond to biotin supplementation over a period of 9–12 months in daily amounts of 5–10 mg (ponies), 15 mg (riding horses) and 30 mg (heavy horses). The strength and conformation of the hoof should also improve. Subsequent supplementation of the diet with 1–3 mg daily should result in maintenance of the improvement (Pl. 4.1) (Comben, Clark & Sutherland 1984).

It is also essential that the hooves are properly shaped and trimmed. Long, unpared hooves exert excessive pressure on the heels and this restricts blood flow and hence impedes adequate nutrition of the foot leading to poor quality and crumbly unsatisfactory growth of walls, sole and frog.

Plate 4.1   Off fore-hooves of (a) 8-year-old Irish chestnut gelding, before (i) and after (ii) receiving 15 mg synthetic biotin per day orally for 13 months and (b) a 5-year-old Thoroughbred gelding hack, before (i) and after (ii) receiving 15 mg synthetic biotin per day orally for 5 months. (Photographs 4.1(a) and (b) by kind permission of Norman Comben, MRCVS; Comben, Clark & Sutherland 1984).

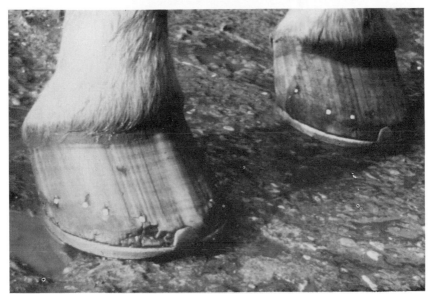

(a)(i)

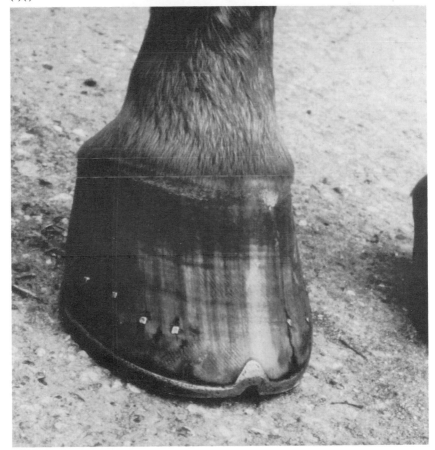

(a)(ii)

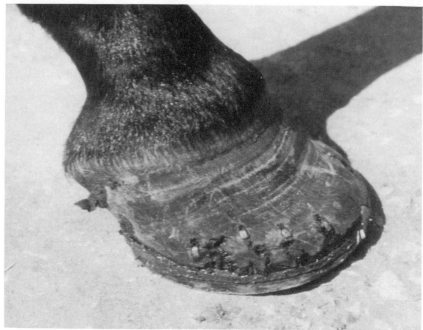

(b)(i)

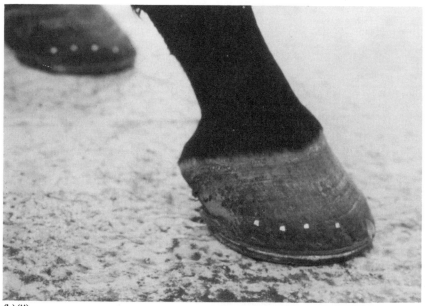

(b)(ii)

Pangamic acid (vitamin B₁₅)

A substance assigned this name has been isolated from apricot seeds. It has been used by human athletes and for horses supposedly to increase the supply of blood oxygen. The evidence for this is dubious and, moreover, the US Food and Drug Administration holds that the substance does not exist as there is no exact chemical formula.

Ascorbic acid

The horse, like other mammals, has a tissue requirement for ascorbic acid. This need is normally met by tissue synthesis from glucose. However, decreased blood and serum concentrations of ascorbic acid are apparently associated in the horse with postoperative and post-traumatic wound infections, epistaxis, strangles, acute rhinopneumonia and 'performing insufficiency' in trotters, perhaps pointing to the value of an exogenous source. Recent studies in Berlin have confirmed that gastrointestinal absorption of ascorbic acid in the horse is slight and comparisons of intravenous, subcutaneous, intramuscular and oral routes of administering 5 or 10 g of ascorbic acid demonstrate that intravenous injection is the only satisfactory route should supplementation be indicated (Löscher, Jaeschke & Keller 1984).

Water

Water constitutes some 65–75 per cent of the bodyweight of an adult horse and 75–80 per cent of a foal's. Water is vital to the life of the animal. The horse also needs to take in water with its food to act as a fluid medium for digestion and for propulsion of the digesta through the gastrointestinal tract, for the useful products milk and growth, and to make good losses through the lungs, skin and in the faeces and urine. In healthy adult horses undertaking light work one estimate showed that the losses of water were distributed such that 18 per cent occurred in the urine, 51 per cent occurred in the faeces and the remaining 31 per cent represented insensible losses (Tasker 1967). Restricted water intake will depress appetite and reduce feed intake.

Equids differ in their ability to conserve body water and to withstand dehydration. Asses from the dry tropics can thwart extreme dehydration because they can conserve water more efficiently than horses do. A rise in environmental temperature from 15 to 20°C increases the water requirement of horses by 15–20 per cent. Work, depending on its severity, will raise requirements by 20–300 per cent above the needs for maintenance through increased losses from the lungs and skin. For obvious reasons, peak lactation can lead to requirements twice those of maintenance (Table 7.3). A high-yielding 500-kg mare may produce 12 kg of water daily in her milk. However, a foal has greater requirements in proportion to its size because it is less able to concentrate urine. A frequent cause of death in neonatal foals is rapid dehydration through persistent diarrhoea, which requires treatment with a physiological salt solution (see Ch. 11).

The horse obtains water for its metabolic needs from three sources – the consumption of fresh water, the water content of natural herbage and metabolic water. Fresh, young, growing herbage may contain 75–80 per cent water so that under many circumstances additional fresh water may not be required, but a source should always be provided. In arid conditions herbage is very different, and horses will seek and consume poisonous shrubs and succulent plants unless water and feed are provided. Metabolic water is that produced during the degradation of carbohydrates, proteins and fats in metabolism and this may amount to the equivalent of 350–400 g/kg of food ingested, depending on its digestibility. Nevertheless, in circumstances of choice, the water intake of horses is highly correlated with intake of dry matter and amounts to between 2–4 litres/kg of dry matter in stabled horses worked moderately.

The horse disposes of excess salts of sodium and potassium and the breakdown products of nitrogen metabolism in its urine. Whereas calcium salts may, in part, be excreted in a solid form there is a limit to which the horse can concentrate the highly soluble salts of sodium and potassium and urea. Thus, where diets are rich in salt, or in protein, more dietary water will be required. Based on evidence from other species, an increase in total dietary salt from 7.5 g to 30 g per kg would be expected to increase the ratio of free-choice water to dry matter in the diet from 2:1 to 3.5:1, other things being equal.

For the maintenance of adult horses in an equable environment, the total water requirement is probably less than 2 litres kg of dry matter intake (about 5 litres per 100 kg of bodyweight). However, strenous effort in hot climates increases the need to as much as 5 to 6 litres per kg of dry matter intake (12–15 litres per 100 kg bodyweight). In this predicament there is also an inevitable loss of relatively large amounts of sodium and potassium chloride in sweat. Excessive dehydration can be fatal. Certain breeds of horse and species of *Equus* (*E. asinus*) (Maloiy 1970) can sustain extensive water loss without apparent discomfort, but horses of temperate breeds may succumb to water losses that amount to 12–15 per cent of their bodyweight (Hinton 1978; Brobst & Bayly 1982). Repletion should be accompanied by balanced electrolytes, although electrolytes may often have to be given first in order to induce drinking when isotonic, or hypotonic dehydration has occurred. Where the horse is fit it should be walked or allowed to graze so that it cools down gradually over an hour before being given substantial amounts of water. Excessive consumption of cold water by hot horses can precipitate colic or founder. During very cold weather, warm water of a temperature between 7 and 18°C should be provided and will be taken more readily than will very cold water. Decreased consumption of water may contribute to the incidence of impaction colic and of depressed performance in racehorses.

The packed cell volume of blood is not a guide to dehydration and water deprivation, because of shrinkage of red cells and changes in the release of red cells from the spleen. Total plasma protein, however, may increase by 10–12 g/litre (say, from 62 to 73 g/litre) with a fluid loss causing a 12–15 per cent decrease in bodyweight. Changes in plasma and urine electrolytes and urea depend on several associated factors. In one study, fluid losses of this extent resulting from dehydration of Thoroughbred geldings increased the concentration of serum and urine urea by 68 and 130 per cent, respectively. The specific gravity of urine reached at least 1.042 and urine osmolality increased 30 per cent to 1310 mosmol/kg when the urine osmolality/serum osmolality ratio increased to 4.14 (Brobst & Bayly 1982).

Where it is feasible, water should be provided from the mains. If mains water is unavailable, then well water, or watercourses, must be free from pollution by sewage or fertilizer seepage. Ideally a new source should be first assessed by a competent analyst.

CHAPTER 5

Ingredients of horse feeds

*Some parts of the kingdom produce no grain so much as oats which probably may
be the reason why they have come to be used as our chief provender.*

William Gibson, 1726

Some of the main chemical characteristics of the ingredients of horse feeds
are given in Appendix C.

Roughage

Hay and physically processed hay

Loose hay

Grasses and forage legumes are cut for hay. Most common species of grass
are suitable, but probably the more popular and productive ones include
ryegrasses (*Lolium*), fescues (*Festuca*), timothy (*Phleum pratense*) and cocksfoot (*Dactylis glomerata*). Many species found in permanent pastures, for
example meadow grasses (*Poa*), bromes (*Bromus*), bent grass (*Agrostis*) and
foxtails (*Alopecurus*), are also quite satisfactory. Among legumes, red, white,
alsike and crimson clovers and trefoils (*Trifolium*), lucerne or alfalfa
(*Medicago*) and sometimes sainfoin (*Onobrychis*) are used. Although the
crude-fibre content of crimson clover (*Trifolium incarnatum*) hay may be
similar to that of other clovers, the fibre tends to be less easily digested
by the horse (forage legume fibre generally is more lignified than that of
grass). For serving as horse hay or haylage crops two reliable seed mixtures
are:

1. 3 perennial ryegrass strains – Melle 5 kg, tetraploid Meltra 15 kg,
hybrid tetraploid Augusta 13 kg per ha, or as a two year crop,

2. tetraploid broad red clover, Hungaropoly 7 kg and tetraploid Italian
ryegrasses: Wilo 15 kg, Wisper 10 kg per ha.

The leaves of forage legumes and grasses are much richer in nutrients than are the stems; they contain about two-thirds of the energy, three-quarters of the protein and most of the other nutrients found in the aerial parts. The leaves of legumes tend to shatter more readily than grass leaves do so that care is necessary at hay making to conserve the nutritional quality of legume hay. Even so, at the same stage of maturity, legume hay contains more digestible energy, calcium, protein, β-carotene and some of the B vitamins, including folic acid, than does grass hay. Horses consuming hay composed predominantly of forage legume tend to produce more urine with a strong ammonia smell and containing deposits of calcium salts. These events are normal physiological responses in healthy animals.

As long as hay is composed of safe, non-toxic nutritious plants, the stage of maturity of the crop at the time of cutting and the weather conditions and care to which the haymaking is subject are much more important characteristics than the species of plant present.

As pasture herbage matures the yield of dry matter per hectare increases, the moisture content of the crop decreases and in the United Kingdom the weather becomes warmer. At Jealott's Hill (ICI Ltd) many years ago the average yield of dry matter from early hay crops was only 57 per cent of that cut at a later date. Even when the aftermath was included, the total yield of the early hay amounted to only 71 per cent of that produced by later cutting. Thus, there is a considerable commercial incentive to produce hay composed of grasses at the late flowering stage. Nevertheless, where hay of good nutritional quality is required for horses, mixtures of grass and clover should be cut before the grass is in full flower when the protein content of the crop lies between 9 and 10 per cent of dry matter and the crop contains high concentrations of calcium, phosphorus and other minerals. Hard, mature grass hays, however, frequently contain between 3.5 and 6 per cent of crude protein, lower concentrations of minerals and more crude fibre (Pl. 5.1). Good-quality hay from pure stands of lucerne or sainfoin is difficult to make when natural drying is relied on because moisture loss from the thick juicy stems is relatively slow and mechanical turning and tedding can result in a considerable loss of leaf, which dries sooner and shatters more readily. For the best product, these legumes should be cut before flowering at the bud stage because after first flowering the crude protein content declines at a rate of some 0.5 per cent daily and the digestible energy declines by some 0.75 per cent daily.

Horses should never be given mouldy hay, so the making of satisfactory leafy hays during inclement weather presents a considerable problem in the absence of a facility for artificial drying. Best-quality hay should be leafy and green, but free from mould dust, weeds and pockets of excess weathering. When ley mixtures are grown for hay, the first cut may contain more weeds, the second cut is generally produced from a faster growth and contains more stem, but the third cut may have the highest nutrient content and leaf, giving a small yield per hectare.

(a)

(b)

Plate 5.1 Hay samples of various types. (a) Hard 'seeds' hay cut when the grass has formed seed heads. The material is clean, low in dust but of low nutritional value. (b) Lucerne (alfalfa) hay, which is similarly stemmy and has been sun dried; bleaching destroys its vitamin-A potency but it adds some vitamin-D potency. Poor harvesting has led to the loss of most of the leaf, so depriving the hay of its most digestible component. (c) Good-quality lucerne hay, which may have been barn-

(c)

(d)

dried. This is 'rich' material and care should be exercised to avoid loss of leaf during feeding (note the leaf particles at the base of the sample). Artificial drying deprives the hay of vitamin-D potency. (d) Meadow hay containing a proportion of timothy (*Phleum pratense*). This sample is of reasonable quality and is free from significant moulding.

Pellets and wafers

Straw, chemically processed with sodium hydroxide or ammonia, will be discussed later in this chapter. Hay is also occasionally processed; it may be ground and pelleted or chopped and wafered. During pelleting molasses and a binding agent are normally added to achieve a satisfactory product. Despite the additional costs of processing, pellets possess a number of advantages:

1. The product is easier to weigh and ration.

2. There is less waste during feeding and particularly with leafy legume or grass material the sifting out of small particles of leaf and their loss in bedding is avoided. This occurs regularly when leafy long hay is consumed.

3. Less storage space is required than for long hay.

4. Transport costs are lower.

5. Those horses particularly prone to respiratory allergies (heaves and broken wind) are less subject to dust irritation when given pelleted hay. Coughing is reduced in normal horses and bleeders are less prone to episodes of epistaxis.

6. Older horses with poor teeth are inclined to masticate long hay incompletely so occasionally precipitating colic through impaction. The introduction of pelleted hay should overcome this risk.

7. Hay belly is avoided.

Pellets do however, have some disadvantages, the principal ones being:

1. Incorrectly pelleted material, or good pellets that are allowed to become wet, may be crumbly and soft so that fines are lost, and wet pellets mould within 18–24 hours.

2. It is difficult visually to assess the quality of pelleted hay.

3. Horses may choke on pellets of about half an inch (12 mm) diameter. The problem is said to be more common when pellets are fed from the hand and might be avoided by placing a large spherical rock (too large to be chewed) in the manger, forcing the horse to eat around it. The scale of this problem is probably exaggerated, however, and episodes of apparent choking are normally overcome without intervention.

4. Wood chewing and coprophagy (faeces eating) are more prevalent where pelleted hay is given without any long hay. The provision of 0.25–0.5 kg of long hay per 100 kg body-weight daily, or good-quality straw bedding, is normally sufficient to minimize these problems. Their incidence seems to be less, but is not eliminated, when wafered hay containing hay particles $1\frac{1}{2}$–2 inches (4–5 cm) long is used.

   There is experimental evidence to suggest a relationship between the caecal environment and the incidence of wood chewing in horses given hay or grain. A higher proportion of propionate in the volatile fatty acids of the caecal fluid and a lower pH seem to be linked with a greater inclination to chew wood (Willard, Willard et al. 1977). A diet of all hay induces a higher proportion of acetate in the VFA of the caecal fluid (see Table 1.2).

5. Although the grinding and pelleting of hay does not affect the digestibility of protein, the digestibility of both dry matter and crude fibre is decreased slightly, possibly because of a decrease in time taken to consume a given amount of feed. From a practical point of view the effect on digestibility is more than offset by the reduction in wastage.

Feeding time can be influenced by the conditions and method of processing the hay. Researchers in Hannover (Meyer, Ahlswede & Reinhardt 1975) recorded that horses took 40 minutes to consume either 1 kg of long hay or 1 kg of hard, pressed wafered hay. They took longer to consume chopped or ground hay but less time to eat soft pressed wafers. Hay of poor quality and high fibre content took longer to eat than better quality hay. Highly digestible chopped maize silage was eaten much more rapidly than hay. Horses of between 450 and 500 kg made between 3000 and 3500 chewing movements in consuming 1 kg of long roughage, but only between 800 and 1200 such movements in eating 1 kg of concentrate. However, ponies of between 200 and 280 kg required twice as long to eat hay and a concentrate meal and between three and five times as long to eat whole oats or pellets. They made between 5000 and 8000 chewing movements in consuming 1 kg of concentrate. The ingesta of horses given chopped hay, or ground hay, passed more rapidly through the stomach than did that of those given long hay and the former led to more fluid stomach contents.

Several other reports have clearly shown that the digesta of horses given ground hay passes more rapidly through the gastrointestinal tract (Wolter, Durix & Letourneau 1974), notwithstanding the evidence that horses masticate roughage to particles of less than 1.6 mm long before it is swallowed (Meyer, Ahlswede & Reinhardt 1975). In one experiment, the mean rate of passage of long meadow hay was 37 hours compared with 26 and 31 hours for ground meadow hay, and ground and pelleted meadow hay, respectively. The decrease in fibre digestibility experienced on grinding almost certainly is a function of rate of passage through the gastrointestinal tract. However, by the same token the faster the rate of passage becomes the greater is the capacity of the horse for feed; but an extension of the time for each meal of pelleted roughage may improve digestibility of fibre.

6. Grinding and pelleting can add up to 10 per cent to hay costs.

Where horses are given a choice of loose hay, wafered hay and pelleted hay, more is consumed of the latter two than of the loose hay. Generally speaking, the horse is a reasonable judge of the quality of loose hay and, among grass hays, well-made ryegrass can be one of the best. Horses prefer grain to either chopped or long hay, and if given a chopped hay–grain mixture they are inclined to sort out the grain. Nevertheless, such a mixture frequently affords a useful function of depressing the rate of grain consumption by a greedy feeder. Recent evidence suggests that the consumption of concentrate feed before hay, rather than after, causes a more intense mixing of ingesta and less variation in the concentration of VFA in the lumen of the large intestine (Muuss, Meyer & Schmidt 1982). This should be an advantage and it may be concluded that advances in processing methods and feeding management should eventually yield more control of digestive health in the horse.

For stabled horses, long hay is given on either a clean area of the floor in the corner of the box, in a hay rack or in a net. The last container should be placed sufficiently high to avoid the possibility of a horse entangling its hoof in an empty net. The amount of hay wasted may be greatest where it is placed on the floor, but this procedure leads to less atmospheric dust.

Dried grass nuts

Grass, clover, lucerne (alfalfa) and sainfoin crops are frequently cut when green and leafy, artificially dried, and preferably chopped and pelleted with a moisture content of about 120 g/kg. In the United Kingdom the product must have a crude protein content of at least 130 g per kg, on the assumption the moisture content is 100 g per kg, to be designated 'dried grass'. High protein grass nuts contain approximately 160 g protein per kg. Dehydrated alfalfa manufactured in the USA contains 150 g to 170 g protein per kg (90 g moisture per kg). These products contain little vitamin D_2, but are rich sources of high quality protein, β carotene, vitamin E and minerals, well suited to horse feeding and of relatively balanced composition. However, where the product is rich in legume forage the Ca:P ratio is frequently too wide, and the protein content too high for it to form the entire diet. It should then be supplemented with a cereal product rich in P.

The artificial drying of green forage yields a product more valuable than hay, as the raw material to be dried is less mature, leaves are not shattered and lost, moulding is avoided and dustiness is minimal. The only disadvantages are the absence of long fibre, β carotene and α-tocopherol contents that are variable and influenced particularly by length of storage of the product. Thus as much as half the vitamin A potency (initially it may be equivalent to 30 000–40 000 iu vitamin A/kg for horses) can be lost during the first 7 months of storage where facilities are not ideal (see pages 89 and 90).

Functions of hay and use of other bulky feeds

Fibre and bulk are useful attributes for part of the horse's diet to have. By diluting more readily fermentable material, fibre suppresses a rapid fall in pH in the large intestine and, by stimulating peristaltic contraction, feed with these characteristics probably aids the expulsion of accumulated bubbles of gas. There are many alternatives to hay as sources of fibre and for horses with sound teeth several are useful where reliable hay cannot be obtained. Best-quality silage and haylage free from moulds can be fed to horses, although an initial reluctance to their consumption is often experienced (see Ch. 10). Good-quality acidified grass silage with a high content of dry matter may replace between one-third and all horse feed; but success depends on its composition, freedom from abnormal fermentation and general quality, on the horse and the skill of the feeder. Compensation should be made for its deficiency, compared with grass, in potency of vitamins A and E. Horses suffering from respiratory allergy should benefit most by changing from hay to silage. Very acid silages should be avoided. Silage with low amounts (less than 25%) of dry matter, and baled or bagged material with a higher content of dry matter and with a pH of around 6 may lead to a greater risk of abnormal clostridial fermentation (see Ch. 10), or may occasionally precipitate explosive intestinal fermentation and colic if given in any but small amounts at each meal. The reason for this may be that the rate of intake of highly fermentable dry matter is much greater in this form than it is in the form of long hay.

Good-quality spring barley or wheat straw in small quantities acts as a source of fibre for horses with sound teeth but is deficient in most nutrients. The inclusion of various wasteproducts in complete diets has been examined, particularly in France and the United States. These products include dried citrus pulp, which is quite satisfactory (Ott, Feaster & Lieb 1979), and such unusual materials as sunflower hulls, almond hulls, corrugated-paper boxes and computer paper. Digestibilities for the last two seem to be about 90 and 97 per cent, respectively, but are much lower for the first two because they are heavily lignified. With increasing competition between domestic animals for feeds and an expanding world population, undoubtedly the search will continue for satisfactory and safe means of sustaining a healthy population of domesticated herbivores by the greater use of wasteproducts of human activities.

During the past decade the chemical treatment of poor-quality roughages, particularly of cereal straws, has improved. Treating straw with sodium hydroxide increases its digestibility to the horse (Mundt 1978) and, with dietary adjustments to its sodium content, the product may become an important supplier of dietary fibre. Ammonia-treated straw also shows promise (Slagsvold, Hintz & Schryver 1979), but results in cattle and horses have been mixed. Both methods of processing are, however, only suitable for exploitation by proficient technicians with appropriate equipment and as yet there has been insufficient practical experience of the direct feeding of either type of material for their general recommendation. The potential digestibility of poor-quality roughages is difficult to predict by chemical analysis as the factors that inhibit the complete digestion of plant cell-wall polysaccharides probably include a difference in structural organization as well as in chemical composition of those structures.

Feeding habits and hunger of stabled horses can vary enormously and succulent roughages are sometimes used to stimulate animals with flagging appetites. One French study showed that the intake ad libitum of a dry feed by a group of stabled horses varied from 8.1 to 19.2 kg daily and the time spent eating ranged from 6 hours 40 minutes to 15 hours 50 minutes. The horses ate several large meals and some small diurnal and nocturnal meals. The night meals represented 30 per cent of the total intake (Doreau 1978). Several factors may contribute to the fastidiousness of finicky eaters, such as environmental stress and nervousness, unpalatability and monotony of ration, nutritional deficiencies, poor health and teeth, lack of exercise and peck order (hierarchy) among group-fed horses.

Many succulent vegetables and fruits (for example, sugarbeet roots, carrots, apples, pears, peaches and plums) are satisfactory as treats for horses. Peaches and plums should be stoned, and hard root vegetables should be sliced into strips to avoid choking, and then mixed with compounded nuts or grain. Carrots contain over 100 mg of carotene per kg and care should be exercised in the quantities used (not more than 0.5 kg of fresh material per 100 kg bodyweight daily) if other large supplements of vitamin A are being given. A similar attitude should apply to all other treats as they represent an unbalanced feed and in large quantities (more than 10% of the total dry-matter intake) can do more harm than good. It should also be realized that succulents, including both root vegetables and fruit, contain 80–90 per cent of water and on a dry-matter basis they may therefore be a very expensive source of energy and protein. Only if they are relatively cheap can succulents be justified and bulkiness restricts their role to that of a supplement to normal rations. Some succulent materials are not very palatable, for example citrus pulp which can be sour. Of all the main flavour groups present in feed, the horse is deterred by sour tastes and attracted by sweet flavours.

Compounded nuts

Compounded feeds have been given routinely for many years to agricultural animals and during the past 20 years have been gaining in acceptance for horse feeding. Several ingredients in a ground form are generally incorporated, among them the common cereal grains, oilseed meals, milling, brewing and distilling byproducts, dried grass and lucerne, fishmeal, and mineral, trace-element and vitamin supplements. Their principal role in this form is to provide a balanced source of all nutrients, but they have to be supplemented with loose hay as a source of long fibre, with water and sometimes with common salt. Different formulations are manufactured for horses with differing needs, so that nuts rich in nutrients and with high digestibility can be supplied to young foals, high-energy nuts can be given to horses in hard work, and low-energy nuts can be provided for adult horses engaged only in light work. The advantages of nuts thus include standardized diets for particular purposes, constant quality, extended shelf life, freedom from dust, palatability, and uniform physical characteristics and density, all of which facilitate routine feeding.

Compounded nuts, particularly high-energy, nutrient-dense formulations, should be introduced gradually to give the horse and its microbial flora time to adapt to the new regime. A too-rapid introduction of nuts, or for that matter of oats, sometimes leads to slightly loose droppings during the first 2–3 weeks, 'filled-legs' and even to colic. Complete nuts are also manufactured for feeding to horses in the absence of hay, but generally speaking, these should be used only for a greater control of dust where horses are subject to respiratory allergies. In the absence of loose hay, wood chewing and some other vices, including coprophagy, may be more prevalent. The preparation of feed in nuts form may have the disadvantage that the user is unable to recognize good-quality from poor-quality ingredients. Products from reputable compounders should therefore always be used for feeding horses, but some indication of the chemical nature of the product can be conjectured by reference to declared analyses required by law in the EEC and found on a ticket attached to the bag. The ticket should give the following information:

1. The percentage by weight of crude oil (lipids extractable with light petroleum, 40/60°C boiling point without prior hydrolysis except in the case of milk products).

2. The percentage by weight of crude protein (the nitrogen content multiplied by 6.25).

3. The percentage by weight of crude fibre (principally organic substances remaining insoluble following alkali and acid treatment).

4. The percentage by weight of total ash.

5. The amounts of added synthetic vitamins A, D and E normally given in international units (iu) per kg.

6. The total selenium content of the product if a synthetic selenium compound has been included. The total is normally given as mg/kg product.

7. If an approved antioxidant is included, this fact must be stated. Table 5.1 gives recommended declarations and some chemical values for compounded horse feeds.

A range of sizes of compounded nuts has been found suitable for feeding to growing and adult horses. However, the optimum seems to be a diameter of 6–8 mm and a length of 12 mm. For very young foals being given a milk substitute nut, a diameter of 4–5 mm and a length of 6–7 mm is probably more suitable. South African work (van der Merwe 1975) indicates that acceptability is not affected measurably by hardness of nut, although most horses dislike nuts that crumble too readily, and those that are excessively hard may occasionally be bolted without mastication. This work revealed that smaller nuts are chewed more slowly and require more time for a given amount to be consumed – a decided advantage.

Where horses are in especially hard work, up to 80 per cent by weight of the total ration can be provided in the form of nuts or grain and

Table 5.1 Recommended declarations and chemical values for compounded nuts and coarse mixtures (assuming 88 per cent dry matter)

	Crude oil (%)	Crude fibre (%)	Crude* protein (%)	Total lysine (%)	Digestible energy (MJ/kg)	Total Ash (%)
Foals						
1 month before weaning to 10 months old	4.0–4.5	6.5–7.5	17–18	0.9	13	7–9
11 months to 20 months old	3.0–3.5	8.5	15–16	0.75	11	7–9
Adults						
Strenous work	3.5–4.0	8.5	12–13	0.55	12	8–10
Light to moderate work, barren mares and stallions	3	14–15	10.0–10.5	0.45	9	9–10
Last quarter of pregnancy, lactation and working stallions	3	9–10	13–15	0.65	11	8–10

* Actual protein concentrations are less important than are the total lysine contents. Note lysine and DE are not normally declared.

supplements, with the remaining 20 per cent composed of hay. Regimens of this nature require considerable skill, 4–5 feeds per day and regular exercise every day. A much more typical regimen for stabled horses in strenuous work is a ration of 50–60 per cent by weight of nuts or concentrate and 40–50 per cent of hay. As the amount of work is reduced, so the proportion of nuts can be decreased, or nuts of lower energy can be used. In stables where horses are given their concentrate measured in terms of the number of bowls per day, the differences in bulk density and energy density of feeds should be recognized. For example, a unit volume of barley is about three times the weight of the same unit volume of wheat bran. Furthermore, the common cereals have different amounts of digestible energy per unit weight. The combined effects of energy density and bulkiness imply that, for example, a unit volume of maize contains nearly double the digestible energy of the same volume of oats (Table 5.2 gives appropriate conversion values).

Some horses in strenuous work have poor appetites and are more likely to 'eat up' when given coarse mixture (sweetfeed) than when given nuts in similarly large amounts. Coarse mixtures have thus gained in popularity during the past 10–15 years, although it is evident that poor appetite for

Table 5.2 Weights of common cereal grains and soyabean meal and average digestible energy (DE) values per unit volume

	Weights		DE
	lb/bushel	kg/10 litres	MJ/10 litres
Oats	27–45	3.4–5.6	51.7
Barley	36–55	4.5–6.9	73.0
Wheat	50–62	6.2–7.7	98.0
Milo (sorghum)	51–59	6.4–7.4	89.7
Maize	46–60	5.7–7.5	93.7
Soyabean meal (solvent extracted, 44%)	47–53	5.9–6.6	83.1
Wheatbran	17–21	2.1–2.6	25.4

nuts results in part from similar volumes and therefore larger weights of nuts being offered, when refusals would be expected. Coarse mixes should be complete, to be supplemented only with loose hay, water and sometimes with common salt. They tend to be more expensive to produce than compounded nuts, but have an advantage in being less dense, and normally contain a proportion of cooked, flaked cereals and oil seeds and expeller oil-seed cakes. Their shelf life is less than that attributed to compounded nuts and their bulkiness demands proportionately more storage space. Storage of these mixtures and of all feeds should be in dry, cool, ventilated conditions where there is little variation in temperature, otherwise moulding can occur.

Cereals

Whereas water is probably the most critical nutrient for the horse's immediate survival, fatness and lack of exercise are its worst enemies. The adequate control of energy intake is the most difficult aspect of optimum and reliable feeding. Cereals (Pl. 5.2) are the principal source of energy in the diet of hard-worked horses and therefore a brief discussion of the characteristics of the common cereal grains and their byproducts is appropriate.

Cereal grains contain from 12 to more than 16 MJ of digestible energy (DE) per kg of dry matter compared with about 8.5 MJ per kg in average hay. Cereal grains embody three main tissues: the husk and aleurone layer, the endosperm and the embryo. The endosperm is a rich store of starch required as a source of energy during the early growth of the plant. Of the nitrogen compounds of cereal grains 85–90 per cent are proteins; these are found in each of the tissue regions but are in higher concentrations in the embryo and aleurone layer. Cereal proteins are not as nutritionally useful as oil-seed and animal proteins because they are relatively deficient in lysine and methionine. The quality of protein (as distinct from the

Plate 5.2 Cereal grains. Maize (corn) grains are the largest and may be given whole to horses with sound teeth, but as they can be very hard cracking is often worthwhile. Barley is smaller and relatively hard and the grains should be crimped or rolled lightly. Oat grains are relatively light and bulky, and crimping or rolling is required only for young horses or for older horses with poor teeth. By comparison, sorghum grains are small, as they are 'naked'. Sorghum is grown in hot dry countries and white varieties are quite satisfactory for horses when coarsely ground, cracked, rolled or cooked. The brown varieties contain large quantities of tannins and are unsuitable for horses.

amount) decreases in the order, oats, rice, barley, maize, milo, wheat and millet. Oat protein contains slightly more lysine than does the protein of other cereals. The oil content of cereal grains varies from about 15 to 50 g/kg, with oats containing slightly more than maize, which in turn contains more than barley or wheat. This oil is rich in polyunsaturated fatty acids, of which the principal one is linoleic acid, that generally constitutes about half the fatty acid composition by weight in the oil. Unsaturated oils, such as these, are prone to rancidity subsequent to the grinding of cereals, unless the meal is compressed into pellets or otherwise stabilized. Cereal grains are deficient in calcium, because they contain less than 1.5 g/kg, but they have three to five times as much phosphorus, principally in the form of phytate salts. These salts tend to reduce the availability of calcium, zinc and probably magnesium in the intestinal tract; those of oat are said to have a greater immobilizing effect than do the phytates of other common cereals.

Grain types

Oats

Under traditional systems of feeding, in which a single species of cereal grain is given, grains of oat (*Avena sativa*) are safer to feed than are the other cereals as their low density and high fibre content make them more difficult to overfeed and the grain size is more appropriate for chewing.

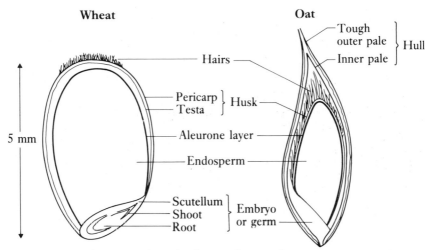

Fig. 5.1 Vertical sections through wheat and oat grains.

They need, therefore, no crimping or rolling for horses aged over 1 year if the teeth are sound. A greater quantity of oats than of the other cereals must be consumed to produce founder or other digestive problems. Nevertheless, they tend to be more expensive per unit of energy than the others as between 23 and 35 per cent of the grain consists of the hull. Figure 5.1 gives a cross-sectional view of oats and in Appendix C their chemical characteristics are listed for comparison with the other common cereals.

Barley

Oats and barley (*Hordeum vulgare*) differ from wheat, maize and grain sorghum in being invested in a hull (botanically known as the inner and outer pales), which the other three cereals have lost during harvesting. All the cereal grains are, however, encased tightly in the thin membrane composed of the less-fibrous fused testa and pericarp (Fig. 5.1).

The hull of barley constitutes 10–14 per cent of the total grain weight and it is relatively smaller and more tightly apposed to a grain, which is larger, compared with the oat hull. Thus, for feeding to horses, barley grain should be crimped or lightly rolled to rupture the case shortly before feeding, but it may be fed as the only cereal after a period of gradual adaptation. This period is necessitated by the higher starch content and bushel weight of barley in comparison to oats. Processes that gelatinize starch grains, such as steam rolling or micronization, are discussed on page 79.

Barley protein is of slightly lower nutritional quality than that of oats being relatively deficient in lysine, and the oil content is quite low, generally being less than 20 g/kg.

Some varieties of naked or hull-less barley, very low in fibre, have been bred. They are comparable to oat groats from which the hull has been removed by processing, again yielding a starchy high-energy, low-fibre product, but their price rarely justifies their use in horse feeding.

Wheat

The grains of wheat (*Triticum aestivum*) are free of hull and relatively small so that they may escape mastication if fed whole. Wheat should therefore be cracked, coarsely ground or steam flaked before use. The two endosperm proteins (known collectively as gluten) are deficient in lysine and can form a pasty impenetrable mass for digestive juices, especially when wheat is finely ground. In the uncooked state, wheat should form less than half the grain fraction of the diet. Moreover, as the starchy endosperm constitutes 85 per cent of the grain, excessive intakes can cause digestive disturbances, particularly if the adaptation period has been short. The bran and germ make up about 13 and 25 per cent of the grain, respectively.

Maize (corn)

Maize (*Zea mays*) is the largest of the cereal grains and is acceptable in any form for feeding to horses. Frequently, however, when the grains are very hard they should be cracked, especially for horses with poor teeth. Maize contains twice the energy per unit volume of oats, but at equal net-energy intakes they will make the horse no more spirited than will oats. They are not a 'heating' feed, as is commonly thought, because the heat produced in fat production is only two-thirds of that produced in horses on oats (see Fig. 6.10). The so-called 'heating' of cereals generally results from their rapid fermentation by intestinal microorganisms (with a fall in caecal pH) and the rapid assimilation of their products of digestation. This causes an abrupt rise in the concentrations of glucose and VFA in the blood, stimulating metabolic rate to a greater extent than occurs after a meal of hay. Precooking of cereals diminishes the fermentative component of this effect and to that extent is fruitful. The grains of maize contain about 650 g of starch per kg, but only 80–100 g of crude protein. The endosperm protein, zein, is deficient in tryptophan and to a lesser extent in lysine, but the protein of the germ in common with that of other cereals is of better nutritional quality.

Yellow and white maize varieties are produced, but the types for stock feeding are predominantly yellow and contain the pigment cryptoxanthin – a precursor of vitamin A. Infrequently some maize imported into the United Kingdom is more than 2 years old when it is a poor source of vitamin A.

Grain sorghums

Sorghums (*Sorghum vulgare subglabrescens*) are the main food grains of Africa and parts of India and China where they are grown on land too dry for maize. They are also a major stock feed in dry areas of the mid-west of the United States. The kernel is naked like that of maize and wheat, but more spherical in shape and smaller than that of wheat. The grain contains more crude protein (deficient in lysine) but less oil than maize and owing to its size should always be rolled, cracked, coarsely ground or steam flaked before feeding to horses. It is a high-energy cereal and, therefore, to avoid digestive disturbances, it should preferably form only a portion of the cereal intake.

Several sorghum varieties are grown, including some for forage use. The grains range in colour from white to deep brown. Only the white or milo varieties (Pl. 5.2) should be used to feed horses as the brown pigmentation is caused by tannic acid in the grain. This produces colic in horses.

Rice

The grain of rice (*Oryza sativa*) is invested in a thick fibrous hull, which is easily removed but which constitutes about 20 per cent of the total weight. The hull is rich in silica and when freed from the grain it is unsuitable on its own for feeding to horses because the sharp edges may cause irritation. Rough rice, that is the grain before the removal of the hull, is more suitable as a horse feed. Rice protein is a reasonable source of lysine.

Millet

The seed of millet (*Setaria* spp., *Panicum miliaceum*), which requires coarse grinding, is small and the hull is not removed during normal harvesting. Its feeding value is somewhat similar to that of oats as it contains 100–120 g of crude protein per kg, 20–25 g of oil and 50–90 g of crude fibre. However, the protein is impoverished of lysine. The grain should be crushed before feeding.

Processing of cereal grains

Specific procedures for processing each species of cereal grain have already been sketched. Where cereals are mechanically rolled the process should be one of kibbling, or bruising, rather than the complete rolling of uncooked material, otherwise the chemical stability of the product is jeopardized, no further increase in digestibility is achieved and greater processing costs are incurred. A similar argument can be made for coarse in contrast to fine grinding. Moreover, finely ground cereal endosperm is floury, unpalatable, dusty and may lead to digestive disturbances. Some other feeds such as bran may occasionally ball up and block the oesophagus if fed dry and therefore are normally dampened or mixed with cut or chopped hay or fed damp with oats. Cereal grains or high-energy nuts should be distributed among as many daily feeds as possible to minimize the risk of colic.

Cereals should be cooked only in the presence of water in order to minimize the risk of heat damage to proteins and oil. Steam pelleting and expansion procedures achieve this objective and, for high-energy materials in particular, have been shown to improve the digestibility of dry matter, organic matter, starch, and the nitrogen-free extract of cereals and nuts without interfering with the digestibility of crude protein. The digestibility of the crude-fibre fraction, however, is either not affected or is only slightly depressed if the food material is ground beforehand. This fraction represents only a relatively small portion of high-energy products. The process of expansion, or popping, relies on the cooking effects of super-heated steam injected into a slurry compressed against a die face by a revolving worm and the subsequent rapid fall in pressure during extrusion. Material is subjected to a temperature of around 120°C for about a minute, but the costs of doing this are rarely justified in the production of horse feed. Processing costs, which constitute more than 10 per cent of the value of the product, are difficult to defend except for young stock and horses under competition rules. But, it is hard to quantify some of the indirect advantages, which, depending on the process, may include reduced storage space, increased stability of product, improved palatability, destruction of natural toxins, insect pests and bacterial pathogens, and the avoidance of high-starch concentrations in the large intestine. The last may be of prime import.

Other cooking procedures include the traditional steam flaking of maize during which the grain is passed through heated rollers, the roasting of oil seeds during the industrial extraction of oil, and the micronization of cereals and vegetable protein seeds. The last procedure, introduced during the early 1970s, represents a somewhat less expensive but safe means of cooking raw materials. A moving belt carries a thin, even layer of cereal grain horizontally beneath a series of ceramic burners that emit infrared irradiation in the 2–6 μm waveband. This results in a rapid internal heating of the grain, a rise in water vapour pressure, during which the starch grains swell, fracture and gelatinize. The product is usually then passed through helically cut rollers and from a cooler to a cyclone. The raw material achieves temperatures ranging from 150 –185°C for between 30 and 70 seconds – for each specific raw material there are optimum values within these ranges. The products are frequently included in coarse mixtures for horses; the process increases digestibility and, for instance, in the case of soya beans, will inactivate antiprotease and other toxic factors.

The alkali treatment of roughages has been briefly described already. Treatment with propionic acid of cereal grains with high moisture content has achieved a certain measure of popularity during harvesting in inclement weather. The acid acts as a mould inhibitor and preservative. Grain treated in this way is only marginally suitable for feeding to horses owing partly to its acidity and more especially to the frequent presence of mouldy patches in the silo. The grain may become infected by the fungus *Fusarium*, which produces the toxin zearalenone known to cause 'poor doing' in all animals and infertility in breeding animals. Furthermore, high-moisture cereals are deficient in α-tocopherol so that supplementation with synthetic forms of vitamin E at a level of about 30 mg/kg of feed is essential.

Cereal byproducts

The industrial use of cereal grains leads to the production of two major types of byproduct: (1) those derived from the milling industries (the seed coats and germ) and (2) those derived from the brewing and distilling industries (principally spent grains, the residues of germinated grains and dried yeast).

Milling byproducts

There are three byproducts of oats: the hull, dust consisting of oat hairs lying between the grain and the hull, and meal seeds composed of hull and the endosperm of small seeds. Oat hull has a crude fibre content of 330–360 g/kg with a digestibility little better than that of oat straw. A combination of oat hulls and dust in the approximate ratio of 4:1 gives oatfeed, which should by legal definition contain no more than 270 g of crude fibre per kg. Each of these byproducts may be fed to horses when appropriately processed and included in balanced feeds in proportions of up to 20 per cent of low-energy diets.

Undoubtedly the major milling byproducts fed to horses in Western countries are those derived from wheat milling. The offals of wheat consist of the germ, bran, coarse middlings and fine middlings, which comprise about 28 per cent of the total weight of the grain and collectively are known as wheatfeed, although in some products a proportion of the germ is marketed separately. The germ contains 220–320 g of crude protein per kg and is a rich source of α-tocopherol and thiamin. This particular byproduct is too expensive for general use but can be of value to sick animals. Bran is derived from the pericarp, testa and aleurone layers surrounding the endosperm, with some of the latter attached. It normally contains between 85 and 110 g of crude fibre per kg and between 140 and 160 g of crude protein. It is sold either as giant, broad or fine bran according to size, or as entire fraction 'straight-run bran'. These grades are similar in chemical composition, although the larger flaked varieties may contain slightly more water. Bran is typically expensive for the nutrients it provides, but it can form, as a mash, a palatable vehicle for oral administration of drugs and it has the capacity to absorb much more than its weight of water. Thus it has a laxative action on the intestinal tract. Bran in particular, but also other wheat milling byproducts, are rich sources of organic phosphorus, as bran contains approximately 10 g/kg, or slightly more, of which 90 per cent is in the form of phytate salts. As bran is deficient in calcium and as phytate depresses the utilization of dietary calcium and zinc, the use of large quantities accelerates the onset of bony abnormalities in young and adult horses.

Coarse middlings are similar to bran, but contain somewhat more endosperm and therefore chemically contain only 60–85 g/kg of crude fibre and about the same amount of crude protein. Fine middlings contain even more endosperm than the coarse, and consequently only 25–60 g of crude fibre.

When adjustments are made for the imbalance in minerals, wheat byproducts are safe feeds as supplements to horse and pony rations, besides being relatively rich sources of some of the water-soluble vitamins.

Maize (corn) byproducts

The byproducts resulting from the industrial production of glucose and starch derived from maize include the protein gluten sometimes sold as gluten meal, a small amount of bran and the germ which, after oil extraction, is occasionally sold as germ meal. These products are similar to the analogous byproducts of wheat and frequently all three are combined for sale as maize gluten feed. This material contains about 180–230 g of protein and 70–80 g of crude fibre per kg. Although the protein is of poor nutritional quality, the feed is quite suitable as a supplement in horse rations as it is a good source of some of the water-soluble vitamins.

Rice byproducts

Large quantities of rice meal or rice bran are produced globally. This byproduct consists of the pericarp, testa, aleurone layer, germ and some of the endosperm of the rice grain, and it has a crude composition of about 110–130 g per kg crude protein and 100–150 g of lipid material composed of a very unsaturated fat. This fat becomes rancid very rapidly and is therefore removed, leaving a product of much better keeping quality. Significant quantities of extracted rice bran are therefore available annually in the United Kingdom. It is a good supplementary feed for horses when used as a component of a mixed ration. This byproduct has a composition of about 15 g of oil, 130 g of crude protein and 120 g of crude fibre per kg. However, frequently as much as 60 g/kg of silica are present and the ash content is normally in the region of 120 g/kg. Extracted rice bran is also a very rich source of organic phosphorus containing 14 g/kg. This again is largely in the form of phytates and care must be taken to ensure that rations in which it is used are appropriately balanced for minerals.

Brewing and distilling byproducts

Three major byproducts are derived from brewing: malt culms, brewer's grains and brewer's yeast. When barley is sprouted for the purpose of hydrolysing the starch, the resulting malt sprouts, which include the embryonic radical (root) and plumule (stem) remain after the malting process. These are removed and dried to form the malt culms. The remainder of the material is mashed to remove sugars leaving the grains, which may be disposed of as a wet byproduct or dried and sold as dried brewer's grains. Malt culms contain about 240–250 g of crude protein of reasonable quality, 12–30 g of oil and 140 g of crude fibre per kg. Although this byproduct is rather dusty, it is a valuable asset in mixed horse feeds in that apart from containing useful protein and some micronutrients, it, like bran, readily absorbs water and stimulates peristalsis. By the same token it should not be fed in large quantities on its own in a dry form. The residual grains after removal of the wort may include maize and rice residues in addition to those of barley, the main constituent. The dried byproduct contains 180–250 g of crude protein and 140–170 g of crude fibre per kg and therefore forms a useful adjunct to mixed horse feeds. The most coveted and expensive byproduct of brewing is, of course, yeast, which in dry form contains 420 g of high-quality protein per kg and is a rich source of a range of water-soluble vitamins and of phosphorus. This yeast is frequently fed to horses in poor condition at the rate of 30–50 g daily, but is too expensive for regular feeding.

The principal residues from the whisky distilling industry are the grains and the solubles. The grains in the malt-whisky industry consist solely of barley residues, whereas grain whisky residues may in addition include those of maize, wheat and oats. After distillation of the alcohol, the spent liquor is spray-dried to yield a light-brown powder known as distiller's solubles. A proportion of the grains is sold wet, but significant quantities are dried and have a composition when derived from barley malt of 260–280 g of crude protein per kg, 100–130 g of crude fibre and 100–120 g of lipid material. The byproduct from grain whisky is generally more digestible, but both are low in sodium, potassium and calcium. Frequently the dried solubles are added back to the dried grains and marketed as dried distiller's grains with solubles, known also as dark grains. This byproduct in small quantities is a valuable supplement to mixed horse feeds and American evidence suggests that dried maize grains with solubles stimulate the digestion of cellulose by microorganisms in the horse caecum.

Molasses

The crystallization and separation of sucrose from the water extracts of sugarbeet (*Beta vulgaris saccharifera*) and sugarcane (*Saccharum officinarum*) leave a thick black liquid termed molasses, which contains about 750 g of dry matter per kg, of which about 500 g consists of sugars. The crude protein in molasses is almost entirely non-protein nitrogen and of minimal value in feeding horses. In beet molasses, a proportion of this is in the form of the amine betaine, which is responsible for the somewhat unpleasant fishy aroma associated with that form of molasses, but cane molasses has a very pleasant smell. The sweet taste of both forms is attractive to horses when used in mixed feeds up to a level of 100 g/kg feed, and in these proportions molasses can act as a relatively effective binding agent in the manufacture of nuts. Cane molasses contains between 5 and 11 g of calcium per kg and the potassium contents range from 20 to 40 g/kg in cane molasses and from 55 to 65 g/kg in beet molasses. Cane molasses is reasonably rich in pantothenic acid and both contain around 16 mg of niacin per kg.

Dried lucerne (alfalfa)

Although not a byproduct but a useful forage, dried lucerne (*Medicago sativa*) is mentioned here as it has been attributed with several indirect effects, possibly caused by unidentified factors. Dehydrated alfalfa meal stimulates cellulose digestion by equine microorganisms and enhances gross-energy digestion of feed. An Eastern European report suggests that lucerne hay may have a protective value in the development of glandular inflammation and may encourage white cell (lymphocyte) and red cell (erythrocyte) production in foals (Romić 1978).

Protein concentrates

Vegetable proteins

The richest sources of vegetable protein fed to horses and ponies are oil-seed residues but other sources include peas, beans, yeast and in the future possibly new sources of microbial protein and, finally, high-quality dried forages, particularly lucerne meal (alfalfa). Soya beans, linseed, cottonseed and, to some extent, sunflower seed after processing are widely used. Groundnuts cannot be recommended because of their frequent contamination with a toxin of the mould *Aspergillus flavus* to which the horse is relatively sensitive.

Two alternative procedures are adopted for the extraction of oil from oil seeds, both of which may be preceded by the removal of a thick coat by a process known as decortication as practised for cottonseed and sunflower. Undecorticated meals contain very much more fibre. Where oil is removed by pressure, this is preceded by cooking at up to 104°C for 15–20 minutes, after which the temperature is raised briefly to 110–115°C. Then pressure is achieved by passing the seeds through a horizontal perforated cylinder, in which a screw revolves and the oil is partially pressed out, leaving a residue containing perhaps 35 g/kg of fat. Expeller cakes, therefore, have the advantage of incorporating more fat than meals derived from the more-efficient chemical extraction process. However, the temperatures achieved during compression can damage the protein, which generally has a lower biological value than that resulting from solvent extraction. In this latter process only material with less than 350 g of oil per kg is suitable so that the feeds are first subjected to a modified screw press, less extreme in its effects than in the expeller process. The seeds are then flaked and the solvent, usually hexane, is allowed to percolate through, effectively removing the oil. The solvent residues are evaporated by heating or toasting, which also benefits some meals by destroying natural toxins.

Oil-seed meals are much richer sources of protein than are cereals and their balance of amino acids is superior. Nevertheless, linseed meal is a poorer source of lysine than is soya, considered the best quality of these proteins. Sunflower seeds are rich in the sulphur amino acids, cystine and methionine, although it is rare for horse diets to be limiting in respect of these amino acids. Oil-seed meals are also relatively reliable sources of some of the B vitamins and of phosphorus, but contain little calcium.

Soyabean meal

Raw soya beans (*Glycine max*) contain allergenic, goitrogenic and anti-coagulant factors in addition to protease inhibitors. The correct toasting and cooking of the beans, as in micronization and well-regulated oil-extraction procedures, destroy these factors without detracting from the protein quality. Reliably cooked products, therefore, may be used as the sole source of a supplementary protein in horse feeding.

Standard hexane-extracted soyabean meal contains 440 g of crude protein per kg. Dehulled meals of uniformly high quality containing 480–490 g of crude protein are also of general commercial availability. Both these meals contain less than 10 g of oil per kg. Full-fat soya flour and cooked soya flakes are much more costly, but the latter is widely used in coarse mixes and both contain 180–190 g of fat and 360–400 g of crude protein per kg. The precise composition varies with the crude-fibre levels, which range from 15 to 55 g/kg.

Linseeds and linseed meal

Linseeds (*Linum usitatissimum*) are unique in so far as they contain a rela-tively indigestible mucilage at concentrations of between 30 and 100 g/kg. This can absorb large amounts of water, producing a thick soup during the traditional cooking of linseed and its lubricating action regulates faecal excretion and sometimes overcomes constipation without causing looseness. The cooking of linseed also destroys the enzyme linase, which, after soaking, would otherwise release hydrocyanic acid from a glycoside present in the seeds, so poisoning the horse. This action implies that the seeds should be added to boiling water rather than to cold water and then boiled, otherwise some enzymatic activity may be initiated. However, HCN is volatile and a proportion of any already present will be driven off by subsequent boiling. Linseeds should not be fed dry because of their water-absorbing propensity, although the contained linase would be rapidly inactivated by the stomach's acid secretions.

The low-temperature removal of oil during the production of linseed meal implies that the product may be toxic if fed as a gruel. In comparison, oil removal by the expeller process normally results in a safe cake, whether it be fed wet or dry. United Kingdom laws state that linseed cake or meal must contain less than 350 mg of hydrocyanic acid per kg, although this takes no account of any linase activity that may be present.

Linseed products are rich in phosphorus and are relatively good sources of thiamin, riboflavin, niacin and pantothenic acid.

Cottonseed meal

Meal from cottonseeds (*Gossypium* spp.) tends to be dry and dusty and has a somewhat costive (constipating) action. Although it contains relatively good-quality protein, it suffers from the presence of toxic yellow pigments called gossypols at concentrations in raw cottonseeds of between 0.3 and 20.0 g/kg of dry matter. Heating during processing partly inactivates the toxin in the raw material, but at excessive temperatures gossypol binds lysine, depressing protein quality. The binding process partially inactivates the toxin as the free form is considered the major risk. However, even the bound form is reported to reduce intestinal iron absorption, partly counteracted by further iron supplementation. Owing to differences in temperatures achieved during manufacture, solvent-extracted meals may contain up to 5000 mg of free gossypol per kg, whereas screw-pressed materials may contain only a tenth of this amount. High-quality cottonseed meal is palatable to horses and can be usefully included in mixed feeds, but varieties with low amounts of gossypol are unavailable in the United Kingdom and mixed feeds containing more than 60 mg/kg of free gossypol are unsatisfactory for horses.

Sunflower seed meal

The imported meals available in the United Kingdom are very variable in composition and generally contain considerable amounts of crude fibre. They are safe feeds for horses but may be unpalatable in large quantities. Undecorticated sunflower (*Helianthus annuus*) meal on average has 320 g of crude fibre per kg dry matter and 200 g of crude protein. The decorticated product contains only 130 g of crude fibre per kg, but up to 400–450 g of crude protein. Some meals with between 250 and 350 g of crude protein per kg are available. Although the protein has a relatively low concentration of lysine, the methionine level is double that of soya protein.

Beans and peas

Within the bean family (Leguminosae or Fabaceae) species belonging to two of the tribes, the Vicieae and Phaseoleae, grow throughout the world and many are important food crops. Winter and spring varieties of field (horse) beans grown in the United Kingdom are members of the species *Vicia faba*, all safe for feeding to horses, especially after cooking. There are, however, importations of bean belonging to the genus *Phaseolus*, especially the lima bean (*P. lunatus*), and the kidney bean, also known as the haricot, navy, pinto and yelloweye bean (*P. vulgaris*). These beans must all be cooked (wet heat) before feeding because they contain several toxic factors, including antiproteases and lectins, which will cause diarrhoea, and many beans of the genus *Phaseolus* also contain a cyanogenetic glycoside identical with that in linseed. Other beans available in parts of the world include the hyacinth bean or lablab (*Dolichos lablab*), horse gram (*D. biflorus*), green and black grams (*Phaseolus aureus, P. mungo*) and chick pea (*Cicer arietinum*), which are widely used in Asia, and lentils (*Lens* spp.) for which India is the chief producer. One lentil is said to induce staleness in racehorses and hunters if given in excess. The Indian or grass pea (*Lathyrus sativus*), at one time imported as an animal feed, and some other members of the same genus cause lathyrism (see p. 92). The horse is particularly and characteristically affected. If beans and peas of unknown origin are used to feed horses, extended cooking will provide a measure of safety from many of the toxins. The field pea (*Pisum aruense*) is safe and palatable.

Although field (horse) beans are safe and palatable without cooking, kidney beans are normally refused by horses unless cooked; if force-fed they will cause colic. The winter field (horse) bean contains on average 230 g of crude protein per kg and 78 g of crude fibre, whereas the spring bean contains 270 g of crude protein and 68 g of crude fibre. Amounts of fat in both varieties are low – about 13 g/kg; like most other seeds they are rich in phosphorus but poor in calcium and manganese. Field-bean protein is of high quality as it is a valuable source of lysine. The bean is normally cracked, kibbled or coarsely ground, but may be given whole to adult horses with sound teeth.

Rapeseed meal

There is a large world production of rapeseed; varieties from two species are grown – *Brassica napus* and *B. campestris*. Western European production has increased rapidly owing to encouragement by the EEC. Unfortunately the product contains quantities of glucosinolates, which occur in many plants and seeds of the family Cruciferae (see p. 94). During digestion these are hydrolysed by the enzyme myrosinase, present in unheated rapeseed and also in gut microorganisms, releasing the two goitrogens – isothiocyanates and goitrin (oxazolidinethione). Although the protein quality of rapeseed meal is good and although heat treatment decreases the hazard by destroying the myrosinase, the intestinal enzyme can still release quantities of the thyroactive substances and so only small amounts of meal of unknown origin are suitable for feeding to horses and many other animals. This predicament led Canadian geneticists to select varieties, predominantly from the species *B. campestris*, known as 'double-low' (low erucic acid, low glucosinolate) varieties and sold as canola meal, which contains less than 3 g/kg of glucosinolates and which may be fed to horses. These varieties may completely replace soya. Unfortunately most European varieties have been based on *B. napus*, that contains on average more goitrogens than the other species. Recently, low glucosinolate varieties have become available in Europe, in particular a variety called 'Duo', and this should not be a cause of goitre in horses. Now under EEC regulations 'double low' seed is eligible for a premium if sold into intervention so that the advantages of a segregated market should be realized.

Several rapeseed varieties contain tannins (averaging 30 g/kg seed), limiting their usefulness and possibly contributing to a slightly lower protein value for the meal in comparison to that of soya, according to several reports. A further discussion of these toxins is given later in this chapter.

Animal proteins

There are only two high-quality animal protein sources suitable as horse feeds – white fishmeal and milk-protein products. They are reserved almost entirely for foals, either in creep supplements or as milk replacers. Small amounts are occasionally given to adult horses in poor condition, but large amounts of dried skimmed milk may cause diarrhoea owing to the presence of lactose (see Ch. 1).

Fishmeals

Two types of fishmeal are recognized under British law, of which the first is a product from the drying and grinding of fish or fish waste of a variety of species. The second, marketed as white fishmeal, is a product containing not more than 4 per cent of salt and obtained by drying and grinding white fish or the waste of white fish to which no other matter has been added. This is a high-quality protein source because it contains abundant lysine, suitable for, but not essential in, the diet of young foals. It is rich in minerals (about 80 g Ca and 35 g P per kg), trace elements (especially manganese, iron and iodine) and several water-soluble vitamins, including vitamin B_{12}. This vitamin is found naturally only in animal products and bacteria. The dietary requirement of the young weaned foal can be met by the inclusion of fishmeal or synthetic sources in its diet. The suckling foal should, however, receive ample in the dam's milk.

During processing the fish waste is dried by one of two procedures. The first and more desirable one is steam drying, either under reduced pressure or with no vacuum applied. The other procedure is based on flame drying, when the temperatures achieved may decrease the digestibility of the protein and decrease the content of available lysine.

About 5 or 10 per cent of white fishmeal in a creep feed or milk replacer is quite satisfactory for foals.

Fishmeals of unknown origin may be contaminated with pathogenic organisms, in particular with *Salmonella* species or other enteric organisms that cause diarrhoea. Good-quality white fishmeal, however, should be a safe feed. Meat meals, meat and bone meals and unsterilized bone flour should on no account be used for feeding to horses because many samples are contaminated with *Salmonella* or are shipped in contaminated bags. The problem of ridding young stock of infection once they contract these diseases is considerable.

Cow's milk

Where liquid milk is used for feeding to orphan foals, it should be diluted with 15–20 per cent of clean water and given in small amounts in as many meals as is practically convenient. Liquid cows milk, on average, contains 125 g of dry matter per kg, 37 g of fat, 33 g of protein and 47 g of lactose. It contains little magnesium and is deficient in iron, a source of which should be provided for the young foal. Whole milk is rich in vitamin A and provides useful quantities of vitamin B_{12}, thiamin and riboflavin. Milk proteins contain abundant lysine.

Dried skimmed milk is widely available and sold commercially as a component of milk replacers and horse supplements. As its name implies, it contains very little fat and therefore practically none of the fat-soluble vitamins. However, the protein quality approaches that of the liquid product if drying has been carried out by the spray process. Roller drying subjects milk to higher temperatures, which result in some loss of lysine availability and in large quantities this product can cause diarrhoea. Any significant quantity of dried skimmed milk should not be fed to horses more than 3 years old owing to their deficiency in the enzyme (lactase) that digests lactose.

Spray-dried skim milk is a useful supplement to feeds of young foals where the mother is providing inadequate quantities of milk to sustain normal growth. Concentrations of 10–15 per cent in the dry diet have proved satisfactory. On the other hand, its use in creep feed for foals approaching weaning may be less satisfactory if the main objective is to encourage the development of a faculty for the digestion of normal horse feeds to be given after weaning. Thus satisfactory creep feeds for use in normal circumstances can be provided as nutrient-rich stud nuts.

Single-cell proteins

During the past 20 years the industrial production of microbial proteins on a commercial scale has been undertaken. These protein sources are of high quality and would be quite suitable for horse feeding but as yet the amounts available are minimal. Several species of bacteria and yeasts have been cultured and their crude protein and fat contents range respectively from 340 to 720 g/kg and 20 to 210 g fat kg for bacteria and 400 to 450 g/kg and 25 to 55 g/kg for yeasts. The relatively high contents of nucleic acids found in bacteria, especially, should be of no particular concern for horses where dietary inclusion rates of up to 50–75 g/kg feed are economically feasible only for foals.

Feed storage

Some organic nutrients and non-nutrients in forages, cereals and compounded feeds deteriorate during storage. Labile, readily oxidized pigments, unsaturated fats and fat-soluble substances are destroyed at differing rates depending on their degree of protection, the environmental conditions, their propensity to oxidation and the presence or absence of accelerating substances. The immediate effects include a reduction in the acceptability of feed to the horse, which is perhaps one of the most discerning and perspicacious of domestic animals over its feed selection. All the fat-soluble vitamins present naturally in feed – that is, vitamins A, D, E and K – are subject to oxidation, together with the unsaturated and polyunsaturated fatty acids. Rancidity of the latter depresses acceptance, although some stability is imparted by natural and permitted synthetic antioxidants, which are respectively present or are used in mixed feed. Added synthetic sources of vitamins A and E are much more stable than are their natural counterparts, but contribute very little antioxidant activity. The critical water-soluble B vitamins are fairly resistant to destruction during normal storage, although riboflavin in feed will be lost where it is exposed to light. Advice given on labels attached to proprietary supplements and feeds should be followed.

Several factors are essential attributes of good feed stores and grain silos. These are a low and uniform temperature, low humidity and good ventilation, absence of direct sunlight, and freedom from rodent, bird, insect and mite infestation. These characteristics imply that feed stores and grain stores should be insulated and without windows, but should be well ventilated and both clean and cleanable. Construction materials should be rodent-proof, stacked feed should be raised from the floor and accessible from all sides, and roofs should be free from leaks. Galvanized bins are generally preferred to plastic bins, which can be gnawed by rodents, but metal bins may be more subject to moisture condensation on the inner surface if the feeds they contain have excessive moisture contents (a maximum of 120 g of moisture per kg should be achieved). Thus the choice of store should rest on the level of general tidiness, whether rat infestation is likely, whether all sides of proposed plastic bins can be reached and whether a uniform temperature can be attained over 24 hours.

An Irish study showed that 14 per cent of both Irish and Canadian oat samples were badly contaminated with fungi, although the Irish samples contained slightly more moisture (MacCarthy, Spillane & Moore 1976). These fungi grow during the maturation of the crop in the field and normally have a minor role in feed stability. However, high concentrations can affect acceptability to the horse and fungal invasion does detract from the stability of cereals during storage. *Fusarium* species and a few others may produce toxins that subsequently affect fertility or other aspects of animal health.

Storage fungi are another matter; these species can grow in environments with relatively low moisture and high osmotic pressure known technically as conditions of low water activity. Such environments cause heating, mustiness, caking, lack of acceptability and eventually decay of stored grains, oilseeds and mixed feeds. Feed stability and nutritional value decline, and fungal toxins may be produced. Species of *Aspergillus* and *Penicillium* are the main culprits and all feed stored where there are moisture contents that give water activities of 0.73–0.78 at temperatures between 5 and 40°C may be invaded by *A. glaucus*. However, *Chrysosporium inops* will spread at moisture contents as low as 150–160 g/kg. Most storage fungi have minimum temperatures for growth of 0–5°C, grow optimally at 25–30°C and do not grow at temperatures above 40–45°C. However, *A. candidus* and *A. flavus* grow vigorously at 50–55°C and *Penicillium* grows slowly at temperatures down to −2°C.

Uninsulated bins and waterproof sacks subject to variations in environmental temperature are particularly prone to moisture condensation on the inner surfaces even when the average moisture content of the product is low, but, of course, the probability of this occurring increases with greater average moisture levels. Once mould growth is initiated, this generates metabolic water and a vicious circle is established.

Insects, for example grain weevils and beetles, and flour mites, not only accelerate deterioration of feed and grain, but generate both heat and metabolic water, and are vectors of fungal spores. Dirty, badly stored grain will be ripe for the hatching of eggs and the multiplication of these insects and mites. Mites will multiply at moisture levels as low as 125 g/kg and a temperature of 4°C. Insects require slightly higher combinations of temperature and moisture. Cleaning and fumigation of long-stored feed and feed stores are, therefore, desirable. Hygiene and the storage of feed and grain at low temperatures and in a dry condition, without pockets of high moisture, are the greatest assurance for the maintenance of feed quality in the long term.

Weevils and beetles can be seen with the naked eye. Mites are extremely small but their presence can be detected in meals by observation for a minute, during which movement of the feed particles should be apparent. There is a characteristic sour smell from mite infestation, whereas, with moulding, discoloration of the grain, dust and a fungal smell can be readily detected.

Rodent infestation not only causes a loss of feed but both rat and mouse droppings introduce to horses the risk of enteric disease, principally salmonellosis.

Mould inhibitors, such as calcium propionate, propionic acid, sorbic acid and hydroxyquinoline, have been recommended but they are really effective only for coating grain and are relatively ineffectual when included in mixed feed. Propionic acid is more effective than the calcium salt.

Toxicants

Feed is sometimes naïvely considered to be a parcel of nutrients, both essential and non-essential. A consideration closer to the truth accepts that natural feeds also contain materials thought either to be inert or which influence the metabolism of other dietary constituents, and substances with nutritional value but which may be present in toxic concentrations. Many natural feedstuffs also contain substances in toxic concentrations with no known nutritive value. Many of these potentially hazardous substances are produced naturally, either by the plants themselves or by organisms infecting them and their products. Finally, there are contaminants that result from human intervention and activity. Table 5.3 gives an arbitrary classification to indicate the extent of the problem, yet the groupings and distinctions drawn are by no means absolute.

Toxicants produced by plants

Those toxicants likely to be consumed by browsing horses will be discussed in Chapter 10. Here our concern is with substances present in seeds used as feeds. Two widely distributed groups of compounds are known as digestive enzyme inhibitors and lectins (previously known as haemagglutinins). The specific compound, its toxicity and susceptibility to destruction by heat, vary among the species of plant within which it is found. Plants producing trypsin inhibitors and lectins include field or horse beans, black grams and kidney, haricot or navy beans. Horse grams, moth bean (*P. aconitifolius*), certain pulses (also containing amylase inhibitors), groundnuts or peanuts (*Arachis hypogea*), soya beans and rice germ also contain these substances. Most rice bran fed to horses has had the germ removed, although some residual activity is normally found.

Trypsin (protease) inhibitors depress protein digestion. Lectins are considered to be more harmful because they disrupt the brush borders of the small intestinal villi, hamper absorption of nutrients but apparently allow the absorption of certain toxic substances. These toxic substances increase tissue catabolism and urinary nitrogen, and thus depress growth in young stock.

The activity of both these groups of substance is destroyed by steam heat treatment. For example, the trypsin inhibitor activity of field beans is reduced by 80–85 per cent during steam heating at 100°C for 2 minutes and by about 90 per cent during treatment for 5 minutes. However, the trypsin inhibitor and lectin activities of kidney beans are very stable because treatment for 2 hours at 93°C is necessary for adequate destruction. Kidney beans are therefore generally unsuitable for non-industrial processing and should not normally be fed to horses.

The Indian or grass pea, after long periods of feeding, causes a condition known as lathyrism, which is exemplified in the horse as a sudden and transient paralysis of the larynx with near suffocation brought on by exercise. This is associated with a degenerative change in the nerves and muscles of the region and profound inflammation of the liver and spleen.

Table 5.3 Classification of naturally occurring toxicants and toxic contaminants of horse feed

---

I. **Naturally occurring toxicants**
   A. Protein or amino acid derivatives
      1. Lectin (haemagglutinins)
      2. Trypsin inhibitors
      3. Lathyrogens
      4. Nitrates/nitrites

   B. Glycosides
      1. Goitrogens
      2. Cyanogens

   C. Miscellanous
      1. Tannins (polyphenols), saponins
      2. Gossypol
      3. Phytin, oxalic acid
      4. Several antivitamin factors

   D. Poisonous shrubs and weeds (principally alkaloids)

II. **Moulds and pathogenic bacteria developing owing to bad harvesting, handling and storage**
   A. Spores causing respiratory allergies
      1. *Aspergillus fumigatus*
      2. *Micropolyspora faeni*

   B. Mycotoxins produced by moulds
      1. Hepatotoxins
      2. Hormonal
      3. Other

   C. Pathogenic bacteria, or their toxins
      1. Pathogenic *Salmonella* spp.
      2. *Clostridium Botulinum* toxin

III. **Dietary allergens absorbed from intestines**

IV. **Dietary contamination during manufacture**
   1. Toxic antibiotics
   2. Dopes
   3. Pesticide and herbicide residues
   4. Nutrients with narrow margin of safety
   5. Heavy metals unlikely to have nutritional value
   6. Breakdown products of feed constituents

Other closely related species, including sweet pea (*Lathyrus odoratus*), wild winter pea (*L. hirsutus*), singletary pea (*L. pusillus*) and everlasting pea (*L. sylvestris*), can also cause lathyrism. Although the whole plant contains the toxin, the seeds appear to be the most potent source and it is only partially destroyed by heat.

Goitrogenic activity is caused by goitrins, which are derived from glucosinolates found in many members of the Cruciferae family, including cabbages, rape and mustard. Goitrins are released by enzymes contained within the plant and the destruction of these enzymes by heat treatment to a large extent eliminates the potential hazard. The effect of goitrins is not counteracted by additional dietary iodine, but further enzymatic metabolism of certain goitrins can release isothiocyanates and thiocyanates. The antithyroid effect of these substances on young horses in particular can be overcome by dietary iodine. The enzymes again would be destroyed by adequate heat treatment. The slight antithyroid effect of uncooked soya beans is said to be overcome by additional iodine.

Certain glycosides, derivitives of α-hydroxynitriles present in lima beans, sorghum leaves, linseed and cassava (tapioca) (*Manihot esculenta*), will generate hydrogen cyanide (HCN) when acted on by specific enzymes that the plants contain. HCN can cause respiratory failure by inhibiting cytochrome oxidase. Again, as the poison is released by enzyme activity, heat treatment will ensure safety, so long as prolonged storage of, for example, moist seeds or cassava roots, has not led to some accumulation of HCN. The suppression of enzymatic activity is another reason for the importance of dry storage of certain uncooked feedstuffs. HCN can also react with any thiosulphate present, producing thiocyanate, which is itself responsible for thyroid enlargement after prolonged feeding.

Tannins, contained in sorghum grains and coloured-flower field beans, depress protein digestion if they are present in sufficient concentration. Autoclaving or pressure-cooking destroys these tannins, but prolonged treatment is required at lower temperatures.

In both its bound and free forms, the pigment gossypol reacts, incompletely, with cottonseed protein to depress appetite and protein digestibility and therefore the efficiency of amino acid utilization; but its toxicity can result in death caused apparently by circulatory failure. The pigment also reacts with dietary iron, precipitating it within the intestines. Fairly large additions of supplementary iron to the diet will then promote further precipitation, which partially suppresses the adverse effects of gossypol.

Several antivitamin factors are present in animal and vegetable feeds, but most are of little significance to horses. A thiaminase present in the bracken fern (*Pteridium aquilinum*) is a cause of bracken poisoning, which is counteracted with large doses of thiamin, and the antivitamin-E factor present in raw kidney beans is partly destroyed by cooking.

High levels of phytic acid, or its salts, present in many vegetable seed products when consumed in large quantities will interfere with the availability of several trace metals, particularly zinc. Oxalates detectable in high concentrations in certain tropical species of grass have been reported to kill cattle and cause lameness in horses, owing to a loss of calcium.

The rapid growth of pasture after a high rainfall and excessive use of nitrogen fertilizers can lead to high concentrations of nitrates in the herbage, and contamination of water supplies through the leaching of soils. Although nitrates are only slightly toxic, they can be reduced to nitrites before or after consumption. High levels of nitrites may accumulate in plants after herbicide treatment and during the making of oat hay, owing to nitrate reduction encouraged by inclement weather. In the body, nitrites convert blood haemoglobin to methaemoglobin, which is unable to act as an oxygen carrier. Large intakes therefore cause death. Pigs are probably more susceptible than are horses, which appear to react similarly to ruminants.

In South Africa, America and Australia numerous species of the legume *Crotalaria* (one species of which causes Kimberley horse disease) have proved very poisonous in horses; lesions are induced in the liver by pyrrolizidine alkaloids similar to those found in ragwort (*Senecio*). Alkaloids in the vegetative parts of certain shrubs and pasture weeds will be considered in Chapter 10 on grazing management and other details are found in Table 10.6. Green potatoes contain the alkaloid solanine. Horses are killed by eating quantities of potatoes, that would not affect ruminants, even when the tubers are not apparently green.

Mould development in feeds

The effects of moulds are of two types. Certain mould spores in large numbers in badly harvested and stored roughages and in cereals can cause a respiratory reaction (allergy) when inhaled by sensitive horses. Many mould species in the appropriate conditions of temperature and humidity produce toxins that have a variety of metabolic effects. There are numerous historical references to the consequences of consuming ergot of rye, namely abortion and other effects of blood vessel constriction. The mould concerned, *Claviceps purpurea*, also infects ryegrasses, and some other grasses, and so can be a hazard in pasture and hay. Several reports, especially from Thailand and the United States, have recorded deaths or brain, heart and particularly liver lesions, and haemolytic enteritis in horses receiving aflatoxin produced by *Aspergillus flavus* at levels of less than 1 mg/kg of contaminated cereals, peanuts and even hay. Horses and ponies seem to be more susceptible to acute aflatoxicosis than are pigs, sheep and calves, as daily intakes of 0.075 and 0.15 mg of aflatoxin B_1 per kg bodyweight are lethal for ponies in 36–39 days and 25–32 days, respectively (Cysewski, Pier et al. 1982). The toxin zearalenone produced by *Fusarium* species causes vulvovaginitis and reproductive failure in females of several domestic species. Whereas aflatoxin usually develops during storage, this toxin may develop pre-harvest. Although no reports of effects in horses are known, it could well cause breeding irregularity in them. Many other fungal toxins with a wide variety of effects and significance also exist.

Dietary allergens

Dietary allergens are not contaminants, but certain horses can react to normal nutritional constituents of specific feeds (Pl. 5.3). These are probably protein in nature and the effects, which include respiratory and skin lesions, are normally overcome according to the author's experience by removal of the offending source from the diet. Problems of cross-reactivity in which related sources of proteins yield similar reactions can, however, pose problems of interpretation.

Heavy-metal and mineral contamination from pastures

Lead is one of the commonest causes of poisoning in cattle, sheep and horses. Signs of toxicity are more frequent in young horses and include lack of appetite, muscular stiffness and weakness, diarrhoea, and, in an acute form, pharyngeal paralysis and regurgitation of food and water. Lead accumulates in the bones and as little as 80 mg of lead per kg diet may eventually cause toxic signs, which are sometimes precipitated by other stresses. Natural feeds with 1–5 mg of lead per kg cause no problems.

The acute lethal dose is 1.0–1.8 g/kg bodyweight as lead acetate or carbonate. The chronic lethal dose depends on many factors but is said on average to be about 12 mg of lead per kg bodyweight daily for 300 days.

The contamination of pasture with lead, cadmium and arsenic – derived from mine workings, dumping or sewage sludge, by aerial dust and water erosion, even by car batteries and lead shot – are local risks. Where the pasture is dense, undoubtedly the greater problem arises from surface contamination of the plants; but where it is sparse, soil, either rich in these heavy metals or contaminated by them, can be consumed in sufficient amounts to cause problems. Lead shot is only slightly hazardous to horses, but, because they close graze, consumption can be greater than might be generally appreciated, followed by some solution in the stomach. Where grass is ensiled, entrapped shot partly dissolves during fermentation and highly toxic levels of 3800 mg of soluble lead per kg dry matter have been detected in England by the author (Frape & Pringle 1984). Whereas lead seems to contaminate the surfaces of plants, cadmium is readily absorbed and accumulated from soils rich in this element. Of common pasture species the daisy (*Bellis*) accumulates 60–80 mg of cadmium per kg (thirty times as much as grasses) from contaminated soils (Matthews & Thornton 1982).

Pastures to the leeward of steel and brick works may amass abnormally high concentrations of fluorine. The horse is probably less subject to fluorosis than are cattle and sheep but damage to its bones and teeth has been induced by this element. However, it will tolerate 50 mg per kg of feed for extended periods.

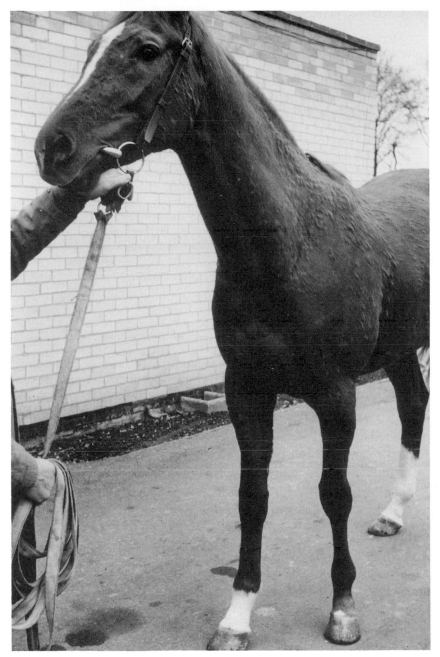

Plate 5.3 A 2-year old Thoroughbred gelding with widespread 'bumps' on the head, neck, shoulders, ribs and flanks. An allergic reaction to bran and oats was detected in the blood serum. The horse recovered over several months when its diet consisted of a high-fibre cubed diet, low in cereal and with water *ad libitum*.

Mercury poisoning expressed as colic and diarrhoea has arisen in horses as a consequence of the mistaken use of dressed seed as a feed. Chance exposure from other sources is unlikely.

The required trace elements – iodine, selenium and molybdenum – may be consumed in toxic quantities following natural accumulation in vegetable materials. Seaweed can be a source of excessive iodine, and certain accumulator plants store large quantities of selenium from selenium-rich soils. When these plants die it is said that they in turn deposit selenium in a form readily absorbed by neighbouring plants. Many selenium-rich areas are sparsely covered and the consumption of soil rich in the element is another source of risk. Molybdenum is readily absorbed by most plants from soils containing excessive amounts. High concentrations of this element in herbage have been shown by the author to depress serum copper in horses, but less strikingly than in cattle, and the ingestion of soil rich in iron and sulphur is also known to reduce copper absorption in grazing animals. Season and the extent of soil drainage can influence the accumulation of several metals in herbage. Concentrations frequently tend to be higher during the winter months.

Pesticide residues

Many normal feeds contain trace amounts of pesticide residues but, excepting gross contamination through negligence, the amounts normally detected are insufficient to cause any problem to horses (after herbicide treatment pastures should be rested for 2 weeks before grazing is permitted).

Feed additives and prohibited substances

Domestic animal feed-additive drugs

Several drugs are used in the feed of farm animals to promote growth, counteract diarrhoea and parasitic infection and to influence the carcass. Most of these drugs have little if any ill effect on horses when present in the diet at normal feed levels, or when horses are mistakenly given feed containing antibiotics and intended for other species. Higher dosages are a different matter. Although framycetin sulphate is sometimes useful in cases of flatulence, or fermentative colic, the persistent use of some antibiotics, especially oxytetracycline, may cause a severe upset to the intestinal flora, possibly including a fungal overgrowth, precipitating acute and intractable diarrhoea, lethargy and lack of appetite. Two other drugs, and one of them in particular, can have severe toxic effects on horses when given at normal feed rates.

Monensin sodium

Monensin is fed to beef cattle for promoting growth and to poultry as a coccidiostat. Poultry feed containing 100 mg/kg of monensin, the normal feed level, has severe toxic effects when fed to horses. At a level of 30 mg/kg in the feed, horses experience a reduced appetite and uneasiness; at a level of 100 mg/kg (about 2.5 mg/kg bodyweight) in a diet fed continuously, it is lethal in a matter of 2–4 days to about half the individuals. Horses present symptoms in the author's experience of posterior weakness, profuse sweating, occasionally muscular tremors, myoglobinuria (dark-brown urine), elevated urinary potassium and an elevated serum level of muscle enzymes. Postmortem examination shows myocardial degeneration and monensin can normally be confirmed by analysis of stomach contents. In the early stages of toxicity, recovery can frequently be achieved by removing the offending feed and dosing the horse with mineral oil, although it may suffer permanent heart damage with increased risk during hard physical exertion.

Lincomycin

Lincomycin is an antibacterial drug sometimes included in pig feed. It is less toxic than monensin in horses, but above dose levels of 80 mg/kg bodyweight daily (5 mg/kg of total diet fed continuously) metabolic symptoms of toxicity and evidence of liver damage have been observed by the author.

Prohibited substances

The list of proscribed drugs embraces a very high proportion of the drugs permitted in livestock feeds by EEC legislation and the detection of any of them, or their recognizable metabolites, in urine, blood, saliva or sweat will lead to the disqualification of a horse subject to the Rules of Racing. Thus, apart from drugs that might be used directly in horses, any antibiotic, growth-promoter or other drug used for feeding to poultry, pigs or ruminants must not be detected in any of the above fluids. Of the proscribed drugs, most are unlikely to be present in feed. However, some drugs acting on the cardiovascular system, some antibiotics and one or two anabolic agents have been detected in contaminated feed ingredients – oats, soya-bean meal, bran – and in feed additives, or they may be present in feeds for other classes of stock mistakenly fed to horses. In practice, those causing chief concern are the xanthine alkaloids – theobromine, caffeine and its metabolite theophylline.

Caffeine is present in tea, coffee, coffee byproduct, cola nuts, cacao and its hull, which is available as a byproduct, and in maté leaves. Tea dust contains as much as 1.5–3.5 per cent of caffeine, whereas coffee byproduct as normally available contains only some 200 mg/kg of caffeine. Small amounts of theophylline are found in tea, but as much as 1.5–3 per cent of theobromine is typically present in cacao beans, and its wasteproduct, the hull, contains as much as 0.7–1.2 per cent. The widespread international traffic in coffee and cocoa beans and in their byproducts constitute a formidable risk through their contamination of the means of transport from ships at one extreme to hemp sacks at the other. These means of transport in their turn put in jeopardy cereals, pulses and other raw materials moved from one place to another leading to infringements of the Rules of Racing through the consumption of contaminated batches of these feeds. Gross contamination and its control in animal feedstuffs has been discussed in a code of practice published by UKASTA (1984).

After an oral dose with caffeine, about 1 per cent appears unchanged in the urine, the excretion of which is almost complete after 3 days. About 60 per cent of the caffeine is excreted in the urine as metabolites, including theophylline and theobromine. Traces of the latter may continue to be excreted for up to 10 days, whereas theophylline excretion is virtually complete after 4–5 days. Thus the inadvertent use of these alkaloids, or of coffee or cocoa wastes containing them, can result in their detection in the urine for up to 10 days and caffeine is moreover demonstrable in urine within 1 hour of an oral dose. Great care must therefore be exercised in the feeding of race and competition horses during such a period. The extent of the excretion curves in urine of various other drugs has been determined, but certainly not those of all antibiotics similarly excreted.

Both caffeine and theobromine are rapidly absorbed from the intestinal tract and soon impart their effects of cardiac and respiratory stimulation and of diuresis. However, tests have shown that their effects on the speed of horses only occur when they are used in high doses.

Further reading

McDonald P, Edwards R A & Greenhalgh J F D (1981) *Animal nutrition.* Longman: London & New York.

Moss M S & Haywood P E (1984) Survey of positive results from racecourse antidoping samples received at Racecourse Security Services' Laboratories. *Equine Vet. J.* **16**, 39–42.

UKASTA – United Kingdom Agricultural Supply Trade Association Ltd (1984) Code of practice for cross-contamination in animal feeding stuffs manufacture. Amended code, June 1984. London.

CHAPTER 6

Estimating the nutrient requirements of the horse

What good receipt have you for a horse that hath taken a surfet of provender. This comes commonly to such horses as are insatiable feeders and therefore it is requisite that they be dieted, especially if they have too much rest, and too little exercise.

Thomas De Gray, 1639

The daily requirements of horses and ponies have been estimated in terms of the amounts of each nutrient – minerals, trace elements, vitamins and amino acids (or more realistically, protein) required per day for the various functions of maintenance, growth, lactation, and so on. The normal vehicle for these nutrients is the daily feed, and if a particular horse were to consume twice as much feed as another horse fulfilling the same tasks, then it might be reasonable to suppose that the nutrient concentration in the diet of the first horse need be only half that in the diet of the second. Thus in order to make useful statements about dietary composition and to facilitate calculation of adequate diets, it is necessary to predict reliably the appetite of a horse, or group of horses, for feed, or more specifically, to predict appetite for dry matter. The appetite and capacity of horses for acceptable feed are regulated by four dominant and related factors: (1) the volume of different parts of the intestinal tract; (2) the rate of passage of the digesta; (3) the concentration of certain digestion products in the intestine; and (4) the energy demands of the horse. The third of these seems to control meal size and the second will be modulated by the physical form of the feed. The first is controlled by the body size of the animal, but to some extent is modified by adaptation.

The most common easily corrected deficiencies in home-prepared feed mixtures are those of calcium, phosphorus, protein, salt and possibly vitamin A. However, in addition to water, the most fundamental, immediate and long-term need of the horse is for a digestible source of dietary energy. Ideally, in situations of moderate work or productivity, the energy demands should just be met by the appetite and capacity of the horse for feed. Not only capacity, but also energy requirement for a variety of functions, is closely allied to bodyweight, although this varies from day to day according to the amount of gut fill. Therefore, a means of estimating weight is fundamental to any rationing system.

In the absence of facilities for weighing horses, the most reliable predictions include a measurement of the heart girth, following respiratory expiration (Fig. 6.1). As conformation changes with age and differs among breeds, the measurement of girth alone is bound to yield only an approximate estimate. Some improvement on this is achieved by inclusion of the length of the horse from the point of the shoulder to the point of the buttocks (Fig. 6.1). If one wishes to include length, then the equation below (from Carroll & Huntington 1988) gives the appropriate relationship:

$$\text{Bodyweight (lb)} = \frac{\text{Heart girth (in)}^2 \times \text{length (in)}}{328.8}$$

or

$$\text{Bodyweight (kg)} = \frac{\text{Heart girth (cm)}^2 \times \text{length (cm)}}{11877}$$

For many, withers height will be a more familiar index of size; Fig. 6.2 shows its approximate relationship to bodyweight for several types of horse and Fig. 6.3 delineates the change in withers height with age during normal growth. Suggested average daily allowances for horses of different live-

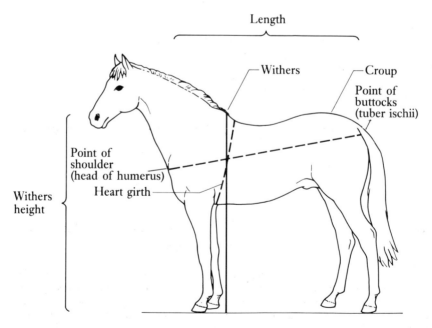

Fig. 6.1 Linear body measurements used in estimating bodyweight of horses and ponies.

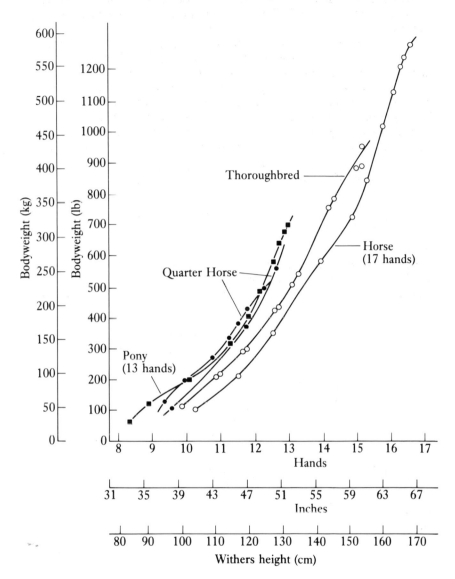

Fig. 6.2 The relationship between bodyweight and withers height in normally growing horses and ponies (1 hand = 10.16 cm). Note that horses and ponies achieve mature height before mature weight so the curves must be concave upwards as maturity is approached (the upper end of the curve). Furthermore, as the curves are not coincident, withers height is generally not a good predictor of bodyweight. (Data from Green 1961, 1969; Hintz 1980a; Knight & Tyznik 1985; and personal communications 1984, R. W. W. Ellis & R. A. Jones).

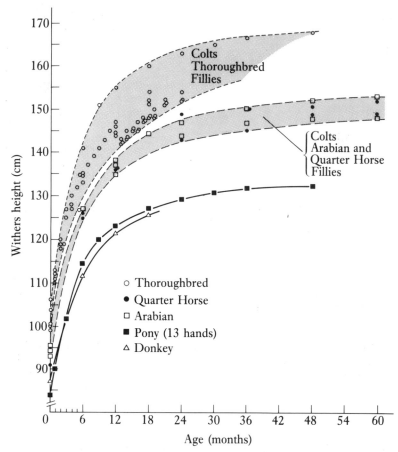

Fig. 6.3 Expected withers height at various ages of normally growing horses and ponies (1 hand = 10.16 cm). (Data from Green 1961, 1969; Hintz 1980a; Knight & Tyznik 1985; and personal communications 1984, R. W. W. Ellis & R. A. Jones).

weights are indicated in Fig. 6.4. The allowances given to idle horses would be lower than those shown, whereas lactating mares will consume more feed. Furthermore, hard-worked animals, such as Thoroughbreds in advanced training for racing, will be entitled to consume amounts near their capacity, although their appetites may decline when vigorous exercise is practised routinely. Observations in the United States showed that among seven racing stables the average daily intake of concentrate by 3 to 4-year-olds was 6.16 kg (4.9–7.5 kg) and that of hay 9.37 kg (6.4–11.9 kg) (Glade 1983a). Comparable observations among 2 to 4-year-olds in Newmarket by the author showed that the concentrate intake averaged 8.15 kg and that of roughage 5.5 kg daily per horse. The lower intake of roughage in the

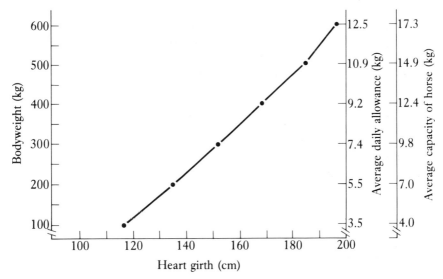

Fig. 6.4 The relationship for horses and ponies between heart girth and average liveweight, average daily capacity for concentrates and forage, and average allowance of concentrates and forage (12% moisture).

United Kingdom may reflect the generally poorer nutritional quality of this feed as supplied to horses. In the American study, the horses were estimated to average 496 kg bodyweight whereas those in the UK were about 480 kg.

Physical form of feed and its density

All horses should receive some form of long roughage, such as fresh grass, hay or silage. A proportion of these may be replaced by succulent green or root vegetables and wet sugarbeet pulp. Hay or grass may comprise the total ration of idle horses and usually forms at least half the ration. The concentrate portion of the daily feed may therefore vary from nothing to 50 per cent and only in exceptional circumstances, or in the hands of experienced feeders, should it rise to the proportions of three-quarters of the daily allowance of dry feed.

Concentrated feeds, such as cereals, cereal byproducts, oilseed meals and the like, are traditionally fed by the bowl, that is by volume. The energy and nutrients that these feeds provide are, of course, much more closely associated with their weight than with their volume and feeding containers should thus be calibrated to show the volume occupied by a unit weight of each type of feed. Table 5.2 contains average conversion values for cereals, although it will be appreciated that the bushel weight of cereals varies from season to season and from crop to crop according to how well they were grown. The energy content per kg of each type of concentrate also differs. Ideally, therefore, feed bowls should be calibrated to indicate the volume giving multiples of 2 MJ of digestible energy for each type in use.

There is often glib talk about the appetite of horses for particular feeds, but apparently they have a true appetite only for water, salt and sources of energy, and if given the choice would not select a balanced diet. Hence it is necessary to induce horses to consume mixtures most appropriate to their needs.

Feed energy

The gross energy of a feed is the heat evolved when it is subjected to complete combustion in an atmosphere of oxygen. Obviously all this energy, measured as heat, is not available to the animal because a portion of the feed remains undigested and is voided in the faeces. In addition, a relatively unknown quantity is lost from the horse as the gases methane and hydrogen, in the main passing out through the anus but also absorbed into the blood and exhaled. Of the products of digestion and fermentation that are absorbed, a proportion of the amino acids is deaminated and the nitrogen incorporated in urea. Much of this is excreted in the urine. The gross energy of a feed, less the energy content of the faeces attributable to it, is the digestible energy (DE), and less the energy content of combustible waste gases voided and urine excreted, leaves us with the metabolizable energy (ME). This is the residue of feed energy that is available to the body for its various processes of tissue repair, the functioning of organs, the physical work of skeletal muscles, growth and milk production. The efficiency of ME utilization depends on the precise chemical form of the nutrients derived from the diet and on which of these functions is performed. This efficiency is measured either as the amount of useful product, or from the quantity of waste heat dissipated. The ME less this heat increment attributable to the feed is the net energy (NE). (Heat increment is the heat loss of a nourished animal in excess of that lost by a fasting animal.) The scheme is summarized in Fig. 6.5. Net energy is used for maintenance, growth, work, reproduction, etc. The energy requirements for maintenance are those leading to a zero change in bodyweight, or more accurately a zero change in body energy content, in a stabled non-working horse.

An idle horse obviously has relatively low energy requirements, yet its appetite should be satisfied. As the horse can consume daily lesser quantities by weight of bulky fibrous feeds than of concentrates, then its appetite

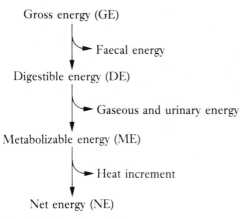

Gross energy (GE)

➤ Faecal energy

Digestible energy (DE)

➤ Gaseous and urinary energy

Metabolizable energy (ME)

➤ Heat increment

Net energy (NE)

Fig. 6.5 Partition of dietary energy.

is more likely to be satisfied with lower intakes of energy when fibrous feeds are used. Conversely, when energy demands are great, concentrate feeds such as cereal grains must form part of the diet if those demands are to be met, simply because the horse can consume larger quantities of dry matter daily when cereals are included and they contain more ME per kg of dry matter.

The energy requirements for maintenance per 100 kg bodyweight decline slightly with increasing bodyweight, so that relative to body size larger horses require slightly less food for maintenance than do ponies in similar conditions. To compensate for this, ponies may develop a greater barrel or appear more pot-bellied. The relationship between bodyweight (W) and the energy requirements for maintenance is depicted in Fig. 6.6. One of these curves is derived from the identity (National Research Council 1978)

$$DE \text{ (kJ/day)} = 649 \, W^{0.75}$$

It has been suggested in the United States recently that this may over-estimate the energy needs for maintenance among boxed animals by about 10 per cent. A simpler relationship of 12.5 MJ of DE per 100 kg body-weight daily has therefore been proposed to reflect more satisfactorily maintenance energy needs (H. L. Hintz, personal communication, 1984), although this does not agree with the extrapolated estimate from the measurements of Anderson, Potter et al. (1983) given in Table 6.1. Individual horses differ in their needs about this mean; some will eventually become fat when subjected to a regime under which others will lose condition.

At the maintenance level of energy intake and expenditure, essentially no work is done by the horse on its surroundings so that NE expended in maintenance (m) is ultimately degraded to heat:

$$ME_m = NE + HE = \text{heat production at maintenance,}$$

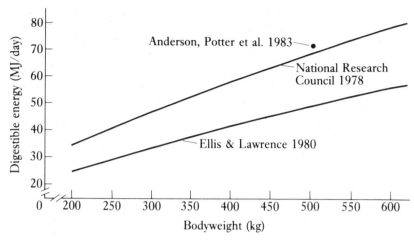

Fig. 6.6 The relationship of bodyweight to daily digestible energy (DE) require-ments for maintenance according to three sources: DE (kJ/day) = 649 $W^{0.75}$ (National Research Council 1978) and DE (kJ/day) = 465 $W^{0.75}$ (Ellis & Lawrence 1980), where W = bodyweight in kg.

Table 6.1   Digestible energy (DE) demands for maintenance plus work on a slope of 9°* at endurance rates (135 heart beats per min, 155 m/min)[†]

Bodyweight (kg)[‡]	400	500	600
Distance travelled (km)	DE per day (MJ)		
1	68	79	90
2	76	88	100
4	89	103	117
6	99	114	127
8	107	121	133
Approximate appetite	108–113	125–130	140–145

*   Work output on an inclined moving belt is in practice greater than that on the level, but in theory no greater.
†   Based upon a quadratic equation relating energy requirements to bodyweight and work in Quarter Horses (Anderson, Potter et al. 1981, 1983).
‡   Average bodyweight 503 kg.

where HE is heat increment or waste heat. The fact that the temperature of the horse's body is normally greater than that of the surroundings, to which heat is continuously being lost, is the expression of this situation. Exposure to a cold, or wet and windy climate accelerates metabolic rate so that the rate of heat production keeps pace with the rate of heat loss in order to maintain a steady body temperature, that is the energy requirements for maintenance rise. Conversely in hot climates, where the environmental temperature is higher than that of the horse, the heat produced must still be dissipated. This is done primarily by evaporation of sweat and of water from the lungs, but also by a rise in body temperature. A physiological stress is induced. Thus, in one environment heat production is a boon and in the other an embarrassment. Can the heat production be manipulated to the horse's advantage?

Waste heat (HE) is a measure of the efficiency of utilization of the ME of feed and it is known to vary between types of feed. Thus, if the NE available represents 80 per cent of the ME, (NE/ME = 0.8), then the remaining 20 per cent is HE. When feeds are selected for use, their difference in heat increment should ideally be considered in the context of the climate and the purpose for which the horse is kept.

Some estimates of the likely efficiency of ME utilization by the horse are given in Table 6.2 and Fig. 6.7. The k values in Table 6.2 subtracted from 1.0 show the proportion of energy lost as waste heat when the feed is used for maintenance or for fat deposition. Thus, 30 per cent of the energy of meadow hay would be lost as waste heat by horses at maintenance, whereas only 15 per cent of the ME of barley would be similarly lost (note that the utilized energy is ultimately degraded to heat also, but more hay would be required for mainenance). During winter ample meadow hay may be a more appropriate feed than in the summer, or than barley, as the greater heat increment of hay may contribute to the maintenance of body temperature when the weather is cold.

The ribs of both breeding and working horses in optimum condition cannot be seen, but can be felt with little fat between the skin and ribs. Acclimatization to cold weather does not necessitate excessive fat deposition but should allow sufficient time for the coat to grow. Horses should therefore be provided with a shelter protecting them from rain, snow and the worst of the wind. In other words, three sides and a roof provide sufficient

Table 6.2 Estimated efficiency of utilization of the metabolizable energy (NE/ME), or k, for various energy sources by the horse

	For maintenance* (k_m)	For fat deposition (k_f)
Mixed proteins	0.70	0.60
Meadow hay	0.70	0.32
Lucerne hay	0.82	0.58
Oats	0.83	0.68
Barley	0.85	0.77
Fat	0.97	0.85

*   These values are higher than those for fattening mainly because the use of these nutrients for that purpose spares the breakdown of body fat

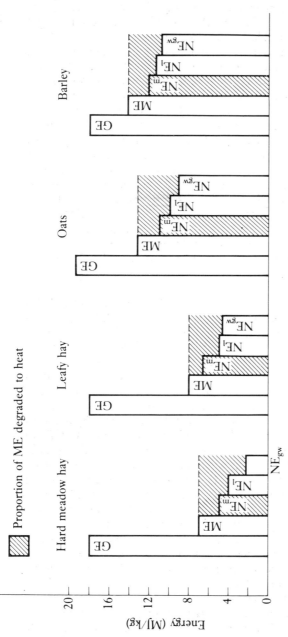

Fig. 6.7 Estimated average utilization efficiency of the gross energy of feeds for several production purposes.

Efficiency of utilization of gross energy as a source of metabolizable energy $= \text{ME/GE} = q$

Efficiency of utilization of metabolizable energy as a source of net energy $= \text{NE/ME} = k$

$= \text{NE/GE} = qk$

where GE is gross energy, ME is metabolizable energy and NE is net energy. The k value for work represents, in the main, the efficiency of glycogen and of depot fat formation. The efficiency of utilization of these sources by muscles in carrying out work is approximately 0.35. The concept of work measured for flat racing, etc. is illusory as the true efficiency can be measured only as a difference in energy expended between exercise on the level and that on a gradient not inclined moving belt. Note in the figure that nearly two and a half times as much hard hay is required for maintenance as would be required of barley so that nearly 25 per cent more heat is produced at maintenance on the hay diet. Other abbreviations: m, maintenance; l, lactation; gw, growth, work, etc.

protection in all seasons for properly fed adult animals. A long hair coat, if dry, and a modicum of subcutaneous fat are an excellent insulation for horses given an ample roughage diet. In the spring when horses are brought in, daily grooming and 2–4 oz (57–114 g) of oil added to the ration each day should accelerate the shedding of the winter coat.

Figure 6.8 gives the DE requirements of horses of various weights at maintenance and when engaged in work of a range of intensities, strenuous work causing a large increase in energy demand. These demands of high productivity, also exemplified by peak lactation, are met from two main sources: (1) the breakdown of body fat and (2) increased feed energy.

Excessive heat increment or waste heat in working horses is an encumbrance and a contributory cause of unnecessary sweating. The data in Table 6.2 show that two-thirds of the ME of meadow hay and 40 per cent of the ME of mixed proteins are lost as waste heat, whereas only 23 per cent of barley grain and 15 per cent of fat ME are similarly lost in hard work or fattening. Concentrated feeds thus have a part to play in the diet of working animals apart from accommodating energy demands more easily. The so-called heating effect of cereals and other concentrates reflects a more rapid rise in blood glucose and metabolic rate after a large meal and the associated feeling of vigour in 'hot-blooded' breeds. In Fig. 6.9 the interactions among the heat increment of feeds, environmental temperature and production of body heat are depicted.

Visual examination of feed reveals nothing about its ME content, but the feed can be weighed. Fortunately, the gross energy (GE) content of most horse feeds is just over 18 MJ/kg of dry matter. This statement is untrue for feeds containing much more than 80 g of ash per kg, or than 35 g of oil per kg. For example, oats on average may contain 45 g of oil and 19.4 MJ of GE per kg dry matter. Figure 6.7 shows how the GE of samples of four different feeds might be utilized for growth or for hard extended work and the data give a revealing and objective comparison of roughages with cereals. The coefficient q represents the approximate efficiency by which the GE of each feed is utilized as a source of ME and as the coefficient k represents the efficiency by which this ME is utilized for the functions of maintenance, growth, etc, $q \times k = NE/GE$, or the overall efficiency of utilization of the 18 MJ for the productive function. Note that $q \times k$ for the hard meadow hay = 0.125, whereas that for barley = 0.593, a value 4.75 × 0.125. The NE in hard meadow hay for growth, or for extended work, is only a quarter of that found in the two cereals, despite a similarity of their gross energies. In energy terms, that is MJ/kg of feed, the losses in the utilization of ME from hay and cereals are not very different (Fig. 6.7) but the k values are (Fig. 6.10), for the reason that the ME values (MJ/kg) differ widely.

It is recognized that roughages are required by horses and ponies, particularly in a long form, in order to maintain general metabolic health and a feeling of well-being. However, are there lessons to be learnt from the above calculations? First, poor-quality roughages can be an expensive

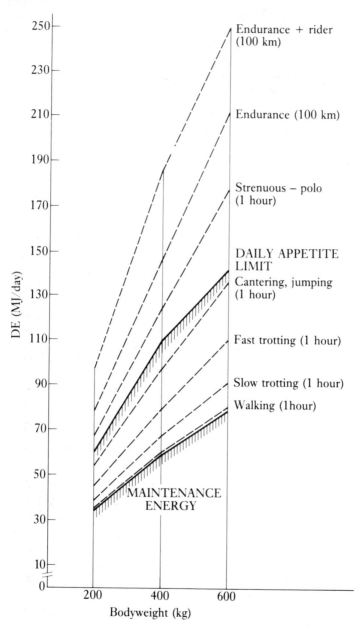

Fig. 6.8 Digestible energy (DE) demands of daily maintenance and work at a constant elevation in relation to appetite of horses of three bodyweights. (Effect of a 67-kg rider of 400-kg and 600-kg horses and 33-kg rider of 200-kg horse given for endurance rides only.)

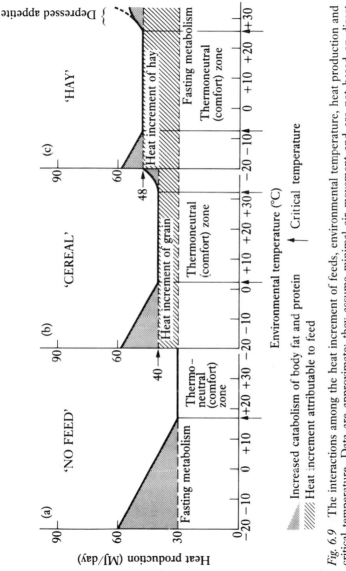

Fig. 6.9 The interactions among the heat increment of feeds, environmental temperature, heat production and critical temperature. Data are approximate; they assume minimal air movement and are not based on direct experimentation. (a) Fasted horse (also with slightly less subcutaneous fat); (b) horse fed 3.5–4.0 kg of grain providing 40 MJ of metabolizable energy (ME) and 30 MJ of net energy (NE) (maintenance level); (c) horse fed 5–6 kg of hay providing 48 MJ of ME and 30 MJ of NE (maintenance level).

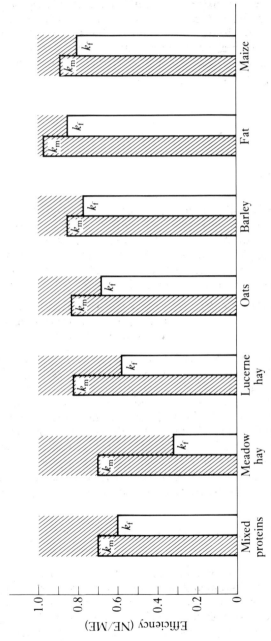

Fig. 6.10 Estimated efficiency of utilization of the metabolizable energy (NE/ME), or *k*, for various energy sources: k_m = NE/ME for maintenance; k_f = NE/ME for fat deposition (later growth), where ME is metabolizable energy and NE is net energy. (Maintenance values are higher than those for fattening, mainly because the breakdown of body fat is spared by the use of these feeds for maintenance.) ▨ Proportion of ME, inevitably degraded to heat.

buy if they are to form a major part of the ration of growing, or hard-worked animals. Second, hard-worked animals can lose condition if poor-quality roughages form a major portion of their diet. Finally, idle horses may put on unwanted fat if too much cereal is included in their diet. Unfortunately there are insufficient data available on the ME content of horse feeds or on their NE for various functions. Therefore it is necessary to follow certain rules based on the above discussion and to use digestible energy (DE) data provided in Appendix C.

Ration formulation

Energy needs

The formulation of a ration requires estimates of (1) total daily dry food intake, and (2) daily energy requirements (Table 6.3 and 6.4). The DE required in MJ/day divided by dry-feed intake in kg/day gives MJ of DE per kg of dry feed needed. The DE contents of the roughage and of the concentrates are given in Appendix C and their required dietary proportions can then be roughly calculated:

DE required (MJ/kg of diet) = roughage (MJ of DE per kg) (proportion, x) + concentrates (MJ of DE per kg) (proportion, $1-x$),

where in most cases the proportion $x = 0.6$ to 0.7 and therefore the proportion of concentrates, $1-x = 0.3$ to 0.4 (see Appendix A). The above equation has been used in deriving the proportions in Table 6.5. The greater the intensity of physical activity the higher the proportion of cereals required. As the speed of the horse increases, the energy expended rises steeply on an hourly basis (Table 6.4). Hence, the types of problem encountered can be quite different in horses undertaking strenuous effort compared with those asked to respond in a leisurely fashion. Furthermore, compared with ponies, large horses tend to require a higher proportion of concentrates in the ration when both are subjected to hard work.

Protein, mineral, trace-element and vitamin needs

The overall principal chemical components of a stabled horse's diet are depicted in Fig. 6.11. Protein and minerals required by working horses are most appropriately expressed in amounts per day, as exercise raises energy needs and may increase voluntary intake. Orton, Hume & Leng (1985) found that by trotting 2-year-old growing horses for 12 km daily, the

Table 6.3 Daily energy requirements of horses for various functions and amounts of hay and concentrates needed to provide the energy

| | Mature bodyweight | | | | |
| | 200 kg | | | 400 kg | |
	Digestible energy (MJ)	Hay* (kg)	Concentrate mixture† (kg)	Digestible energy (MJ)	Hay* (kg)
Mature horse‡, maintenance	34.4	4.2	—	58.2	7.1
Mares, last 90 days of gestation	38.8	3.2	1.1	67.4	5.3
Lactating mare, first 3 months	69.7 (12)§	2.8	4.1	103.2 (18)§	4.8
Lactating mare, 3 months to weaning	55.4	3.7	2.2	86.2	6.2
Stallion:					
Breeding	54.2	3.0	2.6	79.9	3.9
Non-breeding	45.5	3.6	1.4	64.1	4.9
Weanling (6 months old)	35.6	1.0	2.4	55.8	1.8
Yearling (12 months old)	35.8	2.0	1.7	58.5	3.1
Long yearling (18 months old)	34.4	2.8	1.0	61.2	4.4
Two-year-old (24 months old excluding work)	34.4	2.8	1.0	61.2	4.4
Maintenance plus 1 h moderately hard work	54.0	2.8	2.7	98.0	5.0

(Based on National Research Council 1978)

* Hay containing 8.2 MJ/kg and 88% dry matter.
† Concentrate mixture containing 11.4 MJ/kg and 88% dry matter. Quantities of concentrates up to 1.5% bodyweight daily may be fed if a minimum roughage allowance of 1 kg/100 kg bodyweight is given.
‡ 2.3 kg extra feed daily should produce 0.5–0.6 kg gain.
§ Assumed peak daily milk yield (kg).

	500 kg			600 kg		
Concentrate mixture† (kg)	Digestible energy (MJ)	Hay* (kg)	Concentrate mixture† (kg)	Digestible energy (MJ)	Hay* (kg)	Concentrate mixture† (kg)
—	69.7	8.5	—	79.5	9.7	—
2.1	80.5	6.2	2.6	89.8	7.2	2.7
5.6	122.2 (21)§	6.0	6.4	141.1 (23)§	7.2	7.2
3.1	103.4	7.6	3.6	120.4	8.7	4.3
4.2	97.2	4.9	5.0	113.7	5.8	5.8
2.1	77.7	6.0	2.5	90.5	7.0	2.9
3.6	67.1	2.2	4.3	74.7	2.3	4.9
2.9	71.4	3.7	3.6	80.4	4.1	4.1
2.2	73.4	5.2	2.7	80.6	5.8	2.9
2.2	71.1	5.2	2.5	82.9	5.8	3.1
5.0	119.0	5.2	6.7	147.0	5.8	8.7

Table 6.4 Digestible energy (DE) demands of maintenance and work* on the flat

Bodyweight (kg)	200	400	600
Approx. feed capacity per day (MJ of DE)	60	100	150
Maintenance requirement per day (MJ of DE)	35	58	79
	Energy requirements for work above maintenance (MJ of DE)[†]		
Walking (1 hour)	0.4	0.8	1.3
Slow trotting, some cantering (1 hour)	4.2	8.4	12.5
Fast trotting, cantering, some jumping (1 hour)	10.5	20.9	31.4
Cantering, galloping, jumping (1 hour)	25.0	50.0	75.0
Strenuous effort, racing, polo (1 hour)	42.0	85.0	127.0
Slow trotting, some cantering (10.4 h, 100 km) calculated from above	43.5	87.0	130.5

* Based on National Research Council (1978) and more recent evidence (see text).
† 1 kg concentrate provides about 12 MJ of DE.

Table 6.5 Effect of a range of required energy densities (MJ of DE per kg air-dry feed) on the cereal content of the daily ration when hays of two energy contents are available

Energy density of ration required	Oats (%)		Barley (%)	
	7.2*	7.8*[†]	7.2*	7.8*[†]
7.5	7	0	5	0
8.0	19	5	14	4
8.5	30	19	23	14
9.0	42	32	32	24
9.5	54	46	41	34
10.0	65	60	50	44
10.5	77	73	59	54
11.0	88	86	68	64

* Energy content of hay (MJ of DE per kg) 7.2 MJ/kg, medium quality; 7.8 MJ/kg, good quality.
† Hay can be assumed to contain 86% dry matter and where haylage of 45% dry matter is to be used it may be substituted for the hay of 7.8 MJ of DE in the proportions 1.8–1.9 kg haylage per 1 kg hay. Similarly, 1.6–1.7 kg haylage of 50% dry matter could be used.

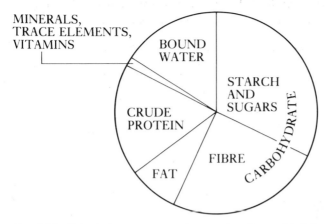

Fig. 6.11 Principal components by weight of a stabled horse's diet (excluding drinking water)

protein concentration of their diet for growth could be reduced from 14 to 8 per cent, owing to increased feed intake, although the surplus protein was probably greater at the higher percentage. Therefore, in stating requirements as fractions of feed weight some average appetite must be assumed. With this assumption the amounts of protein, calcium and phosphorus needed per g of ration are given in Table 6.6 and the levels in the

Table 6.6 Nutrient concentration in diets for horses and ponies expressed on the basis of 90 per cent dry matter

	Crude protein (g/kg)	Calcium (g/kg)	Phosphorus (g/kg)
Mature horses and ponies at maintenance	80	3.2	2.0
Mare, last 90 days of gestation	100	4.5	3.0
Lactating mare, first 3 months	125	4.5	3.0
Lactating mare, 3 months to weaning	110	4.0	2.5
Creep feed	160	8.0	5.5
Foal (3 months old)	160	8.0	5.5
Weanling (6 months old)	145	6.0	4.5
Yearling (12 months old)	120	5.0	3.5
Long yearling (18 months old)	100	4.0	3.0
Two-year-old, (light training)	90	4.0	3.0
Mature working horse, light to intense work	80	3.2	2.0

(Based on National Research Council 1978)

concentrate, hay mixture can be calculated by multiplying their protein contents (g/kg) etc. (see Appendix C) by their proportions and summing. Thus, for hard hay and oats: 55 g × 0.6 + 96 g × 0.4 = 71.4 g of protein per kg of mixture. If the protein requirement is 100 g/kg, then the discrepancy of 28.6 g can be made good by substituting some soyabean meal (other suitable protein sources may be substituted in inverse proportion to their lysine contents) for some of the oats, but attributing the soya (or other protein) with a protein content which is the difference between its content and that of oats, that is 440 g − 96 g = 345 g of protein (see Appendix C). Then the proportion of soya to include and of oats to remove per kg is 28.6/345.0 = 0.083, that is 8.3 per cent or 83 g/kg of mix. For simplicity, the effects of this substitution on the energy, calcium and phosphorus contents of the mix can be ignored. A similar calculation using Appendix C and detailed in Appendix A will have been adopted for calcium and phosphorus but the deficits can be made good by adding limestone flour and/or dicalcium phosphate without making any substitution as for soya/oats. A calcium source of this type will normally be essential unless large amounts of lucerne (alfalfa) or other legume forage are provided.

Table 6.7 gives the requirements for vitamins, trace elements, sodium, potassium and magnesium, supplements of which should be unnecessary if a commercially prepared mixed feed is used at recommended rates. Where compounded feeds are not given a proprietary mixture of trace elements and vitamins should be used, because in high concentrations such nutrients are toxic and the normal horse owner is unlikely to have facilities for their proper handling and weighing. Sometimes compounded feeds intended to form the entire concentrate portion of the ration are used as supplements to oats so diluting their effects in so far as protein, minerals, vitamins and trace elements are concerned. Thus a pellet providing 4000 iu of vitamin A per kg, mixed 50:50 with cereals and in turn fed 50:50 with hard hay will provide approximately 1000 iu of vitamin A per kg of total diet. Dilution of this kind is frequently a cause of incorrect Ca:P ratios in rations.

A discussion of the electrolytes − sodium, potassium and chloride − is given in Chapter 9, where problems of hard training are tackled. The potassium and magnesium needs of normal activity should be automatically met where good-quality roughage is available. The sodium needs (Table 6.7) can be met by providing common salt, for simplicity ignoring the contribution made by natural ingredients. Sodium comprises 40 per cent of common salt so that the common-salt allowance should be two and a half times the sodium allowance, that is 8.75 g of salt per kg provides 3.5 g of sodium. In order that any excessive salt intake is satisfactorily counteracted, clean drinking water, free from contamination with salts, should always be available (Table 6.8), excepting the restrictions to its use discussed in Chapter 4.

The dietary allowances for trace elements, as distinct from those for the major minerals and electrolytes, do not change appreciably per unit of feed with a change in the function of the animal from growth to work or

Table 6.7 Minerals and vitamins per kg diet adequate for horses

	Adequate levels	
	Maintenance of mature horses	Mare last 90 days of gestation and lactation and growing horses
Sodium (g)	3.5	3.5
Potassium (g)	4.0	5.0
Magnesium (g)	0.9	1.0
Sulphur (g)	1.5	1.5
Iron (mg)	40	50
Zinc (mg)	60	80
Manganese (mg)	40	40
Copper (mg)	15	25
Iodine (mg)	0.1	0.2
Cobalt (mg)	0.1	0.1
Selenium (mg)	0.2	0.2
Cholecalciferol (μg*)	10 (400 iu)	10 (400 iu)
Retinol (mg†)	1.5 (5000 iu)	2.0 (6666 iu)
D-α-tocopherol (mg‡)	30	30
Thiamin (mg)	3.0	3.0
Riboflavin (mg)	2.2	2.2
Pantothenic acid (mg)	12	12
Available biotin (mg)	0.2	0.2
Folic acid (mg)	1.0	1.0

(Based on National Research Council 1978)

* 1 iu (international unit) is equal to the biopotency of 0.025 μg of cholecalciferol (vitamin D_3) or ergocalciferol (vitamin D_2).

† 1 iu is equal to the biopotency of 0.3 μg of retinol (vitamin A alcohol). Grass carotene has 0.025 of value of vitamin A on a weight basis.

‡ 1 iu vitamin E is the biopotency of 1 mg of DL-α-tocopheryl acetate. Where 50 g of supplementary fat of average composition are added per kg feed, the requirement rises to 45–50 mg of α-tocopherol per kg, equivalent to 79–88 mg of DL-α-tocopheryl acetate. Also see p. 56 for working horses.

Table 6.8 Characteristics of a good water source

	mg/litre		mg/litre
Ammonia (albuminoid)	<1.0	Calcium	50–170
Permanganate value (15 minutes)	<2.0	Lead	<0.05
Nitrite, N	<1.5	Cadmium	<0.05
Nitrate, N	<1.0	pH 6.8–7.8	
Total dissolved solids	<1000		

to various phases of reproduction. The secretion of iodide in mare's milk may slightly increase her basal needs for iodine, but the margin given in Table 6.7 should satisfy all needs. The difference between this allowance and the toxic level (the margin of safety) existing for selenium and iodine (the elements for which the difference in absolute quantities is probably least) is adequate in the hands of responsible individuals. However, it is unwise and potentially dangerous for the normal horse owner to handle pure forms of trace elements.

The dietary allowances for the fat-soluble vitamins A, D and E per unit of concentrate feed (Table 6.7) again are not varied much per unit with a change in the function of the individual, bearing in mind also the capacity of the horse to store them. A reappraisal of the situation would be necessary if there were any radical change in the basic raw materials traditionally used for feeding horses, for example from dry cereal grains to root vegetables, or to high-moisture conserved cereals, the elimination of conserved pasture, and so on.

Appendix B gives examples of dietary errors encountered by the author in practice. Such errors cover the whole range of dietary attributes.

Reproduction and lactation

The dietary requirements of the breeding mare can be arbitrarily divided into those for: (1) the first 8 months of gestation; (2) the last 3 months of gestation; and (3) lactation (this may coincide with 0–4 months of gestation). Gestation length for most Thoroughbreds is in the range 335–345 days and for other breeds 322–345 days. Lactations of 110–130 days are typical of many husbandry systems, although the non-pregnant mare would produce milk for much longer if given the opportunity.

The first 8 months of gestation have no practical impact on the nutrient needs – that is they do not raise requirements above those for the barren mare, nor do they increase the already high requirements of the lactating mare. Thus, after weaning, this mare's energy requirements approximate those of maintenance until 8 months of gestation have been completed. Most of the foetal growth occurs during the last 90 days of gestation, even so the nutrient drain incurred then to sustain normal foetal growth is much less than that for lactation. The approximate energy contents of the foetus and the other products of conception at term compared with the energy content of mare's milk over a 4-month lactation are given in Table 6.9. By assuming that all foetal growth occurs during the last 90 days, the approximate DE required daily above maintenance to meet these needs can then be calculated (see Table 6.10). For comparison the much greater demands of milk production are also given in Table 6.10.

Table 6.9 Approximate energy contents of the foetus and other products of conception at term compared with the energy content of mare's milk over a 4-month lactation

Mare's weight (kg)	200	400	500	600
Products of conception at term (MJ)	110	200	240	270
Lactation of 17 weeks (MJ)	1700	2840	3400	3900

Table 6.10 The digestible energy (DE) required daily to meet the needs of foetal growth and lactation, but excluding the maintenance requirements for energy of the mare

Mare's weight* (kg)	400	500
		DE (MJ)	
Products of conception at term	5.0	6.0
Average milk production	40	50

* The mare's weight should increase by 15% during gestation so that her maintenance requirements rise proportionately.

As the foetus occupies an increasing proportion of the mare's abdominal cavity, her capacity for bulky feed declines during the period in which nutrient requirements increase. This may correspond to an increase in the quality of grazing (see Ch. 10), but where mares are given hay and concentrates the quality of the forage should be improved during the last 3 months of gestation. The energy and nutrient requirements of the breeding mare (Table 6.3) increase from period (1) to (2) and from (2) to (3) of the cycle given above. (Note that if the mare is lactating during part of period (1), her requirements then will exceed those of period (2).) However, the values are averages and the amounts of feed and therefore DE given to individual pregnant mares should be adjusted to avoid either obesity or poor condition (see also Ch. 7). Workers at Hannover (Meyer & Ahlswede 1976) estimated that a 500-kg mare requires 71, 75, 75 and 84 MJ of DE per day, respectively, during months 8, 9, 10 and 11 of gestation. This demonstrates an increasing energy need right up to parturition and during the eleventh month it is equivalent, in energy terms, to a 0.33:0.66 mixture of oats and hay of nearly 10 kg daily. (Part of the demand in the eleventh month is to sustain udder development.) On average, these values differ little from the flat-rate recommendations of Table 6.3. Feeding studies have not established whether birthweight of the foal is generally influenced by deviations from those rates. One investigation (Goater, Meacham et al. 1981) demonstrated an increase of both 1.5 kg in birthweight and 0.24 kg in daily weight gain during the first 30 days of life as a result of providing the mare with 120 per cent of the gestation rates. Other experiments in which Quarter Horse and Thoroughbred mares were restricted to 55 per cent of the rates and Arabian mares to 85 per cent of the rates led to weight losses by the pregnant mares without affecting birthweight of the foal in comparison with mares receiving the recommended rates. Clearly, healthy mares possess the capacity to adapt without the foal incurring any significant handicap.

The most critical nutrients for breeding mares given traditional feeds are protein, calcium and phosphorus. Mares kept during the last 90 days of gestation entirely on reasonably good-quality pasture or high-quality conserved forage containing some 30–40 per cent leafy clover, lucerne or sainfoin require no other source of calcium, and if the forage contains 10 per cent protein per unit of dry matter, no supplementary protein. The phosphorus requirements, however, amount to 3 g/kg (0.3 per cent) of the total dry diet. This would be provided by good pasture (see Table 10.2), but as both grass and legume hays given to horses in the United Kingdom normally contain less than 2 g/kg (0.2 per cent), a supplement of dicalcium phosphate, steamed bone meal or wheat bran will be required. The discrepancy would be met daily by a supplement of 60 g of dicalcium phosphate or 1.5 kg of bran for horses and 40 g or 1 kg, respectively, for ponies. Where grass hays are used, as assumed in Table 6.3, supplementary protein will be required as well. If the hay contains 7 per cent of protein and it constitutes 70 per cent of the diet, whereas the diet as a whole should contain 10 per cent protein, then the concentrate must contain 16–17 per cent of protein. This is the level found in most commercially prepared stud nuts that would also provide the necessary phosphorus, vitamins and trace elements.

Abundant good-quality pasture will meet the energy, protein, calcium and phosphorus needs of lactation, even though the minimum dietary protein requirement will have risen to 125 g/kg of dry feed (12.5 per cent). Responses in milk yield have been obtained from Quarter Horse mares by providing mixed feeds containing up to 170 g of protein per kg. However, grass and clover proteins are of high quality and it is unlikely that economic responses would be obtained by raising the protein level of the spring grazing diet. If the stocking density is high or good-quality pasture is otherwise moderately limited, supplementation can be provided by horse and pony nuts (see Table 7.1) or a mixture of these and cereals. If pasture is more scarce, a stud nut or an equivalent mix containing 16–17 per cent of protein should be given to lactating mares. Any conserved forage provided should be leafy hay containing a mixture of clover and grass, or should be well-conserved haylage. Typically in the United Kingdom grass hays of only moderate quality are fed when grazing is limited. These contain only 40–80 g of crude protein per kg and so do not meet immediate needs.

It is unlikely that one could overfeed lactating mares with roughage, except that large quantities of poor roughage would limit their capacity for concentrates leading to a decrease in milk yield. When typical grass hays are given, a satisfactory milk yield is obtained only if at least 50 per cent of the dry feed is provided as a stud nut, or equivalent 16–17 per cent protein mix. This mix may be based on oats or barley and soyabean meal, or an equivalent proprietary protein concentrate containing 440 g protein per kg. Horse mares would require 1.75 kg of soya daily (12–14 per cent of the total grass hay and cereal-based ration).

The reason why a 16–17 per cent protein mix is sufficient for lactating horses, as well as for late pregnancy despite a higher protein requirement in lactation, is that the mix forms over half the ration in lactation, to meet the greater energy needs, whereas it forms only 30 per cent of the prefoaling diet.

This grass hay, cereal, soya diet must be supplemented with a mineral mix composed of 35 g of dicalcium phosphate, 65 g of limestone and 70 g of sodium chloride when the total daily intake of dry foods is 14 kg. Proportionately less will be required for smaller rations. A proprietary mixture of trace elements and vitamins should also be given. The latter should include vitamin A for horses with no access to pasture. Where large amounts of silage or haylage are used, supplementary vitamins D and E will also be necessary at levels respectively of 7000 iu and 250 mg daily. Whereas the trace-element content of mare's milk, and therefore the adequacy of the foal's diet, is affected by the supplementation of the dam's diet, a deficiency of water, energy, protein, calcium or phosphorus will ultimately bring about a decrease in milk output, without altering its composition.

Growth

As horses grow they do not simply increase in weight and size – they also display what is termed development. Various tissues and organs of the body grow at different rates. In proportion to body size the rate of weight gain of the body as a whole, if permitted by the feed allowance, is very much greater in the younger than in the older animal. In fact, from the suckling period onwards the rate of gain per 100 kg of bodyweight declines continuously, but the rate of growth of long bone and muscle declines at an even faster rate. An increasing proportion of the gain constitutes fat, which has much higher demands for feed energy. These trends are fundamental to a formulation of requirements for protein, calcium and phosphorus in particular, which decline fairly rapidly as a proportion of total diet with increasing age of the foal and yearling (see Tables 3.2 and 6.6). Further details of growth and the way in which we should guide it, and indeed of the breeding mare, are given in Chapters 7 and 8.

Appetite

There is much yet to be learnt about factors that influence the appetite of horses, but reluctance or eagerness to eat can be assessed either as the amount of dry feed consumed consistently per 24 hours, or the amounts consumed in single meals. The 24-hour feed consumption depends on: (1) volume of gastrointestinal tract, (2) energy demands, (3) rate of passage of ingesta, and (4) over a period of several days, the energy density of the feed.

The ingesta derived from very coarse, poorly digested, long-fibrous feeds, if present in significant amounts, will linger in the large intestine and depress the daily intake of dry matter. Although this is generally a disadvantage, it can occasionally help in containing the appetite of overfat animals. The proclivity of a horse to start eating energy-yielding feeds depends, according to the most recent evidence (Ralston & Baile 1983), upon a fifth factor: the relative absence of products of digestion, including glucose, in the small intestine. To a lesser extent concentrations of blood glucose may play a small part. Under conditions of natural grazing, individuals will feed during perhaps fifteen or twenty periods throughout the 24 hours. A series of small meals reflects not only the low capacity of the stomach, but probably more directly the switching-off mechanism of digestion products in the small intestine. High caecal concentrations of volatile fatty acids (VFA), especially of propionate, have an immediate but small depressing effect on appetite by extending the interval between meals in ponies fed *ad libitum* and by reducing meal size at the time, yet they have no sustained effect over 24 hours. Lower increases in caecal VFA may even stimulate appetite. Thus, many factors influence the capacity for feed but an estimate of the daily appetite of average healthy animals for leafy hay and oats can nevertheless be given (see Fig. 6.4).

A true appetite apparently exists not only for energy but also for water and salt. Thirst will sometimes drive a horse to consume poisonous succulent plants, as a source of moisture, in an otherwise arid landscape. The appetites for water and for salt are interrelated. Thirst depends to a considerable extent on dehydration, or increased osmotic pressure of body fluids. When blood is hypertonic, horses will normally drink; if much salt has been lost through sweating the body may be dehydrated, the fluids will be hypotonic and thirst is then engendered by giving salt. The appetite for salt varies among horses – deprivation causes a greater craving in some than in others, despite a more uniform essentiality. On the other hand, an extreme thirst probably takes precedence over appetites for both salt and energy.

Feed in an unacceptable physical condition, and more importantly with an unfamiliar or unattractive aroma, will also be a deterrent to the initiation of feeding. Stale feed, which has incurred the oxidation of many of its less-stable organic components, is unlikely to encourage a horse to eat. Occasionally spices are used to cover, or camouflage, inadequacies of general acceptability and although horses can become 'hooked' on these additives they are generally to be discouraged.

In the stable, horses are fed generally by the bowl, and overfeeding, or apparent loss of appetite, can frequently reflect lack of recognition on the part of the groom of differences in the bulkiness and energy density of feeds (see Table 5.2). For example, taking both these characteristics into account, many of the better, coarse feeds currently available provide 25 per cent more energy per bowl than is provided by a similar volume of crushed oats. Failure on the part of the horse to eat up may then simply mean that it has been given 25 per cent more energy in its hard feed than it is familiar with or requires.

Rate of feeding

A too-rapid consumption of cereals and concentrates by stabled horses is sometimes a problem needing attention. Although the consumption of long hay is a relatively slow process, cereals and other concentrates are normally eaten first at a particular meal, but the form in which the concentrate is given can have some impact on the rate at which it is consumed. German workers (Meyer, Ahlswede & Reinhardt 1975) showed that 1 kg of feed in the form of oat grain or a pelleted concentrate took horses weighing between 450 and 550 kg about 10 minutes to consume, but meal took longer. Feeding time and the number of chewing movements were, however, increased by 30 to 100 per cent if 10–20 per cent of chopped roughage was added to the oats or concentrate. Finely ground meal mixes took longer to eat than crushed grains or pelleted feed; the addition of chopped roughage to fine mixes speeded up intake. Poor quality retarded consumption rate suggesting a place for good-quality barley straw in occupying the time of greedy horses.

Many horses develop vices, such as wood chewing, faeces-eating (coprophagy), or less-frequently cribbing (forced-swallowing gulps of air) (see Pl. 11.1). Boredom in isolated boxes is a contributory factor in their initiation and the provision of long hay or good-quality straw does help to circumvent this, but does not eliminate their occurrence.

Sound teeth enable horses to grind roughage to a small particle size, but decaying teeth, or molars with sharp points abrading the cheek and tongue (see Fig. 1.1), encourage bolting and poor mastication with the consequences of rapid intake, choking and sometimes colic. Dental treatment is clearly indicated, together with the provision of coarsely and freshly ground cereals and roughage chopped in short lengths. These may be given mixed together in a wet mash, which in turn may reduce wastage. The dampening of feed is, however, generally unnecessary except when large quantities of bran and beet pulp are given, in which case time for the thorough absorption of water should be allowed so that their true bulkiness is realized. Linseed also swells considerably on soaking and boiling, the latter, of course, is to be recommended as described in Chapter 5.

Where feeds are dusty, damping down lays the dust, so decreasing the likelihood of irritation to the respiratory tract in susceptible individuals. The damping of feed may also inhibit the segregation of its components. If feeds are too dry minerals, for example, may sift to the bottom and be left unconsumed in the manger.

Processing of feed

Preferred methods for processing raw cereal grains are coarse grinding, cracking, rolling, crimping (passing between corrugated rollers), steam flaking and micronizing. The overall objective of each is to improve digestibility and acceptability, but the last two may accelerate the onset of satiety during a meal. The processing of oats and barley is generally difficult to justify if the costs exceed 5 per cent of the cost of the raw material, but it is necessary shortly before feeding for the small grains, milo and wheat. Steam flaking or micronizing tends also to extend shelf life and destroy most heat-labile antinutritive substances. Normal cooking is, however, inadequate for the destruction of some mould toxins found in poorly harvested crops (see also Ch. 5).

Feeding frequency and punctuality

Horses, like most other animals, are creatures of habit and their reactions are in part affected by an inner clock. Thus the wise groom sticks to regular feeding times, week in and week out, and with equally regular exercise metabolic upsets and accidents can be avoided, and damage to stable doors can be decreased. Designs for feed mangers and water bowls recommended by the French Ministry of Agriculture are shown in Figs. 6.12 and 6.13.

Fig. 6.12 Designs for feed mangers (units in mm): (a) unbreakable manger; (b) manger embedded in concrete. (From Ministère de l'Agriculture, Service des Haras et de l'Equitation, Institut de Cheval, Le Lion d'Angèrs *Amènagement et Equipment des Centres Equestres, Section Technique des Equipments Hippiques*, Fiche N° CE.E.13).

(a) Cast ductile aluminium alloy or similar

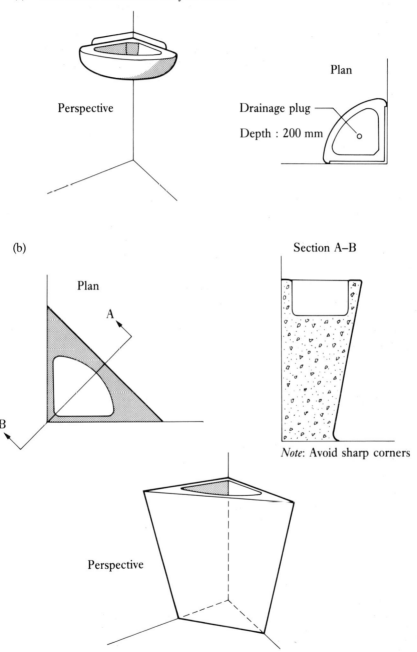

Perspective

Plan

Drainage plug

Depth : 200 mm

(b)

Plan

A

B

Section A–B

Note: Avoid sharp corners

Perspective

(a)   Cast ductile aluminium alloy or similar

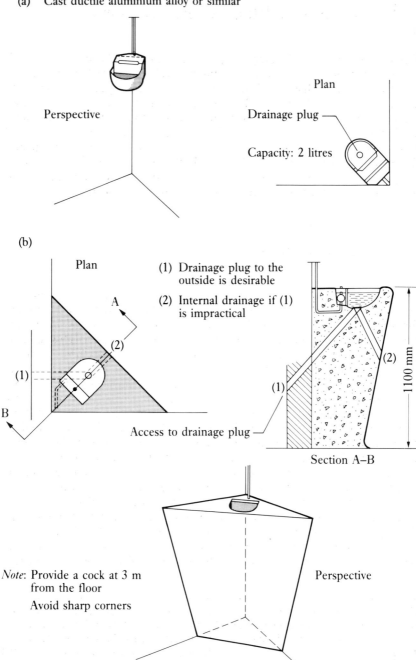

Perspective

Plan

Drainage plug ⎯

Capacity: 2 litres

(b)

Plan

(1) Drainage plug to the
     outside is desirable

(2) Internal drainage if (1)
     is impractical

A

(2)

(1)

B

Access to drainage plug ⎯

(1)

(2)

1100 mm

Section A–B

Note: Provide a cock at 3 m
       from the floor

       Avoid sharp corners

Perspective

Cereal costs and their energy contents

Cereals, and more particularly oats, of low bushel weight provide less DE per kg than do cereals of higher bushel weights. Digestive upsets are minimized, when changing from one feed or cereal type to another, by ensuring that the DE intake does not simultaneously rise. A bowl of an average sample of oats may provide only 56 per cent of the DE of a similar bowl of average wheat (see Table 5.2). As a result of its bulkiness oats tend to be a safer feed than the other cereals although, through careful and expert feed management, no differences in this respect should arise. Therefore, in assessing the feeding value of the various cereals, account should be taken of the purchase price per unit of DE (see Appendix C).

Traditional feeds

For some considerable time many stables in the British Isles have given their horses a diet based on grass hay, oats and a little cooked linseed. Before the advent of the tractor-drawn baler, hay was probably made for horses with greater care than it is today when the horse in many situations no longer reigns supreme in its claims to conserved forage and grazing privileges. In terms of nutrients, the diet now leaves a little to be desired, but the adult horse has some capacity to adapt to its culinary shortcomings because it may take a considerable period of time for a large non-productive animal to be depleted of essential nutrients. Many of these nutrients are made good where summer grazing is available and nutrients provided by productive pasture in summer will serve as a reservoir over the ensuing winter until the spring grass again refurbishes losses. Lactation of the mare and early growth of young stock rely traditionally almost entirely on the availability of grazing, any deficiencies of which are soon revealed.

A diet of oats, hay and linseed is alien to the natural inclinations of a browsing animal so that there is no good reason why other equally strange feeds should not be used if economies can be achieved and deficiencies rectified. There are now better protein sources than linseed and there are oils less prone to peroxidation than linseed oil as these other oils contain

---

Fig. 6.13 Designs for water bowls (units in mm): (a) unbreakable water bowl; (b) water bowl embedded in concrete. (From Ministère de l'Agriculture, Service des Haras et de l'Equitation, Institut du Cheval, Le Lion d'Angèrs *Aménagement et Equipment des Centres Equestres, Section Technique des Equipments Hippiques*, Fiche N° CE.E.14).

much lower concentrations of linolenic acid and a higher proportion of linoleic acid. Stabilized forms of groundnut, soya, sunflower and corn oils are widely available today. By understanding the necessity for protracted periods of adaptation and by recognizing the differences in bulk and energy density among the various cereals, the careful feeder has a wide choice of feeds. The principal advantage of oats is to the novice feeder but this advantage could easily be supplanted (with a probable saving in costs per unit of energy) by the use of a variety of cereals and byproducts. However, the traditional feeder knows by long experience how many bowls of oats a performance horse will consume without refusal. Without careful weighing the quantities of alternative mixtures required can be misjudged and over-feeding or underfeeding can result.

Complete diets

Many horses and ponies receive diets based on compounded nuts and long hay. In normal circumstances this hay should be preferred to a source of ground fibre if the quality is good and the price acceptable. Compounded feeds described as complete pellets are for consumption by adult horses required to carry out moderate work where no other feeds are to be given. These pellets have the advantage of providing a reasonably balanced diet relatively free from dust for horses subject to respiratory irritation and allergy. The absence of long hay, or straw is, however, likely to cause a higher incidence of vices, such as wood chewing and coprophagy, but these annoyances are undoubtedly preferrable to the exacerbation of a respiratory problem. Rationing is simplified by the constant bulk and energy densities of the feed and storage space is minimized, but some problems of boredom require ingenuity to overcome.

Shelf life of feeds

The grinding of cereals increases their surface area exposed to the atmos-phere and deterioration proceeds continuously, retarded only by the pres-ence of natural antioxidants. Staleness detracts from feed acceptability. Legislation permits the addition of limited amounts of synthetic antioxidants to products for consumption by animals. The most commonly used of these are butylated hydroxyanisole (BHA), butylated hydroxytoluene (BHT) and ethoxyquin. These substances are safe, have the effect of extending shelf life and acceptability of feeds, and they cause no infringement of the Rules of Racing.

Some nutrients are particularly susceptible to destruction by light. Hence, concentrated feedstuffs should be covered during storage as the products of light destruction are also harmful to the stability of certain other nutrients. Atmospheric oxygen brings about the gradual destruction of fat-soluble vitamins and unsaturated fats. Among the more critical water-soluble vitamins thiamin (vitamin B_1) and pantothenic acid are somewhat unstable. The loss of these nutrients and other unstable compounds detracts from the value of feedstuffs. However, whole grains in an uncrushed or unground form remain relatively stable for several years with only a slight decline in feeding value, although this, of course, assumes good storage, freedom from pests, moisture levels of less than 135 g/kg (the lower the better) and absence of cracked and moulded grains. Discoloration of cereal grains may indicate only superficial microbial damage during ripening as observed when one compares good-quality Scotch oats with their counterparts from Canadian or Australian sources. However, discoloration may be an indicator of more profound internal damage through fungal infection, which will seriously impair feeding value and stability in storage. Thus, bright and bold grains are to be preferred in the absence of other information. Other aspects of storage have already been discussed in Chapter 5.

Feed contaminants

Dressed seed corn or cereal grains that have been exposed to toxic pesticides should not be used for feeding to horses, although several pesticides leave quite harmless residues.

Further reading

The National Research Council (1978) *Nutrient requirements of domestic animals*. No. 6. *Nutrient requirements of horses*. National Academy of Sciences: Washington.

CHAPTER 7

Feeding the breeding mare

You should wean your foals at the beginning of winter, when it beginneth to grow cold, that is about Martinmas, or the middle of November, and wean them three days before full moon, and hang about their necks upon a piece of rope seven or eight inches of the end of a cow's horn, to catch hold of them upon occasion, after which bring them all into your stable, with racks and mangers pretty low set

Sieur de Solleysel, 1711

The oestrous cycle and fertility

The natural season for maximum breeding activity in both the mare and the stallion in the British Isles is from April to November, but the breeding season can be shifted by artifically changing daylight length and by manipulating the diet; a geographical move to the Southern Hemisphere can, of course, have comparable effects. During the season the normal mare expresses consecutive oestrous cycles of approximately 22 days long; within each cycle there is a period of oestrus of varying intensities that lasts on average 6 days. The fertility of the oestrus is low at the start of the season, but the creation of large follicles that ovulate and generate corpora lutea can be stimulated by the extension of day length and dietary adjustment. High fertility tends to coincide naturally with the flush of grass in late spring – it was noted in the very dry summer of 1976, when many pastures became non-productive and bleached, that breeding mares stopped cycling; they went into anoestrus when the ovaries became small and inactive.

An extension of daylight to 16 hours and an increase in the plane of nutrition in December will stimulate the onset of normal cycling 2 to 5 weeks later during the first months of the year. Successive oestrous periods will be of increasing fertility in healthy, barren or maiden mares and individuals that are increasing in bodyweight are more likely to conceive. Therefore, by starting with a lean individual in November and December, this objective is more likely to be achieved. It has, however, been suggested that forcing barren mares in December and January enhances the probability of twins; yet if early conception is desirable such a procedure is obligatory.

Pregnant mares should be kept fit but not fat as this reduces foaling difficulties and provides greater freedom for controlling milk secretion by feeding during lactation. The foal heat occurs within 14 days of foaling and subsequent heats occur at 22-day intervals in unbred mares. The recommended rate of feeding of lactating mares is given in Table 6.3, yet it has not been clearly established whether this is the optimum rate for maximum fertility of the foal and subsequent oestrous periods. Milky mares nevertheless have a greater tendency to resorb fertilized eggs at first oestrus. This could be the reason for a putative association between overfeeding during the last 3 months of pregnancy and a reduced subsequent fertility. Unfortunately the experimental evidence to support this assertion is conflicting. Jordan (1982) noted that no reduction occurred in conception among pony mares losing 20 per cent of body weight during gestation but allowed to gain weight during lactation. Heneke, Potter & Kreider (1981) reported that mares in thin condition at foaling had reduced conception rates, longer postpartum intervals and more cycles per conception. Conception rates of mares in good condition at foaling, but who lost weight during lactation, were as good as those of mares in good or thin condition at foaling that maintained or gained weight in lactation. Recent preliminary evidence in the USA indicates that mares foaling in a fat condition should be allowed to hold their weight, rather than lose it, and that thin mares should gain weight during lactation in order to maximize the pregnancy rate at 90 days post-foaling. It is concluded that thin mares should be fed well in lactation to stimulate fertility.

Experience with dairy cows might suggest that if mares are fed too liberally through gestation and given inadequate feed during lactation they are more prone to a fatty liver condition, known to reduce fertility in the dairy cow. Observation of both horses and ponies shows that various stresses during late pregnancy and early lactation, accompanied by an inadequate and impoverished diet, predispose the mare to an extreme metabolic upset associated with loss of appetite, abnormal reactions, diarrhoea, hyperlipidaemia and eventual death. This probably represents a breakdown in energy metabolism with liver fat accumulation, as happens in the dairy cow. Any imposed weight reduction in obese pregnant mares should therefore take place before the last 3 months and, for preference, the fatness should be corrected before breeding is instigated. This may be achieved by providing the mare with good-quality hay, but no concentrated feed.

The pregnant mare

The gestation period of the mare commonly lasts for 335–345 days but may continue for 1 year. The period in part depends on the month of breeding. In the Northern Hemisphere early bred mares (that is those conceiving before the end of April) normally have a gestation period exceeding 350 days and up to 365 days. Those bred in May normally foal after 340 to 360 days, and those bred in June and July generally foal after 320 to 350 days. The critical factor may be day length during the last 3 months of pregnancy, as when the photoperiod was artificially extended to 16 hours in late gestation of Quarter Horse mares (Hodge, Kreider et al. 1981), the gestation period was shortened by 11 days and the interval from parturition to first ovulation was decreased by 1.6 days in comparison with mares subjected to natural day length.

The feed requirements of pregnant mares are given in Chapter 6. The condition of an overfat pregnant mare can be improved by gradually reducing the cereal component of the ration while the protein and mineral mixture is maintained at the previously determined level of intake.

Clearly, the mare is capable of considerable adjustment to a variety of situations. But in the extremes, excesses or deficiencies of energy will lessen her reproductive efficiency. During winter or summer the ribs of a mare should not be seen, but should be detectable by touch with no appreciable layer of fat occurring between them and the skin. Where diets are grossly imbalanced in terms of protein and minerals, especially calcium and phosphorus, the foal will be adversely affected at birth and reduced milk yield and infertility will ensue.

Parturition

In the 24 hours before birth of the foal, the mare should be fed lightly with good-quality hay and a low-energy cereal mixture including bran, or proprietary horse and pony nuts (10–11% protein, 3% oil, 14–15% crude fibre; Table 7.1), with access to restricted quantities of warm water. The first feed after parturition can effectively be a bran mash and the second can include some bran with small quantities of good-quality proprietary stud nuts (16–17% protein, 3% oil, 8% crude fibre; Table 7.1) or a cereal protein mixture. Obese mares tend to be less active and so poorer muscle tone may lead to birth difficulties and delayed expulsion of the placenta, which should be passed during the first hour after birth. The rate of concentrate feeding up to day 10 should be restricted in order to avoid excessive milk secretion and digestive disturbance in the foal. However, inadequate amounts of energy may contribute to the metabolic abnormalities outlined in the previous section. Recommended allowances are given in Table 6.3.

Perhaps 5 per cent (Rossdale & Ricketts 1980) or 10 per cent (Jeffcott, Rossdale et al. 1982) of foals may be lost through perinatal mortality, including stillbirths and postnatal mortality. Of these, significantly more are male. Although nutrition is a vital factor, the significance of it in this statistic is

Table 7.1 Composition of foal milk replacer and stud concentrate mixture to be given with hay and water

Mare's milk replacer (see footnote for mixing*)	(%)		Stud mixture† (%)	Horse and pony mix (%)
Glucose	20.0	Oatfeed	—	25.0
Skimmed-milk powder (20% fat)	5.0	Oats	46.0	33.0
Spray-dried skim milk powder	40.0	Wheat bran	15.0	20.0
Spray-dried whey powder	32.7	High-protein grassmeal (16% crude protein)	15.0	10.0
High-grade fat‡	1.0	Extracted soyabean meal	15.0	4.0
Dicalcium phosphate	1.0	Molasses	5.0	5.0
Sodium chloride	0.2	Fat‡	1.0	1.0
Vitamins/trace minerals§	0.1	Limestone	1.0	1.0
		Dicalcium phosphate	1.1	0.9
		Salt	0.75	0.5
		Vitamins/trace elements§	0.1	0.1
Total	100.0	Total	100.0	100.0

* Disperse in clean water at the rate of 175 g/litre (for 2 days after colostrum at 250 g/litre) or 1.75 lb/imperial gallon. Also can be pelleted and mixed with stud mixture as a weaning feed for orphan foals.

† This mixture is satisfactory as a creep feed and post-weaning diet. However, a mix specifically for young weaned foals to be fed with grass hay could to advantage contain an extra 5% soyabean meal replacing 5% oats.

‡ High-quality tallow, lard including dispersing agent. Stabilized vegetable oil could alternatively be added at time of mixing.

§ To provide vitamins A, D_3, E, K_3, riboflavin, thiamin, nicotinic acid, pantothenic acid, folic acid, cyanocobalamin, iron, copper, cobalt, manganese, zinc, iodine and selenium.

entirely unknown. Birthweight is a crucial characteristic in determining the prospects of foals and, despite the influence nutrition can have on this, size of the dam is a major controlling influence so that an acceptable minimum weight must depend on the breed and the purpose intended for the individual offspring. In Thoroughbreds, an early rapid growth rate is normally expected and required for work at an early age. For this, foals of less than 35 kg probably should not be kept. Where twins are born, their total weight approximates that of single births with a mean in the region of 55 kg for Thoroughbreds, implying that it is practical to retain only the heavier of the two.

Colostrum and acquired immunity

The mare must pass adequate passive protection to her foal through the colostrum, and she therefore should have been situated in the foaling area for, at the very least, 2 weeks and preferably a month before foaling. This means she will confer some immunity to the strains of microorganism peculiar to her environment, for example those causing scours, joint-ill and septicaemia.

Newborn foals will normally first suck within 30 minutes to 2 hours of birth. Colostrum is rich in protein (particularly immunoglobulins), dry matter and vitamin A. If foals are deprived of colostrum, an injection of about 300 000 iu of vitamin A is in order. Immunoglobulins do not pass through the dam's placenta and can be absorbed efficiently through the intestinal wall of the foal only during the first few hours of life. The major causes of colostrum deprivation in the foal are premature birth and delayed sucking, small intestinal malabsorption, premature leakage of milk through the teats or death of the mare. The immunoglobulins are concentrated by the mare in her udder within the last 2 weeks of gestation, when their level in the mare's serum falls. There is, therefore, a selective concentration of this protein fraction in the mammary gland.

If the foal suckles normally, the concentration of the immunoglobulin fraction in the colostrum, 12–15 hours after birth, is only 10–20 per cent of the initial value. It is known that the protein content of mare's colostrum is around 19 per cent during the first 30 minutes after parturition but by 12 hours the level falls to about 3.8 per cent and after 8 days it reaches a fairly constant level of 2.2 per cent (Ullrey, Struthers et al. 1966). The foal absorbs gamma-globulin as intact undegraded molecules throughout the first 12 hours of life so that the level of this protein fraction in its blood approaches a value somewhat lower than that of the dam's at parturition. Amounts of specific antibodies so acquired by the foal's blood decline from 24 hours of age; by 3 weeks the values are halved and by 4 months the titre of specific antitoxins provided by the mother is barely detectable.

The foal's own system for building active immunity in the form of autogenous gamma-globulins first provides detectable products at 2 weeks of age in the blood of colostrum-deprived foals and at 4 weeks in those reared normally. By 3–4 months of age the gamma-globulins have attained adult plasma concentrations. Up to this age, therefore, the foal is more susceptible to infection than is an adult in the same environment, particularly when it has received an inadequate quantity of colostrum, or colostrum at the wrong time.

If the mare has ejected much of her colostrum before foaling then it will be necessary to give the foal colostrum from another mare, preferably one accustomed to the same environment, or failing this, cow's colostrum rather than milk. The foal should receive about 500 ml (1 pint) of colostrum by nipple or stomach tube every hour for three or four feeds before 12 hours of age. After 18 hours the colostrum has little systemic immune value, although it does have some beneficial local effects within the intestinal tract.

Where the concentration of immunoglobulins in the foal's plasma is less than 400 mg per 100 ml the foal may be given blood plasma from another horse at the rate of 22 ml/kg bodyweight over a period of 1–2 hours, that is an amount totalling approximately 1 litre per foal. This should raise the antibody titre of its blood to about 30 per cent of the donor level. Although giving plasma orally will be of value in the prevention of enteritis, it should normally be given aseptically into a vein. A simple field test has been developed in which the turbidity of plasma is assessed following the addition of zinc sulphate; the results correlate well with concentrations of blood globulin in foals indicating whether sufficient antibodies have been absorbed in the neonatal period. This subject is discussed further under the section on orphan foals below.

Neonatal problems in the foal

Hygiene is generally outside the scope of this book, yet the importance of cleanliness in the foaling area cannot be overemphasized. It is essential that the foal receives colostrum to provide it with some protection (passive immunity) from potentially harmful organisms in the environment. Nevertheless the consumption of excessive quantities of milk can overload the digestive capacity of the foal and the milk may then become a substrate for rapid bacterial growth in the intestines. This situation can precipitate diarrhoea despite the consumption by the foal of normal quantities of colostrum.

There are two immunological problems related to the feeding of the foal, yet unrelated to disease caused by microorganisms. The foal's blood differs immunologically from that of its dam and on rare occasions it may react with the dam's immune system causing the production by the mare of isoantibodies to the foal's red cells. These antibodies are transmitted to the colostrum and the suckled foal may absorb sufficient to initiate a considerable destruction of red cells, precipitating an anaemia and jaundice, known as haemolytic icterus. In severe cases, the foal's urine will be discoloured with haemoglobin. If a mild attack is detected before icterus has occurred, the foal should not be allowed to nurse its dam for 36 hours. Where the mare has previously produced foals with the condition, it may still carry similar antibodies. In this case the foal should automatically be given colostrum from another mare at the rate of 500 ml (1 pint) every 1–2 hours for three to four feeds, followed by milk replacer until 36 hours, when it may be returned to its dam. In the meantime the mare should be milked out by hand.

If the problem is anticipated, blood samples can be taken from the foal and the red cells allowed to settle. The abnormality causes the plasma to appear pink rather than the normal straw colour. In these circumstances the foal may be severely anaemic and red cells from the mother may be slowly infused into a vein after syphoning off as much plasma as is conveniently possible. For preference, however, the source of red cells should be three or four geldings that have not previously received transfusions so the risk of immunological reactions is minimized. A simple procedure has been proposed as a means of precluding damage *before* colostrum is taken by foals that are considered to be at risk. One drop of umbilical blood is mixed with four drops of saline and five drops of mare's colostrum on a clean microscope slide, checking after several minutes for agglutination reactions.

At birth much of the large intestine, including the caecum and rectum, contains a substance, the meconium, which is normally completely voided within the first 2–3 days of life. Sucking usually sets up a reflex, promoting defaecation of this material. If this does not occur the normal passage of colostrum and milk may become blocked so that the gases formed during their fermentation cause distension and pain to the foal. It may then go off suck, act abnormally, crouch, lift its tail and flex its hocks in an effort to pass the offending material, or roll over in pain. Eventually a yellowish milk dung reaches the rectum, the meconium is cleared and the symptoms subside. The problem is treated conservatively by administration of a lubricant through a stomach tube, plus one or two enemas of soap and water, or liquid paraffin, and the injection of pain-relieving drugs. If the foal goes off suck for an extended period it should be given a fluid feed by stomach tube or appropriate intravenous solutions. Intravenous feeding of glucose and an isotonic electrolyte solution (see Table 9.2) is a life-saving procedure in cases of severe enteritis with consequent dehydration. In normal circumstances the foal will eat a quantity of its mare's faeces. In so doing it introduces beneficial microorganisms into the intestinal tract, which compete with pathogens present in the general environment.

Lactation

At a given stage of lactation the composition of mare's milk is remarkably similar among the various breeds of horse. The composition changes rapidly during the first days of lactation and then more slowly (see Figs 7.1–7.4 and Table 7.2).

Milk contains about 2.1 MJ of gross energy per kg. Recent work with eight Thoroughbred and two Standardbred mares receiving a diet of concentrates and hay showed that they achieved yields of 16, 15 and 18 kg daily (3.1, 2.9 and 3.4% of bodyweight daily, or 149, 139 and 163 g/kg$^{0.75}$) at 11, 25 and 39 days postpartum, respectively (Oftedal, Hintz & Schryver 1983). A group of Quarter Horses lactating for 150 days attained their maximum milk yield after 30 days. The average yield in early lactation was 11.8 kg daily and 9.8 kg during late lactation. Milk yields are markedly influenced by the mare's innate ability, by feed consumption during the latter stages of pregnancy, and, more importantly, by water availability (Table 7.3) and intake of energy and nutrients during lactation. Experimental work with Quarter Horse and Thoroughbred mares has shown that a reduction in the energy intake to 75 per cent of that recommended for lactation by the US National Research Council does not lead to a parallel decrease in foal weight at 75 days (Banach & Evans 1981a, b). Undoubt-

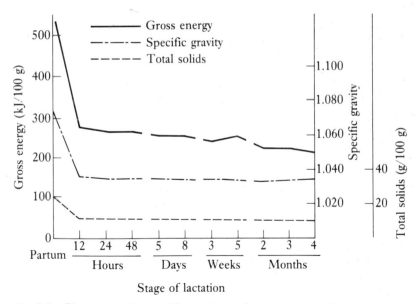

Fig. 7.1 Changes in the specific gravity and concentration of gross energy and total solids in mare's milk at various stages of lactation. (After Ullrey, Struthers et al. 1966).

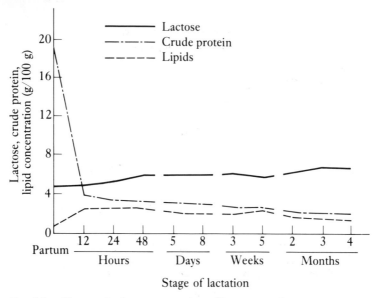

Fig. 7.2 Changes in the concentration of lactose, crude protein and lipids in mare's milk at various stages of lactation. (From Ullrey, Struthers et al. 1966).

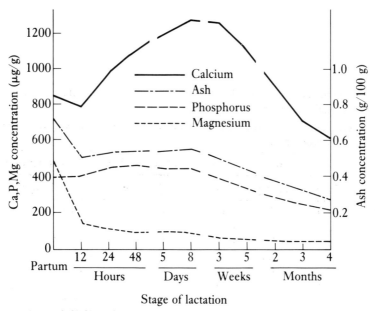

Fig. 7.3 Changes in the concentration of ash, calcium, phosphorus and magnesium in mare's milk at various stages of lactation. (From Ullrey, Struthers et al. 1966).

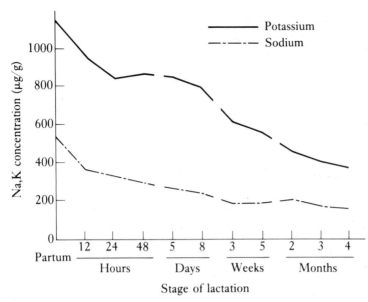

Fig. 7.4 Changes in the concentration of potassium and sodium in mare's milk at various stages of lactation. (From Ullrey, Struthers et al. 1966).

Table 7.2 Nutrient content of milk in Dutch warmblooded saddle horse, Quarter Horse and Thoroughbred mares

	Composition (%)				
	Dutch			**Quarter Horse**	**Thoroughbred**
	Initial	**28 days**	**196 days**	**2–150 days**	**24–54 days**
Total solids	13.0	11.2	10.5	10.5	10.5
Fat	2.7	1.1	—	1.3	1.3
Protein	3.3	2.0	—	2.1	1.9
Ash	0.53	0.20	—	—	—
Calcium	0.12	0.07	—	—	—
Phosphorus	0.1	0.04	—	—	—
Sugar	—	—	—	—	6.9

(From Bouwman & van der Schee 1978; Gibbs, Potter et al. 1982; Oftedal, Hintz & Schryver 1983)

Table 7.3   Water requirements in the stud

(a)   Water provided by pasture herbage (kg) per kg of herbage dry matter and total (including herbage) minimum water needs (kg) per kg of feed dry matter consumed

Water content of herbage per unit of dry matter (DM)			Water requirement of horses per unit of dry matter (DM) consumed		
	DM	Water		DM	Water
Spring growth	1	4	Last 90 days of gestation	1	3
Dry summer	1	2.5*	First 3 months of lactation	1	4
Mild winter	1	3	Breeding stallion	1	3
			Weaned foal	1	3
			Barren mare	1	2

*   When 'burnt' and bleached the proportion of water may be as low as 0.15–0.3.

(b)   Minimum supplementary water requirements in litres (and imperial gallons) per mare daily assuming average milk yields

Liveweight of mare: (kg)	200	400	500
(lb)	440	880	1100
Last 90 days of gestation†:			
In stable	12.7(2.8)	22.3(4.9)	26.4(5.8)
On pasture‡	0	0	0
First 3 months of lactation:			
In stable	27.7(6.1)	41.8(9.2)	49.6(10.9)
On pasture‡	7.3(1.6)	10.9(2.4)	12.3(2.7)

†   The breeding stallion's needs are similar to those of the pregnant mare.
‡   These amounts will be insufficient for mares on parched pasture in environmental temperatures exceeding 30°C (86°F) where shade should be provided.

edly the thriving mare has considerable capacity to adapt within limits to a restricted diet. Work with Dutch warmblooded saddle horses weighing from 533 to 676 kg, revealed that their maximum milk yield reached 23 kg/day with a mean initial yield of 14 kg rising to an average of 18 kg during the first 10 weeks and then falling to 11 kg by week 23 (Bouwman & van der Schee 1978). Mares weighing between 400 and 550 kg may be expected to yield 5–15 kg daily during the first few weeks, 10–20 kg daily in the second and third month, falling to 5–10 kg by the fifth month; but normal yields assume an ample supply of water (Table 7.3). The observations with the Dutch saddle horses indicated that sucking frequency averaged 103 times per 24 hours in the first week, falling to 35 times per 24 hours by week 10, and on each occasion the foal suckled for 1.3–1.7 minutes. (Small amounts taken frequently are unlikely to cause digestive disturbances.) At birth, the foals weighed 57 kg on average and the weight doubled in the first 37 days of life.

Suckling Thoroughbred and Standardbred foals have been shown to take in daily milk dry matter equivalent to 3.1, 2.1 and 2.0 per cent of bodyweight at 11, 25 and 39 days post partum, when weight gain averaged 1.14 kg daily (Oftedal, Hintz & Schryver 1983). The comparable daily intakes of gross energy were 39, 32 and 37 MJ, respectively.

Some high-yielding dairy cows suffer from a condition known as milk fever, or parturient paresis, which is probably caused by a sudden draining of blood calcium into the milk after parturition without an equivalent mobilization of bone calcium. Successful treatment entails giving cows a low-calcium diet from 2 to 4 weeks before parturition and then providing a diet relatively rich in calcium during lactation. A regulated, but high dose of vitamin D given 8–10 days before parturition has also been beneficial in some instances, but the dose has to be carefully calculated as a gross excess can be toxic. Pony and horse mares with a history of tetany associated with depressed blood calcium in lactation might well benefit from being given a diet containing 1.5–2.0 g of calcium per kg and approximately 3000 iu of vitamin D per kg throughout the last 2–3 weeks of gestation. It is difficult to predict parturition with sufficient accuracy 10 days before foaling and therefore single doses of large amounts of vitamin D at that time cannot be recommended. The diet should, of course, be adequate in other respects and during lactation the total diet of such mares should contain 5–6 g Ca per kg. Horses that are suffering from this form of tetany are slowly given calcium gluconate intravenously while cardiac action is continuously monitored.

Creep feeding of the foal

Foals will start to nibble hay and concentrates at 10–21 days of age and if the milk supply of the dam, or the amount of grass, is inadequate the provision of creep from this time may well enable a normal growth rate to be achieved. However, the principal objective of creep feeding is to accelerate the anatomical and physiological maturation of the gastrointestinal tract of the foal so that ultimate weaning presents no particular stress or hazard and digestive disturbances caused by abnormal fermentation of ingesta are prevented.

Where the mare secretes minimal quantities of milk, a creep feed based on dried skimmed milk provided from about 2 weeks of age is recommended. The composition of this feed should be changed gradually to that of a stud nut (16–17% protein, 3% oil, 8% fibre) or a growing foal diet (17–18% protein, 3% oil, 7% fibre) (see Tables 7.1 and 8.6) from 10 to 14 weeks of age. The use of milk pellets as a creep feed before weaning is contraindicated as it defeats the prime objective.

Foals that are growing well with mares on pasture have no particular need for additional dry feed until 2 months before weaning, say at 2–3 months of age. Here the functions are to compensate for a waning milk production in the dam, to redress the effects of a decline in pasture quality and probably more importantly to accustom the foal to the dietary regime it must expect after weaning. Thus the feed should be a concentrate of the type given in Tables 7.1 and 8.6 on which the foal will subsequently be maintained throughout the forthcoming winter and spring. Supplementary creep feeds should be restricted in quantity to 0.5–0.75 kg per 100 kg bodyweight. Such a restriction will give a measure of control over the incidence of growth-associated ailments, including epiphysitis and contracted tendons (flexural deformity) (Pl. 7.1). Where evidence of either of these is at hand a restriction of growth rate, imposed by cutting the supplementary feed and by reducing the mare's feed for a period of 3–4 weeks, should not prejudice ultimate mature size if carefully regulated. The extent of restriction must depend on how serious the problem is. Foals that stand high on their toes at birth should be exercised regularly and allowed to grow at a submaximum rate if the condition is to be contained. Overtly contracted tendons at birth are relatively untreatable and possibly result from intrauterine malpositioning. If therapy is possible, splints or extension shoes are used and the foals are exercised regularly with restricted access to feed until the abnormality is satisfactory.

The vertical growth of normal foals is very rapid throughout the first 3–4 months of life. Access to creep feeders should be controlled by regulating the width of the entrance rather than by restricting its height. Foals of breeds with mature weights of 550 kg can be weaned easily when consuming nearly 1 kg of creep feed and 0.5 kg of hay (or equivalent grass) daily by which time the foal should be in excess of 140 kg. These restrictions normally ensure that growth will not falter during the first postweaning week and by the end of the second week the rate of gain will be at least 1 kg daily in healthy stock. Several days before weaning it is appropriate to remove the mare's daily concentrate allowance and access to hay and pasture may also be restricted.

Epiphysitis (Pl. 7.2), probably more correctly called metaphysitis, is not uncommon in faster growing, larger, fine-boned foals of Thoroughbred, Saddlebred or modern Quarter Horse breeding. It is encountered particularly in the fetlock joint at the end of the metacarpus and in the 'knee' joint at the distal end of the radius. Where it is slight, the foal will probably aright matters, but where severe, supplementary feed should be restricted to good-quality roughage, the milk intake should be limited and the animal should be boxed until the worst of the 'bumps' subside. A restriction in the rate of weight increase allows joint maturation to continue without the stress of excessive pressure on the joints. Light exercise must then be undertaken daily and the normal feeding regime gradually reinstated. Exercise may, however, be damaging in severe epiphysitis, and in this case no analgesics should be used. Problems of this nature can arise in less than

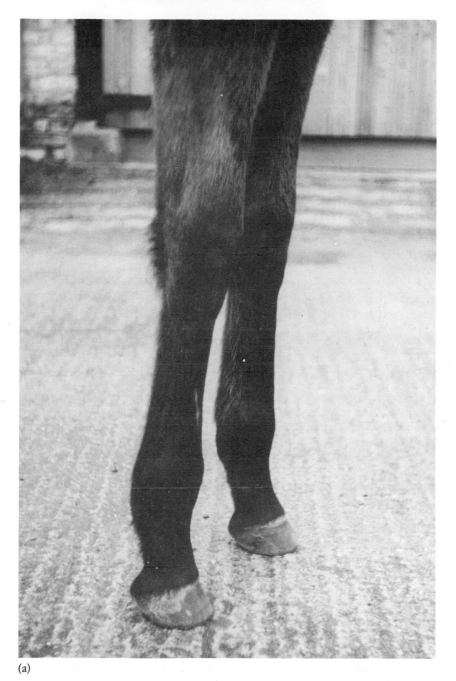

(a)

Plate 7.1 Chronic 'contracted tendons' in a yearling showing enlarged fetlock joints and upright stance. Later improvement was achieved by desmotomy of the superior check ligaments. Before surgery (a) and after surgery (b).

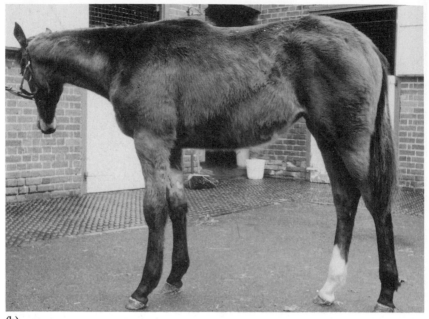

(b)

a week and may be complicated by angular deformities in one limb together with epiphysitis in the opposite limb. Successful treatment is contingent on immediate action. Attention by the farrier to the hooves should allow small misalignments of the limbs to be corrected during growth.

Glade, Gupta & Reimers (1984) have advanced an intriguing explanation of the relationship between epiphysitis and the energy and protein consumption at each meal. Excessive intakes suppress normal postprandial hyperthyroxaemia (raised plasma concentration of the thyroid hormone thyroxine, T_4), because an intense insulin secretion stimulates T_3 (triiodothyronine) formation from T_4, in turn inhibiting the thyroid-stimulating hormone (thyrotrophin, TSH) and thus T_4 secretion. Hypothyroidism is known to cause skeletal manifestations similar to epiphysitis and osteochondrosis, as T_4 is required for bone maturation, whereas insulin stimulates the formation of immature cartilage. Each of these diseases is characterized by enlarged growth centres, failure of bone formation from cartilage, occasional cartilage necrosis and cyst formation. The solution would seem to lie not only in the control of dietary energy and protein and the correction of errors in mineral and trace-element nutrition, but also in raising the number of daily feeds and decreasing their individual size. The logical extension of this may be to change to a system of earlier weaning and feeding foals *ad libitum* a complete mix described on page 166. This will encourage nibbling and avoid large postprandial surges in blood glucose and amino acids that stimulate insulin secretion.

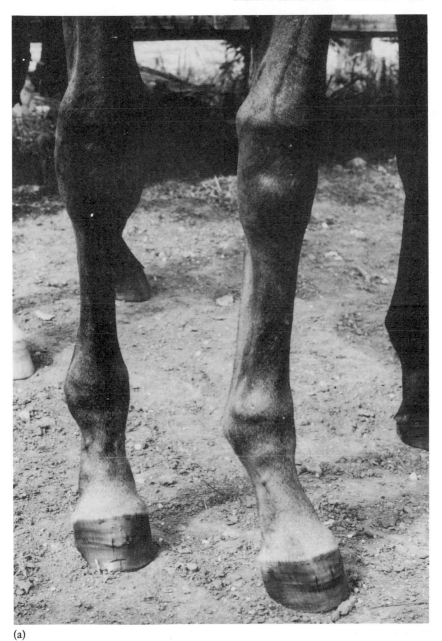

(a)

Plate 7.2   Epiphysitis in the lower (distal) end of the metacarpus and the upper (proximal) end of the proximal phalanx (fore fetlock) (a) and (b); and spavin of the hock (c) in foals. (Photograph 7.2 (c): Dr Peter Rossdale, FRCVS).

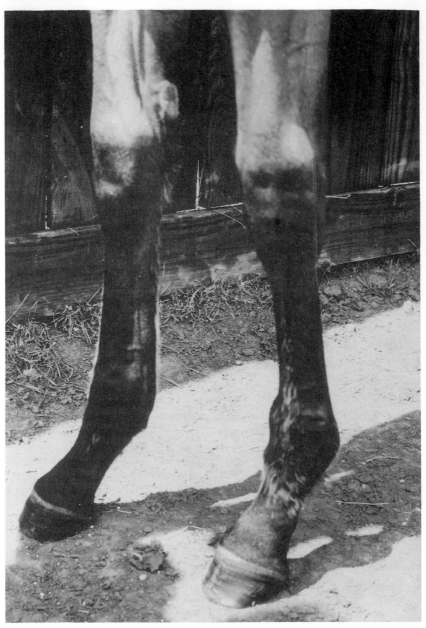

(b)

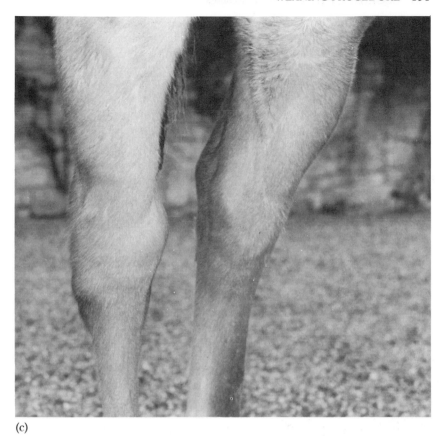

(c)

Worming of the foal should not coincide with weaning, but the first dose of anthelmintic may be given at 2–3 months of age or ideally at 4 months of age and thereafter at 6-weekly intervals.

Weaning procedure

Restricted feeding of the mare limits milk secretion, but after weaning the udder should not be milked out. Some breeders rub camphorated oil into the udder.

The psychological attachment of the foal to its dam is greatest between the second and the twelfth weeks of lactation with a peak around the third week, at which time separation leads to the greatest agitation of both. Risk of injury to the foal at weaning as a result of excitement induced by separation, is a major factor to be contained. On large stud farms, three alternative procedures are practised. These are:

1. All the mares are removed at the same time from a year's crop of foals.

2. One or two mares are removed at a time, starting with the first foaled, or most dominant mare, and allowing a few days to elapse before the next is removed.

3. Separate the foals for increasing time periods, such that the foals are allowed to nurse three times per day, twice and then only once on successive days keeping mares and foals in sight of each other.

Procedures (1) and (2) may require access to another farm to ensure that weaned stock are out of sight, sound and smell of their mothers. Method (2) may lead to some foals being kicked by more aggressive mares when searching for milk. The last method protracts the drying-up procedure, is more laborious, and not without risk.

It is, however, recommended that unless there are other mitigating circumstances mares should be abruptly removed from foals, starting with that having the largest and most independent foal, or with the most dominant mare likely to cause problems to other foals. Several days should elapse before the next most dominant mare is removed, leaving the foals in familiar surroundings out of sight, sound and smell of their mothers. It is helpful to leave a gentle dry mare with the foals and any foal having a cold or other debilitating condition, should not be weaned until it has regained health.

Foals at first may become frantic and it is important to ensure that all have company, that there is ample space for play, that the pasture has a clean water source and a shelter, and that the shelter and fences are free from protruding nails, splinters and loose wire. The pasture should also be of good quality, without a worm burden, and free from flints and rabbit or other holes that might cause leg injury.

The orphan foal

The artificial rearing of relatively small foals should not be lightly attempted. Thoroughbreds of less than 40 kg are normally destroyed. If the mare is lost after the first 24 hours the prospects of foal survival are greatly enhanced as sufficient colostrum should have been sucked. Where this is not the case the maintenance of a bank of frozen colostrum is an asset, but great care should be exercised to ensure that no bacterial contamination has occurred, otherwise organisms will proliferate on thawing. Thus, minimum quantities, sufficient for each feed, should be thawed at one time and consumed immediately on warming. Where colostrum is unavailable, one can use blood plasma, preferably from a donor gelding horse or unrelated mare that has never received a blood transfusion and which has been on the farm for some time. The dose is about 20 ml per kg bodyweight, given intravenously. If the foal trembles, the rate of dosing should be reduced and rapid recovery will soon follow. Oral dosing with plasma helps in cases of enteritis.

Orphan foals are deprived of the warmth of the dam and in cold weather should be covered with lightweight quilted material. Normally a nurse is desirable, but in the meantime bottle feeding or, for preference, bucket feeding, using human baby food or a calf milk replacer reconstituted with water is suitable. The initial concentration should be 22 per cent of dry matter for the first 1–2 days dropping by 1 per cent daily until a normal concentration of 14–15 per cent of dry matter is reached and maintained until weaning. If there is diarrhoea the milk can be diluted, or preferably replaced, for a short period with a glucose–electrolyte solution, which provides sodium, potassium, chloride, organic base and glucose in particular (see Table 9.2).

Where the foal is to be trained to a bucket, the head can be drawn into it with a finger in the mouth – initially this may require the assistance of another person, but soon the foal will adapt to the procedure. All feeding utensils should be clean before each feed. The intake of liquid milk replacer, or a 50:50 mixture of skimmed milk and whole cow's milk, should be at the rate of 280 ml (10 fluid ounces) every $1\frac{1}{2}$ hours so that the daily energy intake amounts to 9–10 MJ of DE. The initial feeding can take place to advantage close by a horse acting as a decoy, but never at the stable door to avoid the association of it with feed. In order to minimize affinity with man, orphan foals should not be fondled.

Liquid milks are normally given at body temperature, but can equally be given cold. Within a few days the daily intake will attain 9–18 litres (2–4 imperial gallons) and if the foal is permitted to drink freely it may reach 36 litres (8 gallons). However, intake should be restricted to a maximum of 18 litres (4 gallons) in a large foal and with any evidence of diarrhoea the quantity should be reduced until the problem has subsided. Once the first few days are over the liquid can be provided in four then three feeds per day and any excess disposed of.

Creep pellets in the form of stud nuts, or a concentrate mix, together with milk pellets and a little best-quality leafy hay from 7 days of age will encourage dry feeding. Access to fresh faeces from a healthy adult horse that has been wormed regularly will provide bacteria of the appropriate kind for seeding the intestinal tract. Any strongyle or ascarid eggs in the faeces should be immature and therefore of low infectivity, and so will passively traverse the gastrointestinal tract of the foal; but the foal's faeces should be removed regularly. If progress is normal, liquid milk can be discontinued from 30 days of age and intake of dry feed will rapidly rise. At this time the foal may be consuming as much as 2–3 kg of dry feed daily, although the consumption of hay will still be rather slight.

Automatic liquid-milk feeders of French design have proved very successful on large studs for groups of foals, and they avoid the problem of humanizing. They are electrically operated; the water is warmed and mixed with milk replacer powder (Table 7.1) at an adjustable rate. Fresh liquid is prepared to replace that used up as the foals drink. The appropriate concentrations of dry matter suggested above should be adhered to as solutions that are either too weak or too concentrated can precipitate looseness or constipation. New foals are rapidly trained by experienced foals in the same yard.

Fostering

The least-troublesome nurse is frequently an old coldblooded mare, especially of piebald breeding, or even a nanny goat; the worst type is a young flighty Thoroughbred. Prospective individuals should be checked for disease, and their milk should be examined. The udder and the tail should be thoroughly washed and disinfected. The mare can be brought to the stable hooded and disorientated by walking around the area. A strong smelling substance such as camphorated oil can be placed on the muzzle and the same substance smeared on both the mare's own foal and the orphan. When these two are held together for a while the mare confuses their sounds.

A fostering gate is a boon for nurse mares, enabling foals to be suckled without being kicked. It will have facilities for feeding and watering but the mare should be allowed out for exercise at regular intervals. The crate should have a gate at each end; the critical dimensions are a length of about 250 cm, a width of 65 cm and a gap at both sides of one end 90 cm × 40 cm, the lower edge of which is 70 cm from the floor for access by the foals to the udder.

Early weaning of foals

Foals can be weaned at 3 days of age when the mare is not imprinted on the foal. A separation of 6 hours without feed eliminates fretting and normal weight gain can be subsequently achieved. The procedures outlined for both liquid and dry feeding of orphaned foals can be followed and, although good growth is possible, the procedure is both labour intensive and an interruption to normal activities on the stud farm. From both a practical and an economic point of view, therefore, early weaning has little to recommend it.

Feeding the stallion

The stallion is subject to the same seasonal influences as affect the breeding cycles of the mare: his fertility is greatest in the summer and least in the winter. Some evidence suggests that improved fertility in the early months of the year may be obtained by following a regime similar to that proposed for mares, in which artificial light and a richer food are provided in late December and January. At no time should the stallion be allowed to fatten and higher fibre, but balanced feeds are quite satisfactory out of the breeding season. Poorer quality hay supplemented with horse and pony nuts should then be satisfactory and they will facilitate the imposition of a rising plane of nutrition as the breeding season approaches when stud nuts or an equivalent concentrate mixture with good-quality hay should be introduced (Table 7.1). The energy requirements of the stallion rise during the breeding season as a consequence of increased physical activity, in particular that of pacing his run or stall, yet space for physical exercise is important to the stallion in all seasons.

There is little evidence to support the use of special supplements to enhance the fertility of stallions, but a diet of 0.75–1.5 kg of cereal-based concentrates plus hay per 100 kg bodyweight daily and clean water (Table 7.3) should suffice.

Further reading

Rossdale P. D & Ricketts S W (1980) *Equine stud farm medicine*. Cassell (Baillière Tindall): London.

CHAPTER 8

Growth of the horse

For it·is easy to demonstrate that a horse may irrecoverably suffer in his shape and outward beauty as well as in strength by being underfed while he is young.

William Gibson, 1726

Birthweight and early growth

Growth proceeds through a process of cell division and enlargement initiated by the fertilization of the egg. Cells differentiate forming the various embryonic tissues. Soon after birth the number of cells in most tissues has reached a maximum and further growth is accomplished by enlargement of the individuals cells; but in some tissues, for instance epithelial tissue, cell replication continues throughout life in order to replace the cells that are sloughed off. Not all tissues, organs and structures increase in size at the same rate so that during growth the shape of the animal changes. The overall rate of growth measured relative to the existing weight is initially slow, it accelerates to a maximum before birth and then declines. However, the potential for a maximum rate of growth measured in kg daily bodyweight gain persists until about 9 months of age in the horse, at which time it also gradually declines and ceases as the adult size and shape are attained. By the seventh month of gestation, merely 17 per cent of the birthweight and only 10–15 per cent of the birth dry matter have been accumulated. Thus the accretion of most of the energy and minerals present at birth occurs during the last months of gestation.

Partly because the mature number of cells in many tissues of the adult animal has been achieved by birth, or shortly afterwards, the maximum adult weight of horses and ponies is, to a large extent, determined by birthweight. As a rough guide, birthweight constitutes 10 per cent of the adult weight; among Thoroughbreds, individuals weighing less than about 35 kg at birth are very unlikely to reach 15 hands in adult life. One study revealed that the proportion of foals with a birthweight of less than 40 kg that actually raced was much smaller than the proportion of those weighing more than 40 kg (Platt 1978). Horses out of small mares by small stallions will be small as adults, but will achieve their mature size slightly sooner than will the products of large parents. Differences between breeds in rate of attainment of mature weight are greater than the differences in rate of attainment of mature height.

Table 8.1 gives data for several breeds. As a general rule, foals attain 60 per cent of their mature weight, 90 per cent of their mature height and 95 per cent of their eventual bone growth by 12 months of age. In the period from birth to maturity, the growth coefficients (the rate constant for the growth of a tissue, or structure, relative to that for the whole empty body) of the three major equine tissues in order of increasing magnitude are bone, muscle, fat, implying that bone is the earliest maturing tissue and fat the latest. This, and the earlier assertion that overall growth rate declines from birth (weight gain per unit empty liveweight) signify that ultimate height is determined in very early life, that early growth demands diets rich in bone-forming minerals, protein and vitamins, and that with increasing age an increasing proportion of dietary carbohydrate is required. The early rapid extension of the long bones of the legs, pronounced in tall breeds, render them subject to deformation through early malnutrition.

Table 8.1 Percentage of mature bodyweight and of withers' height attained at various ages of horse breeds

Age	6 months		12 months		18 months	
	Weight	Height	Weight	Height	Weight	Height
Shetland pony	52	86	73	94	83	97
Quarter Horse	44	84	66	91	80	95
Anglo-Arab	45	83	67	92	81	95
Arabian	46	84	66	91	80	95
Thoroughbred	46	84	66	90	80	95
Percheron	40	79	59	89	74	92

(From Hintz 1980a)

The nutrition of the foal at birth is affected, not only by the feeding of the mare but also by the physiological efficiency of the uterine environment. This may to some extent be affected by the age of the mare, as indicated by the data in Table 8.2. Differences in birthweight brought about by nutritional deviations in the pregnant mare can, however, be proportionately lessened by nutritional adjustments in early postnatal life.

Initial and ultimate weights and heights differ as between colts and fillies (Table 8.3). The differences are small and limited studies in England (Green 1969) failed to detect sex differences in linear measurements or any between early and late-born foals. Nonetheless, some substantial evidence suggests that foals born late in the season are heavier and taller than early

Table 8.2   Effect of age of Thoroughbred mare on bodyweight and height at withers of foals

Age of mare (years)	30 days		540 days	
	Foalweight (kg)	Height (cm)	Weight (kg)	Height (cm)
3–7	93.0	108.0	393.7	152.4
8–12	97.5	110.5	401.4	153.7
13–16	98.0	110.5	396.9	153.0
17–20	95.3	109.2	391.0	152.4

(From Hintz 1980a; data based on 1992 foals)

Table 8.3   Effect of sex on growth of Thoroughbred foals

Age (days)	Bodyweight (kg)		Withers height (cm)	
	Colts	Fillies	Colts	Fillies
2	52.2	51.3	100.3	99.7
60	136.5	134.7	118.3	118.1
180	244.9	235.9	134.6	133.0
540	435.5	401.4	154.3	152.4

(From Hintz 1980a; data based on 1992 foals)

Table 8.4   Effect of month of birth on bodyweight and withers' height of Thoroughbreds

Month of birth	Age			
	30 days		540 days	
	Weight (kg)	Height (cm)	Weight (kg)	Height (cm)
February–March	95.3	109.2	396.9	153.7
April	97.5	110.5	402.8	153.7
May	100.7	111.1	403.7	153.7

(From Hintz 1980a; data based on 1992 foals)

born foals (Table 8.4) despite a somewhat shorter period of gestation. As previously indicated (Ch. 7), gestation length seems to be a function of daylight length in late gestation.

Although birthweight has a major impact on ultimate size for both genetic and environmental reasons, weaning age, in conditions of good management, has little influence. Foals weaned after receiving colostrum can achieve growth rates equal to that of those weaned between 2 and 4 months of age. These in turn may grow faster than foals weaned at 6 months of age. The appropriate weaning age, therefore, for any particular stud farm turns on the most convenient and reliable management practice.

Later growth and conformational changes

On the whole, differences in growth rate after the neonatal period have little influence on mature size. Although maximum height may be approached soon after 12 months of age, this may be delayed without a reduction in the ultimate measurement by reducing the rate of feeding slightly. Similarly, 90 per cent of mature weight may be achieved at 18 months, but delayed until 24 months by the same restriction. In Fig. 8.1 the changes in height at the withers over the first 12 months of pony and Thoroughbred foals are compared (Hintz 1980a). Although the foals achieve very different heights the pattern of growth is similar. Height at the withers largely reflects linear growth of the long bones in the front legs. Long bones increase in diameter or thickness throughout their length, but no increase in length occurs in the shank, or diaphysis, after birth (Figs. 8.2 and 8.3). They increase in length by growth in a metaphyseal plate at both the near

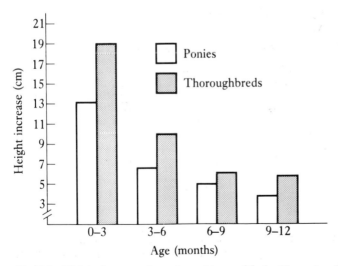

Fig. 8.1   Height increase (in cm) at withers of foals: □, ponies; ▨, Thoroughbreds. (From Campbell & Lee 1981).

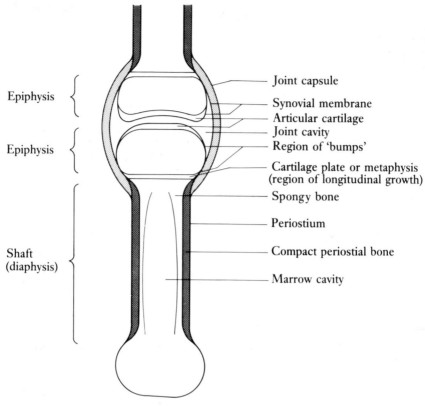

Epiphysis

Epiphysis

Shaft
(diaphysis)

Joint capsule

Synovial membrane
Articular cartilage
Joint cavity
Region of 'bumps'

Cartilage plate or metaphysis
(region of longitudinal growth)

Spongy bone

Periostium

Compact periostial bone

Marrow cavity

Fig. 8.2 Long bones and their articulation in late growth (joint cavity has been expanded for purposes of visualization). Note the regions of growth: cartilage at either end of diaphysis; cartilage of epiphysis; and periostium of shank.

and distant ends (proximal and distal) from the body. The rate of growth at each end is different; Table 8.5 gives values recorded for crossbred ponies (Campbell & Lee 1981). Correction of distortions in bone growth, owing to bad feeding practices and other causes, is possible during the phase of rapid growth of the end in question. Thus, for the distal radius or tibia, such correction could be imposed up to 60 weeks of age, whereas fetlock distortion, a fairly common condition, requires treatment by 3 months. In either case, temporarily restricted growth will either not, or only slightly, influence ultimate size.

The balanced adult horse of good conformation has a height at the withers that equals its length from the tip of the shoulder to the point of the buttocks (see Fig. 6.1). By contrast, a foal is taller than it is long (Pl. 8.1) so that it inevitably trots wide behind. Fillies at birth tend to be fairly level across the top, but may be as much as 5 cm higher over the croup than

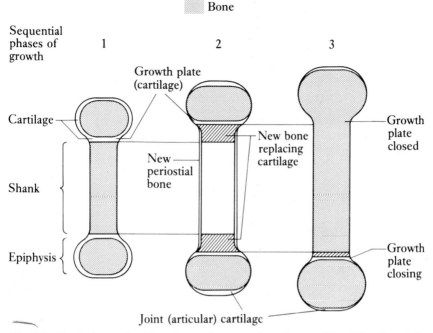

Fig. 8.3 Typical growth of long bone, for example radius or tibia. (After Rossdale & Ricketts 1980).

over the withers at 1 year and then balanced again by 5 years. The length of their bodies tends to be greater than their height at the withers. By contrast, colts are generally higher over the hip at birth but level by 3 years. Some horses that have grown poorly in the front legs are lower over the withers than the croup at maturity and, as nearly 60 per cent of the bodyweight is normally carred by the front legs, this unusual conformation can force additional weight and stress on the forequarter, increasing the risk of damage.

When fed liberally, the growth of ponies up to 9 months of age may increase to nearly 1.5 kg/day, although by 12 months the rate may have fallen to half this amount. Under hill conditions, where young stock can

Plate 8.1 (see page 162) Grey colt (51 kg at birth) by Vitiges out of Castle Moon at Derisley Wood Stud, Newmarket: 5 days old (a) and 82 days old (b). Note that the height of the young foal exceeds its length, a characteristic that changes as adult proportions are attained. Over a period of as little as 77 days the length of the foal has increased considerably so the body proportions are already more like those of the adult. The height of the croup tends to increase faster than that of the withers during the first few months, but by 3 years of age in the Thoroughbred the withers height has caught up, through differential growth of the leg bones.

Plate 8.1

(a)

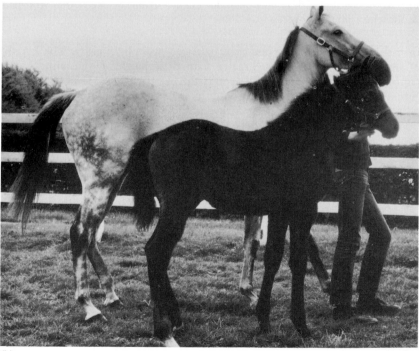

(b)

Table 8.5 Growth of leg bones in three male and three female crossbred ponies

Bone	Length of bone (cm)		Increase at each end (%)			Age at closure of growth plates (weeks)		
	0–7 days	2 years	Proximal	Distal		Proximal	Distal
Femur	21.6	30.9	24	20		55	55
Tibia	21.3	29.1	22	17		60	60
Radius	20.7	28.1	12	24		54	69
Humerus	16.0	22.6	31	11		80	70
Metatarsal	22.7	23.7	5			40–44	
Metacarpal	18.4	19.4	5			40–44	
Phalanges 1	5.9	6.6					
Phalanges 2	2.3	2.5					
Phalanges 3	5.6	6.2					

(From Campbell & Lee 1981)

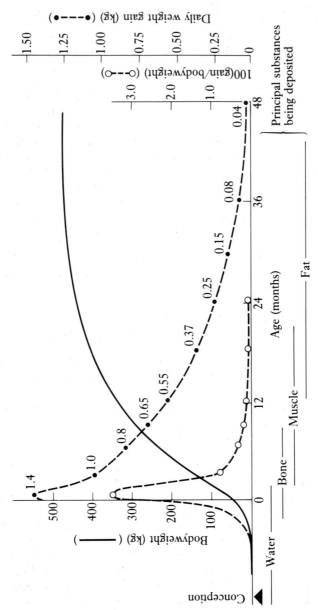

Fig. 8.4 Normal growth curve for the horse (mature weight of example, 500 kg).

receive little in the way of concentrates, growth can be strikingly depressed. Morgan horses on range were shown to be between 50 and 100 lb (23 and 46 kg) lighter and 1 inch (2.5 cm) shorter at 5 years of age than their contemporaries in the same environment receiving supplementary cereal-based concentrates (Dawson, Phillips & Speelman 1945). With only moderate restrictions, such horses may attain normal mature height at 5 years, whereas with more liberal feeding they can reach it by 3 years (Kownacki 1983). New Forest filly ponies, fed from 6 months of age throughout the winter at a rate that only maintained bodyweight, grew faster during the following summer on rough pasture than their counterparts permitted to gain 0.4 kg/day in the winter (Ellis & Lawrence 1978, 1979, 1980). Although bodyweight may remain constant in such deprived circumstances, parts of the skeleton grow differentially so that the restricted fillies in this example were tall, thin and shallow-bodied. With a delay in the closure of growth plates these differences are partly corrected during later growth. Catch-up growth of this kind is not without risk among potentially tall breeds. Foals weaned at 4 months of age and restricted to gaining weight at 0.3 kg/day and then allowed free access to feed, were observed to develop a condition similar to contracted tendons (Hintz, Schryver & Lowe 1976). Thus a smooth growth curve should be sought for Thoroughbreds, Saddlebreds and the like (Fig. 8.4). The period of maximum risk of skeletal abnormalities occurs when the rate of gain (per unit of bodyweight achieved) is high.

The amounts of supplementary winter feed necessary will depend, first, on the quality and, second, on the amount of grazing available. Winter pastures have little value as a stimulus to growth in young stock, but will provide reasonable roughage to complement cereal-based concentrates. Until the spring flush is available, stock at this age should receive from 1.25 to 1.5 kg of concentrates containing 16 per cent protein and appropriate vitamins and minerals per 100 kg bodyweight (see Figs 6.2–6.4 for height equivalents). With full access to good summer pastures, yearlings require no supplementary feed (that is assuming the land has no specific trace-element deficiencies).

When exercise is limited by in-wintering, nuts or other concentrate feeds should be given to young stock in two meals daily to avoid intestinal upsets that might cause oedema in the legs. Stabled, weaned horses may be given free access to good-quality hay, and when they are receiving concentrates at the rate of between 1.25 and 1.5 per cent of bodyweight, they will consume hay or other forage dry matter at a rate of about 0.5 –1.5 per cent of bodyweight. This should allow for a normal growth rate during the winter and early spring. In the study of New Forest and Welsh ponies, the maintenance requirement of the filly foals amounted to between 14.0 and 14.6 MJ of digestible energy (DE) per 100 kg bodyweight, equivalent to 1.75 kg of average hay or 1.1–1.2 kg of average nuts per 100 kg bodyweight (Ellis & Lawrence 1980). At the end of the first winter, that is when yearlings are 11 or 12 months old, colts and fillies should be separated. The first oestrus should occur before May, unless winter conditions and feed have been poor.

From 90 per cent of mature-weight young horses and ponies can be fed most economically on maintenance rations as for mature horses. This should include good-quality hay and a concentrate offered at a rate of 0.75–1.0 per cent of bodyweight and providing cereals, a concentrate protein source, minerals and vitamins. However, when these growing stock are out on good-quality pasture continuously they should need no other source of feed, apart from a trace-element mixture should a specific deficiency exist.

Feed mixtures for growing horses

Complete mixtures of concentrates and chopped roughage are given *ad libitum* routinely to growing cattle and to suggest that they should be given to groups of growing Thoroughbreds and other horses may seem an affront, or at the very least inappropriate. When a number of such stock are reared together through both the first and even the second winter months, there is much to recommend such a procedure. It does require sheltered '*ad lib*' feed hoppers, as by ensuring that the feed is always present, risk of colic or of laminitis (founder) is non-existent in foals. Complete feed mixtures

Table 8.6 Feed mixture for growing horses for providing with chopped forage or grass hay or as a supplement to poor pasture

	Concentrate mixtures (%)		
	Weaned foal	Yearling	Presale of long yearling
Oats	41.1	44.7	59.0
Wheatbran	15.0	15.0	10.0
High protein grassmeal	15.0	15.0	10.0
Extracted soyabean meal	18.0	15.0	10.0
Molasses	7.5	7.5	7.5
Feed-grade fat	1.0	1.0	2.0
Limestone	1.2	0.7	0.5
Dicalcium phosphate	0.5	0.5	0.4
Salt	0.5	0.5	0.5
Vitamins/trace elements*	0.2	0.1	0.1
Total	100.0	100.0	100.0

* See Table 7.1

Table 8.7 Daily feed allowances (kg) for growing horses (500 kg mature weight)

Foals			Yearlings		
Bodyweight (kg)	Concentrates	Hay (clover/grass)	Bodyweight (kg)	Concentrates	Hay
100 (4–5 weeks)	0.5	—	310–360	3.5–4.5	3.0–3.7
130–180	2.2–3.2	1.3–1.9	360–410	3.0–4.2	3.6–4.1
180–230	2.9–3.9	1.8–2.3	410–460	3.0–3.8	4.0–4.5
230–270	3.6–4.8	2.2–2.8			
270–320	4.0–4.6	2.7–3.2			

should initially contain a cereal-based concentrate and good-quality chopped hay in a ratio of 2:1, gradually falling as the growth rate declines to a ratio of 1:1. Stock will normally consume a total amount of dry feed daily equivalent to 3–3.5 per cent of their bodyweight. The physical make-up of the mix and particle size should be so regulated that no segregation between the hay and the concentrate arises. Molasses is often a good material to include in the mix as an aid in preventing segregation and some ingenuity is required to ensure that no bridging occurs in hoppers.

Thoroughbred yearlings are normally prepared for the autumn sales by excluding them from pasture. A diet based on 1.5 kg of good-quality stud nuts or other concentrate, plus 1.0–1.5 kg of hay per 100 kg bodyweight is a typical practice. Traditional concentrate mixtures contain 75–90 per cent of oats, plus bran, soyabean meal, dried skimmed milk and sometimes a vitamin/mineral premix. There is no particular reason, however, why these horses should not be provided with a complete mixture of concentrates and chopped hay as advocated for younger stock, but in a ratio of 0.5:0.5 concentrates to hay. Table 8.6 gives proposed feed mixtures for growing horses and Table 8.7 rates of feeding (on the basis of 90 per cent dry matter). The composition of these diets conforms with the principles of equine growth previously propounded in this chapter.

CHAPTER 9

Feeding for performance and the metabolism of nutrients during exercise

But if you intend the next day to give him an heat (to which I now bend mine aim) you shall then only give him a quart of sweet oats and as soon as they are eaten, put on his bridle, and tie up his head, not forgetting all by ceremonies before declared.

Thomas De Gray, 1639

A major role of feed for working horses is the conversion of the chemical energy of feed into locomotion at speeds varying from 160 m/min to over 900 m/min (6–35 mph) for distances varying from 1 to 150 km or more. This enormous range may superficially lead to equal fatigue in fit horses, but quite different processes of nutrition physiology are involved at the two extremes of distance and speed. At the one extreme a flat race of 6–8 furlongs (1.2–1.6 km) would *theoretically* increase the day's energy needs by a mere 4 per cent (Pl. 9.1), a barely perceptible effect, whereas at the other they would be increased by five- to sixfold. Training regimes recognize both this and the different responses of breeds to contrasting forms of work. These distinctive training procedures induce profound and dissimilar physiological changes in the attainment of fitness. Diets must be formulated to conform with these changing needs, but appetite may flag in the process. First, an adequate and optimum nutrition for a particular purpose implies an optimum supply of nutrients to each tissue and cell and the efficient disposal of waste products.

The way in which food and nutrients are utilized is affected by the well-being and training of the horse, the environmental temperature, the extent of nutrient depletion and therefore by the adequacy of diet. The most critical feeds are those that provide energy, water and electrolytes. Electrolytes are those elements that in solution have an electrical charge; they include sodium, potassium, magnesium, calcium, chloride and phosphate. It must be clear therefore that training and nutrition should work together like hand and glove.

Plate 9.1   Rathgorman, ridden by Chris Bell, cantering 8 furlongs (1.6 km) on the all-weather track at Dunkeswick, West Yorks, watched by trainer Michael W. Dickinson. As part of the training for National Hunt racing the horses canter 5 days per week. If a horse races every 7 to 21 days it has an easy canter on the day before and for several days after a race. A regular work programme is maintained during the racing season but with some variety in location, scenery and type of work.

There have been few investigations on the optimum dietary supply of nutrients to horses in work, but there is an increasing understanding of the metabolism of nutrients and of the production of readily available chemical energy at the level of the muscle cell during aerobic and anaerobic respiration. In sprint races, horses obtain much of their muscular energy from anaerobic pathways whereas in extended work, such as endurance competitions, energy is derived almost exclusively through aerobic respiration. (Anaerobic respiration is the breakdown of organic nutrients in the absence of oxygen but with the release of energy captured by ATP.) A day's hunting, with episodes of hill climbing, cantering and jumping, and periods of waiting and walking combine both processes and rates of energy expenditure. The expenditure during a long day by a hunter carrying a heavy huntsman exceeds the average daily consumption of feed energy by several-fold. The processes of energy metabolism are summarized here as a greater understanding of them should allow more rational feeding and it should foster comprehension of each future development in the feeding of working horses.

Training and energy expenditure

The purpose of training is to modify muscular action and indeed the whole metabolism of the horse so that it functions at maximum efficiency with minimum fatigue at the speed and over the distance at which the sights are set. In this process feed energy is converted into work and Table 6.4 gives the approximate amounts of feed energy required by horses of different weights undertaking work of a variety of intensities. In all but one instance the work covers a period of 1 hour, although it is appreciated that the most strenuous effort could not be sustained continuously for such a period. Nevertheless, the method allows comparisons of different degrees of effort. On the one hand, walking creates practically no further demand on the requirements for maintenance, whereas at the other extreme, racing – and more particularly lengthy endurance work – create an energy demand that exceeds the capacity of the horse for the immediate replenishment of the losses through feeding. Endurance work is at the rate of 154–224 m/min (5.75–8.35 mph) over variable slopes whereas flat racing is at about 940 m/min (35 mph). For comparison, the stage-coach horse, pulling a heavy load, covered about 40 km (25 miles) at speeds similar to those required of endurance horses – 200–214 m/min (7.5–8.0 mph). Strenuous effort tends to depress appetite so recovery from extreme effort requires several days for reserves to be refurbished. The data in Table 6.4 show that the capacity of larger horses for feed is proportionally less than that of smaller horses so that after undertaking strenuous work they may require a longer period for recovery, especially if it is assumed that the weight of the rider is proportional to the weight of the horse.

The information in Table 6.1 is based on experimental results and although there has been some extrapolation it does indicate greater demands for feed energy than the theoretical estimates for endurance work given in Table 6.4. The reason for the difference is unclear. It may be the stress of, rather than the inclination of, the belt. Blaxter (1962) calculated that the energy expenditure for vertical effort in horses, in addition to any horizontal effort, amounted to seventeen times that expended in horizontal movement above the costs of energy metabolism at rest. A 400 kg horse is calculated to expend 0.67 kJ/m in horizontal work and 11.4 kJ/m vertically. Exercise over uneven and hilly ground can therefore be much more arduous than that on the flat. Sprint work on the flat must be considered a special case as it is almost entirely confined to young horses that may be still growing. Extended work for which there is more published experimental evidence is generally undertaken by older horses.

Work entails an increase in nutrient supply to the muscles and other organs of the body, and in converting principally glucose or free fatty acids to high-energy phosphate compounds (ATP and creatine phosphate, CP), which the muscles use as an immediate source of energy, there is an increase in the rate of production of wasteproducts, more particularly carbon dioxide and heat. The supply of nutrients and the effective disposal of wastes entail large physiological adjustments, which training seeks to encourage.

Sources of energy to muscles and other tissues

Most potential energy for muscular work is absorbed from the intestinal tract as glucose, volatile fatty acids (acetic, propionic, butyric and smaller quantities of related acids), longer chain fatty acids, neutral fats and amino acids. Absorbed glucose, propionate and glucogenic amino acids are potential sources of blood glucose and of liver glycogen, a storage starch, whereas absorbed long-chain fatty acids, neutral fats, ketogenic amino acids and particularly acetate and butyrate are potential sources of blood fats and fatty acids, storage fat and acetyl coenzyme A (CoA). Blood glucose and its precursors are also, of course, potential sources of body fat through the key substance acetyl CoA.

The liver, while storing energy in the form of glycogen, fat (and also as protein), serves the vital role of maintaining normal levels of blood glucose through the breakdown and re-storage of the glycogen. Muscle cells form high-energy phosphate compounds necessary for muscular relaxation and contraction by drawing on blood glucose and fatty acids as fuels. The complete release of chemical energy from them requires a supply of oxygen reaching the muscle cells through the arteries. However, an immediate and rapid release of energy, particularly important in short sprint races, can be achieved through the process of glycolysis, in which glucose is broken down to pyruvate in the muscle cell without the consumption of oxygen, and also by the release of energy from previously stored ATP and CP. The further and complete breakdown of pyruvate and of fatty acids, demands the presence of oxygen and this process takes place exclusively in the mitochondria, through the agency of what are known as β-oxidation of fatty acids and the tricarboxylic acid (TCA) cycle.

The anaerobic production of energy by glycolysis would soon be halted by an excessive accumulation of this intermediary pyruvate. The cunning mechanism has therefore evolved in which pyruvate is reversibly converted to lactate as an intermediate 'waste' product. This reduction even more importantly oxidizes reduced NAD^+, essential for triggering an important glycolytic step. Therefore, in sprint racing, lactate diffuses into the blood from the muscle cell and there accumulates until sufficient oxygen is available for its hepatic conversion to pyruvate. This mechanism allows more energy to be obtained by glycolysis than would otherwise be possible. As the oxygen debt has eventually to be repaid, the yield of energy per unit volume oxygen consumption, owing to anaerobic processes, is only half that achieved by aerobic processes. However, during recovery from sprint races oxygen is abundant so that the advantages of the rapid availability of chemical energy outweighs the minor disadvantage of interest payment.

The operation of the TCA cycle in the mitochondria requires the diffusion of oxygen to these organelles and the removal of carbon dioxide to the blood. During light work this process occurs quite smoothly, and with each turn of the cycle one unit of oxaloacetate is produced that is required for the subsequent metabolism of acetyl CoA in the presence of oxygen. However, when larger quantities of oxaloacetate are present, the metabolism of acetyl CoA is likely to proceed more rapidly. When larger quantities of fatty acids are dissimilated to acetyl CoA, the cycle must turn at an accelerating rate. This is also achieved by the provision of extra quantities of oxaloacetate from outside the mitochondrion. There must, therefore, be adequate quantities of pyruvate requiring an ample supply of glucose, lactate, glucogenic amino acids or even glycerol. During extended work, the trained healthy horse finds no difficulty in breaking down fatty acids to carbon dioxide as sufficient oxygen is taken in by normal respiration. In fact, such work leads to an accumulation of glycerol, signifying that it is not called on to form pyruvate units in any great quantities. If the utilization of fatty acids was interrupted, this would probably be expressed as a retarded metabolism of acetyl CoA and there would be a build-up of blood ketones (acetoacetate and 3-hydroxybutyrate). Our work in Newmarket (Frape, Peace & Ellis 1979) shows that these ketones accumulate only after work stops (Fig. 9.1), implying that no bottleneck occurs to the complete combustion of fats, but that the signal for slowing down the dissolution of storage fat is delayed. Hyperlipidaemia recognized in starving horses possibly reflects a blocking to fat metabolism in animals relying overwhelmingly on residual fat stores in relative absence of carbohydrate substrates. The activity of enzymes required in this fat breakdown may also be depleted in an associated deficiency of dietary protein. Figure 9.2 gives a brief account of the pathways by which energy sources are metabolized in horse muscle cells and liver.

An examination of the Fig. 9.2 will reveal that the changing demands of the horse for energy are monitored by a number of endocrine secretions, or hormones. Where rapid changes are necessitated, the signal for secretion by the appropriate glands is provided by the involuntary action of the autonomic nervous system, reacting to environmental stimuli, which in turn brings about other essential changes in cardiac muscle and the smooth muscles of arteries, intestines etc. Endocrine secretions to a considerable extent mediate their effects by switching on and switching off some of the enzymes that regulate the chemical reactions in energy metabolism. In addition, such hormones as insulin seem to influence the permeability of

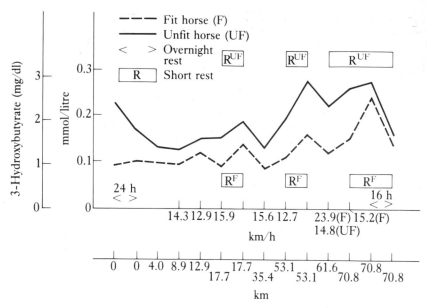

Fig. 9.1 Effect of rest periods during an endurance ride on plasma 3-hydroxybutyrate in a fit and an unfit horse. (Data from Frape, Peace & Ellis 1979).

cell membrances, for instance by increasing the cellular uptake of glucose so that the blood concentration is decreased.

Blood, the supply of which to muscles is greatly increased during work, is the principal vehicle by which the wasteproducts of energy metabolism, water, carbon dioxide and heat, are removed from muscle cells. In the absence of their efficient disposal, pathological cellular changes would result. Training augments the blood supply to skeletal and cardiac muscles.

Blood-glucose concentration is the expression of a dynamic balance of glycogen breakdown and glycogen synthesis and the production of glucose from other sources – amino acids, lactic acid and propionate (gluconeogenesis). The resting level is higher in horses highly trained for sprint races. This state is brought about by the stimulation of the two systems leading to the formation of glucose and by increasing the efficiency of fatty acid utilization, sparing glucose. Blood glucose, however, also fluctuates throughout the 24 hours, showing a peak value 3–8 hours after feeding when pyruvate also tends to increase, whereas fatty acids and glycerol are in lower concentrations. In one experiment with ponies receiving a pelleted diet, blood-glucose concentrations, after appetite was satisfied, reached values of between 85 and 115 mg/100 ml (4.7–6.4 mmol/litre), whereas after 3 hours of fast the levels fell to between 50 and 90 mg/100 ml (2.8–5.0 mmol/litre) (Ralston, Van den Broek & Baile 1979). The fluctuation between peak and resting levels varies with the type of diet, in that

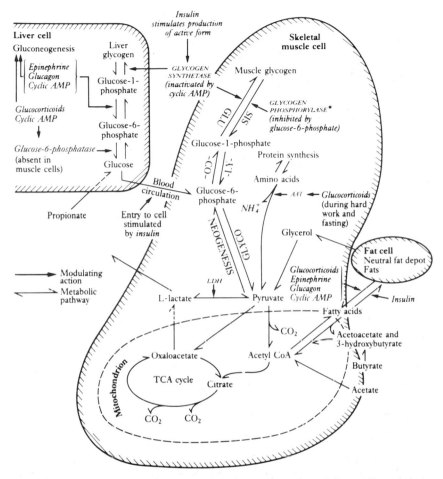

Fig. 9.2 Major pathways of energy metabolism in muscle and liver cells and their modulation by the endocrine system (intermediate steps have been excluded).

*Bursts of muscular action with the secretion of glucagon and epinephrine and nervous stimulation of the release of Ca^{2+} initiate the formation of cyclic AMP and then the phosphorylation and activation of glycogen phosphorylase.

foods containing more grain and less roughage lead to higher peaks and lower troughs. Horses subjected to a rougher diet develop a greater faculty for gluconeogenesis and can therefore resist a depression in blood glucose more readily during a fast, but are unable to meet the demands of an excessive rate of exertion. Furthermore, breeds such as the Thoroughbred with higher insulin sensitivity, experience a greater fluctuation in blood glucose than do ponies given the same diet at similar times of the day.

The anaerobic lactic acid system in unconditioned Thoroughbreds is probably untaxed until work rates exceed around 600 m/min (22 mph) on the flat (Williamson 1974). The anaerobic threshold of Standardbred trotters is said to be 300–400 m/min (11–15 mph) on the flat (Milne, Skarda et al. 1976), although a striking increase in blood lactate of Standardbred trotters was not observed until speeds exceeded 600 m/min (Lindholm 1974), or 684–750 m/min (25–28 mph) (Lindholm & Saltin 1974), and Williamson (1974) recorded similar blood responses in these breeds undertaking the same types of exercise. The threshold of exercise-induced blood lactate accumulation may be taken to be 4 mmol/litre.

Cantering for up to 22 km in one study by workers at the University of Glasgow caused no stress on nutrient reserves in the blood in that immediately after the exercise blood glucose, glycerol and free fatty acids (FFA), as well as pH, rose slightly (Snow & MacKenzie 1977b), which indicated an adequate rate of their mobilization and a sufficient irrigation of muscle cells with oxygen. Snow & MacKenzie (1977a) found that at a maximum work rate of 864 m/min (32 mph) over 3×600 m, with two 5-minute intervals between the gallops, anaerobic respiration was necessary in that blood glucose, glycerol, FFA and lactate were all elevated and there was a fall in pH. The increased fund of blood glucose was associated with a considerable rise in adrenocorticosteroid secretion; training for 10 weeks before the gallops caused a further increase in blood glycerol and lactate with smaller changes in FFA and pH than was the case with the untrained horses. As corticosteroid secretion was even greater after training, an explanation may be proferred that training confers a more efficient mobilization of reserve fat, and both it and glucose are used to a greater extent by trained horses resulting in faster times. The lesser fall in blood pH indicates that training also leads to better ventilation and oxygenation of the muscles, as this effect of training coincided with a greater, rather than a smaller, increase in blood lactate.

The accelerated use of glucose is accommodated by a stimulation to glycogen deposition in the muscles of adequately fed trained horses. Training for a period of 10 weeks can increase these reserves by approximately 30 per cent (Guy & Snow 1977). Blood glucose cannot, however, be maintained indefinitely during work, and endurance rides of up to 150 km indicate a gradual decline and an exhaustion of muscle glycogen. In fact, faster horses generally have lower blood-glucose concentrations on completing these rides. In one 50-mile (80 km) ride, blood glucose on average fell by 40 per cent whereas FFA rose eightfold (Hall, Adrian et al. 1982). The glucose effect is in accord with the recent observations of Essén-Gustavsson, Karlström & Lindholm (1984) that blood-glucose concentration is not affected by endurance rides of 50 km (31 miles) but it decreased 23 per cent during a ride of 100 km (62 miles).

A medium-paced canter is normally accomplished with aerobic respiration so that the ratio of lactate to pyruvate in the blood is unchanged. But during galloping, the ratio rises sharply and an effect of training is to induce smaller changes in blood lactate and in the lactate to pyruvate ratio.

Blood-lactate concentration may nevertheless be unrelated to racing speed, but as it is a relatively strong acid it tends to be correlated with blood pH. Blood-lactate concentration is partly an expression of adaptation to training and Persson (1983) proposes that the estimation of blood lactate at a work-induced heart rate of 200 beats/min could be used to gauge fitness in training where lower lactate concentrations down to 2 mmol/litre would reflect greater fitness. Heart rate would be measured by telemetry on the track and the results would be unaffected by track conditions that would affect speed. In addition to dehydration and electrolyte depletion, blood acidity is a dominating factor in determining exhaustion (Table 9.1) and serum levels of the enzyme creatine kinase (CK) provide an indication of the severity of exercise and metabolic acidosis. In one study where horses raced at the rate of 700 m/min (26 mph) serum AAT rose 50 per cent, serum calcium rose 13 per cent, but serum CK rose 227 per cent (Williamson 1974). It is thought that the increase in serum concentration of muscle enzymes after exercise is explained by an increase in the permeability of muscle cell membranes owing to hypoxia (lack of oxygen). Thus, inadequate training or severe work loads will induce a greater increase in the serum concentration of CK.

Table 9.1   The effect of racing 1900–2500 m on concentrations of blood lactate and acidity and their relationship to exhaustion

	Blood lactate (mequiv/litre)			Blood pH		
		After race			After race	
	Before race	Immediately post	15 min post	Before race	Immediately post	15 min post
Exhausted	0.58	23.14	24.37	7.379	7.086	7.105
Not exhausted	0.63	17.99	16.92	7.379	7.164	7.213

(From Krzywanck 1974)

As the intensity of training increases, the demand for energy release by way of a particular metabolic pathway rises, so increasing the need for the appropriate enzymes to catalyse the reactions. Sprint racing imposes greater demands on anaerobic metabolism and longer races call predominantly on aerobic processes. The latter implies a more intense use of the TCA cycle in the mitochondria. Training for sustained work is therefore seen to bring about an increase in the number of these cellular organelles and their associated enzymes. Training for sprint racing, however, increases glycolytic activity and a twofold increase in the activity of the enzymes aldolase and alanine amino transferase (ALT) has been observed. The latter promotes the formation of the amino acid alanine from pyruvate so lessening the conversion to lactate and therefore probably lessening fatigue by moderating a fall in pH. The activity of enzymes depends on the presence of the necessary cofactors and an increasing demand for these enzymes implies an increasing need for the cofactors. These cofactors include magnesium and zinc together with forms of the vitamins thiamin, riboflavin, niacin and pantothenic acid, all of which play major parts in carbohydrate and fat metabolism. The horse derives these B vitamins from its diet and by microbial synthesis in its intestine. As the intensity of work increases, the composition of the diet and the amount of feed consumed change as a consequence of the increased consumption of starchy cereal grains. This will alter not only the dietary supply of B vitamins, but also the intestinal synthesis of these substances, and it is an open question whether the rate of their absorption is exceeded by the tissue demand when horses are in intensive training.

The microbial fermentation of starch yields a higher proportion of propionate in the VFA. Metabolism of this acid requires cyanocobalamin (vitamin B_{12}) and work with ruminants has revealed that such diets may create a dietary requirement for this vitamin (Agricultural Research Council 1980), the lack of which causes an accumulation of propionate, depressing appetite. Uncontrolled observations of horses in training by the author have shown that the blood concentrations of vitamin B_{12} in them is lower than in many other horses and that the palates of those with flagging appetites may be whetted by supplements of the vitamin. A reasonable inference is that an increased consumption of cereal grains by horses increases propionate production and hence the dietary requirement for vitamin B_{12}.

It can be concluded that exhaustion during long-distance work is an expression of nutrient depletion whereas during sprint work it is principally the result of raised blood lactate and a consequential fall in blood pH.

The endocrine system (hormones)

The activity of many enzymes involved in energy metabolism is regulated by specific hormones, which in turn may react to nutrient status as in the case of insulin, or to the outside environment through the medium of involuntary nervous reactions, as applies to the catecholamines epinephrine and norepinephrine (adrenaline and noradrenaline) (see Fig. 9.2).

Insulin

Insulin is secreted in response to the incoming tide of glucose after a meal. It retards the breakdown of glucose and favours its cellular uptake and storage as glycogen or fat. Horses accustomed to a high starch diet possess a high insulin activity and therefore are more inclined to hypoglycaemic shock when subjected to a fast than are those normally fed hay diets and accustomed to deriving blood glucose from other sources. The insulin status of a horse can be determined by measuring glucose tolerance, which is the quantity of glucose that can be absorbed without causing glycosuria. A horse with a decreased tolerance exhibits a more marked rise in blood glucose following a dose and a slower subsequent rate of fall in the blood concentration. Such a condition can arise from diabetes mellitus, fasting, or low starch diets.

In Thoroughbreds, blood glucose normally peaks between 3 and 8 hours from the start of feeding. The peak value may be 150 mg/dl (8.3 mmol/litre) at 4–5 hours and then there is a gradual decline to a postabsorptive level of 80 mg/dl (4.4 mmol/litre). Ponies on roughage diets may maintain normal levels 25 mg/dl lower than this. Blood-insulin activity tends to fall during work because of the catabolism of glucose. However, as insulin appears to increase the permeability of muscle cells to glucose, the fall in the blood activity may be very slight, even after a trot of 40–50 km. Nevertheless, on completion of endurance rides when glycogen reserves are much depleted, blood-insulin activity is considerably depressed.

Glucagon

In order to sustain concentrations of blood glucose within normal limits the influence of insulin is counteracted by that of glucagon. Whereas the former promotes uptake of glucose in all cells of the body, glucagon appears to focus its effects primarily on the liver and adipose tissue. It achieves an increase in blood glucose by stimulating those enzymes that cause a breakdown of liver glycogen (Fig. 9.2) and by encouraging gluconeogenesis. In this latter function it works in concert with other hormones discussed below, a particularly important task in roughage-fed animals. These other hormones are the glucocorticosteroids produced by the adrenal cortex and epinephrine and norepinephrine secreted by the adrenal medulla.

Adrenal hormones

The rapid initiation of intense work necessitates an immediate response in terms of energy mobilization. This is brought about by sympathetic nervous activity, which not only causes splenic release of red cells but also stimulates the adrenal medulla to secrete epinephrine. The extent of this reaction depends on the intensity of the work load – that is the faster the horse is running the greater is the secretion. The medullary hormones affect several tissues, increasing the mobilization of fatty acids from adipose tissue and stimulating the production of glucose with a rapid rise in the blood concentration by the breakdown of liver glycogen and by amino acid metabolism (Fig. 9.2).

The glucocorticoids secreted by the adrenal cortex are somewhat slower in responding to work demand and their secretion depends on a hormonal signal from the pituitary. Moreover, they stimulate a rise in blood glucose and the accumulation of liver glycogen by promoting gluconeogenesis, through the inhibition of protein synthesis, and they provoke the breakdown of depot fats to FFA and glycerol. Synthetic analogues of these secretions, when given repeatedly, have a comparable effect and cause muscular wasting. They also initiate bone problems through an inhibition of calcium absorption from the gut. The stimulation from glucocorticoids of amino acid mobilization is expressed by excitation of transferase enzymes raising serum (Codazza, Maffeo & Redaelli 1974; Essén-Gustavsson, Karlström & Lindholm 1984; Sommer & Felbinger 1983) and muscle (Guy & Snow 1977) activities of ALT and AAT (see Fig. 9.2) observed after exercise.

In an analogous fashion to the response of glucagon, a reduction in blood glucose triggers the secretion of the glucocorticoids and their circulating level increases with agitation, trauma and psychological stress. Training leads to a greater adrenocorticoid response under such stress. This applies to sprint, endurance and other training so that normal concentrations of blood glucose are maintained more effectively in all circumstances. In contrast to the response of medullary hormones, amounts of plasma cortisol seem to be uncorrelated with the intensity and speed of work.

The heart and lungs

The heart is simply a pump with the function of circulating blood to the various tissues and organs of the body so that nutrients and oxygen are delivered, wasteproducts collected and heat redistributed. To facilitate these objectives during increased work the capillaries of the skeletal and cardiac muscles become dilated while those in the visceral region contract. This in turn diminishes the digestion and absorption of nutrients and gives credibility to the tenet that horses should not be fed before hard work. Heart rate is linearly related to speed of the horse and varies between the approximate limits of 30 and 240 contractions per minute. After a gallop at 700–800 m/min (26–30 mph) the rate may be achieving 240 beats per minute with an output from each ventricle of 3–4 litres/s. Blood volume of horses is about 9.7 per cent of bodyweight so that a horse weighing 560 kg (1235 lb) would contain around 51.2 litres (specific gravity 1.06). The volume of blood discharged per beat (stroke volume) in that horse at rest would be 1.2 litres and as blood volume is proportional to bodyweight, cardiac output per ventricle ranges between 56 and 75 ml per (kg × min) at rest. The need for such a flexible system can be appreciated when it is realized that oxygen consumption of skeletal muscles can increase a hundred times in strenuous exercise. Slight pulmonary bleeding seems to be a normal consequence of strenuous work and reflects no dietary abnormality but simply the physiological stress of a massive increase in nutrient and gaseous irrigation of muscle tissues.

Heart rate, particularly after work, is a good indicator of fitness. In the pre-ride checks of endurance horses it is agreed that pulse rate should fall within the limits of 36–42 beats per minute and respiration rate between 8 and 14. Both are higher in unconditioned horses. After an endurance ride and 20 minutes' rest the pulse rate should have fallen to less than 55 and the respiration rate to 20–25 per minute. In exhausted horses, the rate of both these is greater (tachycardia and hyperpnea) and the occurrence of muscular spasms more likely (Figs 9.3 and 9.4). The ratio of pulse to respiratory rate should fall within the limits of 2:1 to 5:1. During heavy exertion and heat stress the pulse and respiration rates have been known to rise to 85 and 170, respectively, a ratio of 1:2, that is an inversion of the pulse : respiration rates. Poorer horses and those suffering from adrenal exhaustion tend to exhibit lower heart to respiratory rates both before and after exercise. After a 20-minute rest during an endurance ride horses exhibiting heart rates exceeding 70, or heart : respiratory rate ratios of less than 2:1 should be eliminated. Hyperventilation of the lungs may simply reflect a shortage of oxygen and normal respiratory acidosis (not commonly

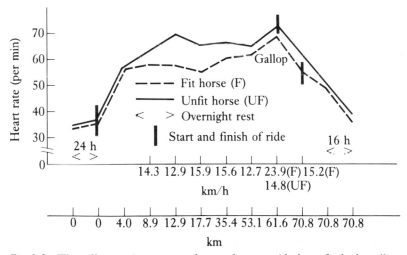

Fig. 9.3 The effect on heart rate of an endurance ride by a fit Arab stallion and a less-fit pony gelding. (Data from Frape, Peace & Ellis 1979).
Note raised heart rate of unfit horse between 53 and 61 km may have resulted from anxiety when partner galloped off.

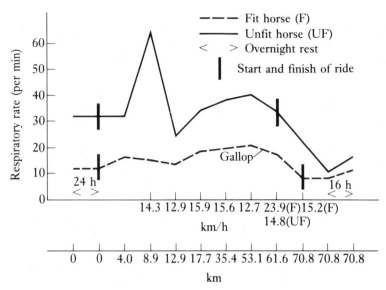

Fig. 9.4   The effect on respiration rate of an endurance ride by a fit Arab stallion and a less-fit pony gelding. (Data from Frape, Peace & Ellis 1979).

seen in endurance horses), or it may indicate an increase in body temperature (easily measured per rectum) brought on by hot weather or inadequate training, or both (Fig. 9.5). Alkalaemia normally follows a raised body temperature (see below).

Blood composition

The blood of the horse is a fluid containing, by volume, about 45 per cent of red cells (erythrocytes), 1 per cent of white cells (leucocytes) plus platelets and 54 per cent of plasma. The red cells have, as a major function, the transport of oxygen from the lungs to the muscles and other tissues. To accommodate an increased oxygen demand a reserve of red cells is held in the spleen. This splenic reserve is very large in Thoroughbreds so that they can increase the numbers of red cells in circulation by between 30 and 60 per cent. Thus, in heavy work, the oxygen-carrying capacity of the blood may increase from rest by as much as 8.8 volumes per cent. Between rest and galloping at 700 m/min (26 mph), the oxygen-carrying capacity changed from 15.9 to 21.4 volumes per cent in one study and from 16.35 to 25.19 volumes per cent, a 54 per cent increase, in another study (Milne 1974) using Thoroughbred, Quarter Horse, Arabian and Standardbred

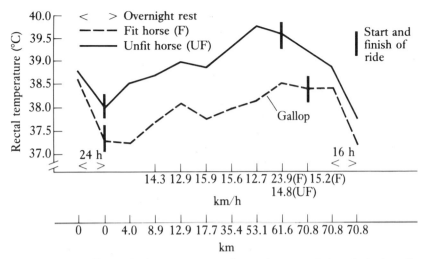

Fig. 9.5 The effect on body temperature of an endurance ride by a fit Arab stallion and a less-fit pony gelding in cool weather (Data from Frape, Peace & Ellis 1979).

horses, the effect of which is augmented by a redistribution of the blood supply to the muscles. The degree of splenic release tends to be proportional to speed; in one set of measurements the packed cell volume (PCV) rose 32 per cent at 350 m/min (13 mph) and 55 per cent at 700 m/min (26 mph) (Williamson 1974) (for other observations, see Figs 9.6 and 9.7). The red cells contain haemoglobin that not only carries oxygen but also acts as a buffer to the lactic acid produced in the contracting muscles. This is probably why the pH of blood falls only during galloping whereas the haemoglobin content rises appreciably during only moderate exercise. To assess the red-cell count and haemoglobin content of blood as affected by dietary inadequacies and other factors, the most consistent results are obtained after splenic release has been accomplished.

In order that blood travels freely through the small capillary bed of muscles, it is essential that it retains its fluidity and where the PCV exceeds 55 per cent there is an exponential increase in blood viscosity. The interest shown in providing horses with additional red blood cells before races can therefore be a misguided activity. The importance of a low viscosity is recognized in horses subject to dehydration in hot climates and during long-distance rides (Figs 9.8 and 9.9). For this reason it may be no chance that Arab horses have a lower PCV than other hotblooded horses and are highly adapted to both hot climates and extended work. A voluminous splenic pool may therefore be a disadvantage for some purposes. Plasma albumin, a reserve protein and a major contributor to blood viscosity, displays a decreased concentration during training for reasons that may be understandable in this context. The dilutions achieved still provide

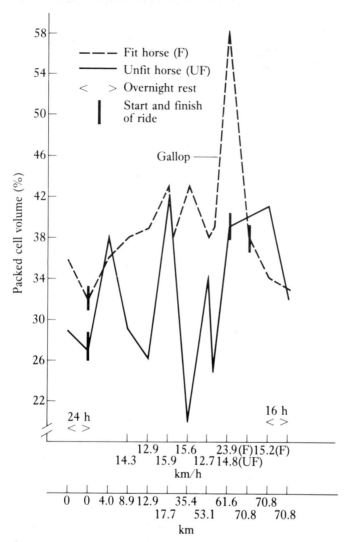

Fig. 9.6   Packed cell volume of the blood during an endurance ride by a fit Arab stallion and a less-fit pony gelding. (Data from Frape, Peace & Ellis 1979).

adequate osmotic pressure and do not necessarily impute a dietary protein deficiency. On the contrary, from recent observations (Lunn & Austin 1983) it could be inferred that dietary carbohydrate excess may be a factor through induction of increased protein catabolism.

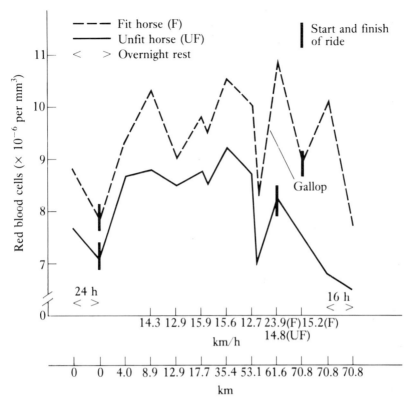

Fig. 9.7 Changes in erthyrocyte count during an endurance ride in a fit Arab stallion and a less-fit pony gelding. (Data from Frape, Peace & Ellis 1979).

Muscular changes in training

Adaptation to hard muscular work entails changes not only in the blood vascular system but also in the skeletal muscles. Increase in size (hypertrophy) of muscle occurs during training. The detailed changes induced vary with the breed of horse and with the type of work. Long-distance riding requires a preponderance of muscle fibres capable of slow contraction but resistant to fatigue, whereas sprinting ability demands the presence of a higher proportion of fast-contracting fibres that happen to be readily fatigued. These fast-contracting fibres are categorized as having myosin with high ATPase activity and a high glycolytic activity. They are subdivided into fast-twitch, low-oxidative (FT) and fast-twitch, high-oxidative (FTH) fibres. Fibres with low myosin-ATPase activity are present in relatively higher proportions in the skeletal muscles of horses more suited to

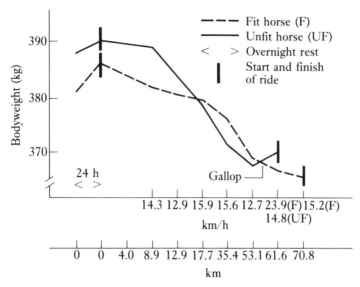

Fig. 9.8   The effect on bodyweight of an endurance ride by a fit Arab stallion and a less-fit pony gelding in cool weather. (Data from Frape, Peace & Ellis 1979).

long-distance riding. These are known as slow-twitch, high-oxidative fibres (ST). Standardbred trotters older than 4 years have 54 per cent FTH, 22 per cent FT and 24 per cent SF fibres in the gluteus medius muscle (Lindholm & Piehl 1974). All muscles have about the same ratio of low-oxidative to high-oxidative fibres, but a mixture of anaerobic and aerobic training increases the proportion of high-oxidative FTH fibres and decreases the proportion of ST and possibly also of FT fibres. Training (Pl. 9.1) increases the glycogen content and enzymatic activity of muscles but, where sprint training is practised, the enzymes involved in glycolysis are preferentially stimulated. In contrast, endurance work tends to favour the enzymes of the TCA cycle.

Muscle glycogen is consumed by exercise, but the net loss is extremely small in light work. Lindholm (1974) observed that Standardbred trotters, for instance, lost 0.3 mmol of glycogen per (kg × min) when trotting at 300 m/min (11 mph), whereas trotting at a maximum rate (750 m/min or 28 mph) led to a loss of 14 mmol of glycogen per (kg × min). Three aggregate minutes of maximal trotting were proved to cause a 48 per cent decrease in muscle glycogen but the decrement was not equally distributed among the fibre types. Maximal work causes a striking depletion of ST fibres in addition to a loss from the other two types. On the other hand, slow trotting leads to a gradual depletion of ST fibres after which the FTH fibres become active and depleted. Thus there seems to be a preferential fibre recruitment with increasing speed and duration.

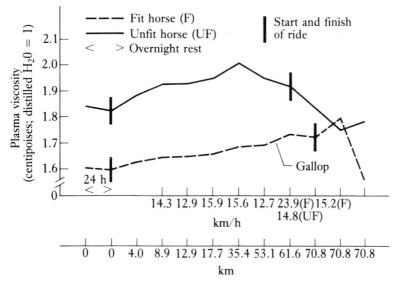

Fig. 9.9   The effect on blood plasma viscosity of breed and of an endurance ride by a fit Arab stallion and a less-fit pony gelding. (Data from Frape, Peace & Ellis 1979).

Preliminary evidence indicates that with high starch diets and regular exercise muscle glycogen overload can be induced but this is not maintained if the starch content of the diet is subsequently reduced. Glycogen loading in man has been practised for long-distance running, although for sprinting this loading can apparently precipitate stiffness associated with intracellular water storage. No clear advantage from glycogen overloading in horses has yet been demonstrated. Such stores *per se* seem not to be the cause of 'tying-up' in horses, which affects mainly the FT and FTH fibres and is considered to be a hypocalcaemic tetany associated with raised muscle lactate and glycogen loss.

Acid–base balance

An acid is a substance, such as lactic acid, which yields hydrogen ions in solution. The acidity of blood or other solutions is expressed as the pH. Acids and bases are produced during the metabolism of nutrients and abnormalities in the acid–base balance result from dysfunction, or overloading, of general metabolism and respiration. The normal pH of arterial blood is 7.5 and that of venous blood 7.4. The blood carries carbon dioxide to the lungs, partly in the form of weak carbonic acid, one of the principle acids of fizzy drinks. It and haemoglobin act as the principal buffers in blood; that is they prevent the pH from shifting appreciably and so prevent death from this cause. In the plasma, carbon dioxide reluctantly and slowly forms carbonic acid, but on diffusing into the red cells this reaction is accelerated 13 000-fold by the presence there of the enzyme carbonic anhydrase. Despite the accelerated change only 1 part in 800 of CO_2 forms carbonic acid. This is described in Fig. 9.10, which shows that nearly all this carbonic acid dissociates into hydrogen and bicarbonate ions. The former are partly buffered by haemoglobin and the latter to a large extent diffuse back into the plasma so that around twenty times as much CO_2 is carried as bicarbonate as remains in the form of dissolved gas. Now, the dissociated form of carbonic acid in blood solution is in a constant (K) proportion to the undissociated form as shown below:

$$\frac{H^+ \times HCO_3^-}{H_2CO_3} = K$$

If acid is produced during muscular activity, or during intestinal colic (see Ch. 11), this increases the H^+ ions in the numerator and these react with bicarbonate forming CO_2 and water (Fig. 9.10). In this process the bicarbonate concentration falls but the H^+ ion concentration does not rise as much as it would in the absence of the bicarbonate buffer, and so the ratio bicarbonate to CO_2 governs the pH of the blood.

As haemoglobin is also a major buffer, blood pH is changed very little for a considerable change in bicarbonate content. However, in performing this function haemoglobin carries less oxygen. In extreme cases of acidosis the horse becomes cyanotic, or oxygen-starved.

The lungs have a vital short-term role in the acid–base balance by providing a route for the excretion of carbonic acid (Fig. 9.10). Although the kidneys also function in this role, quantitatively they are insignificant as the lungs dispose of 200 times more of the acid. However, the kidneys fulfil a long-term role in disposing of non-volatile acids and bases from the diet, to be discussed below under cation–anion balance. In this balance, and in that of acid–base, an important principle is disclosed. This is the principle of electroneutrality in which all urinary and lung excretion and in all movement across normal cell membranes in the horse no electrical charge can accumulate, as in a battery. The number of anions has to correspond with the number of cations moving in the same direction. This may lead to pH changes of various bodily fluids.

Like most laws, this one is only approximately true as it may conflict with that of iso-osmolarity (see Glossary), in which all bodily fluid compartments (among which water is exchangeable through semipermeable membranes) approach isotonicity. In achieving this a small voltage gradient occurs as, for example, between intracellular fluid in the muscles and extracellular fluid. These principles play a central part in muscular ailments, such as 'tying-up' (see Ch. 11).

In so far as their acid–base status is concerned, horses are subject to

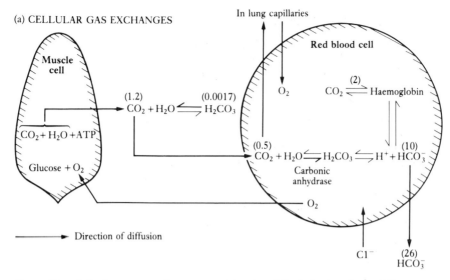

(a) CELLULAR GAS EXCHANGES

Note: • Numerical values are approximate concentrations [mmol/litre] of various forms of CO_2
• Total concentrations of CO_2 $[CO_2]$ = dissolved $[CO_2]$ + $[HCO_3^-]$ + protein-bound $[CO_2]$
• The enzyme carbonic anhydrase has a vital influence on CO_2 carrying capacity of blood, which is therefore a function of the number of red cells

(b) PLASMA REACTIONS

(1) **Metabolic acidosis**   *Note*: O_2 debt

$$CO_2 + H_2O \rightleftharpoons H_2CO_3 \rightleftharpoons \uparrow H^+ + \downarrow HCO_3^-$$

Increased H^+ causes increased
ventilation rate and decreased
blood CO_2

$H^+ + $ lactate$^-$

(Tendency to be exchanged
K^+ in functioning kidney tubules
for potassium ions)

(2) **Metabolic alkalosis**   *Note*: HCO_3^- generated
replacing Cl^- lost
in sweat

$$\uparrow CO_2 + H_2O \rightleftharpoons H_2CO_3 \rightleftharpoons \downarrow H^+ + \uparrow HCO_3^-$$

Decreased ventilation rate

K^+ lost in sweat is
replaced in renal
tubular filtrate by H^+

(3) **Respiratory acidosis**   *Note*: Plasma $[H^+]$ tends
to rise

$$\uparrow CO_2 + H_2O \rightleftharpoons H_2CO_3 \rightleftharpoons H^+ + HCO_3^-$$

Respiratory insufficiency causes
increased ventilation rate

(4) **Respiratory alkalosis**   *Note*: A decrease in $[H^+]$
and possibly $[HCO_3^-]$
results from
respiratory changes

$$\downarrow CO_2 + H_2O \rightleftharpoons H_2CO_3 \rightleftharpoons H^+ + HCO_3^-$$

Overheating or pain causes increased
respiratory minute volume

⟷ Increased direction of reaction

$\uparrow\downarrow$ Initial changes in plasma concentration of metabolite

Fig. 9.10 The effect of metabolic state on carbonic acid in blood plasma and associated relationships in muscle cells and red blood cells.

Table 9.2 Composition (mmol/litre) of various electrolyte solutions for intravenous use*

	Na	K	Ca	Mg	Cl	Glucose	Bicarbonate	Lactate	Acetate	Gluconate	Propionate
0.9% NaCl	154	—	—	—	154	—	—	—	—	—	—
5% Dextrose	—	—	—	—	—	278	—	—	—	—	—
Hartmann's solution	131	5	2	—	112	—	—	28	—	—	—
Ringer's lactate	130	4	3	—	109	—	—	28	—	—	—
Normosol R†	140	5	—	1.5	98	—	—	—	27	23	—
Dilusol R‡	140	5	—	1.5	98	—	—	—	27	23	—
Normosol M†	40	13	—	3	40	278	—	—	16	—	—
Balanced electrolyte solution§	137	5	3	3	95	—	—	—	27	—	23
Solution to treat acidosis¶	137	20	—	—	97	—	60	—	—	—	—
5% NaHCO₃ (hypertonic)	600	—	—	—	—	—	600	—	—	—	—
5% Dextrose saline (hypertonic)	154	—	—	—	154	278	—	—	—	—	—

(From Rose 1981)

Note: Intravenous solutions should contain a bicarbonate precursor (lactate, acetate, gluconate or propionate) as the sole use of fixed bases such as chloride can cause metabolic acidosis, hypokalaemia and hyperchloraemia.

* Give only 1–2 litres/h of solutions containing 5% dextrose but up to 3–5 litres/h for other solutions, all of which should be at 37°C.

† Abbot Laboratories, Illinois, USA.

‡ Diamond Laboratories, California, USA.

§ Merritt 1975 in Rose 1981.

¶ Rose 1979 in Rose 1981.

four types of abnormal metabolism. Only those conditions relating to physical work will be discussed here.

Metabolic acidosis

In strenuous exercise an oxygen debt with a tissue accumulation of lactic acid will occur to varying degrees. It is also possible that this form of acidosis may exist in combination with a B-vitamin deficiency, causing an incomplete metabolism of pyruvic acid, but this proposition has not yet been substantiated. The build-up of H^+ ions causes hyperventilation and thus an increase in respiratory minute volume with a fall in blood CO_2. This shifts the carbonic acid equation to the left (Fig. 9.10), which brings about some respiratory compensation of the acidosis and prevents an excessive fall in pH. In an analogous situation, the raised H^+ ion concentration and pain of laminitis will have a similar respiratory effect so that in its advanced state respiratory alkalosis supervenes.

Sodium bicarbonate at a rate of between 250 and 300 g (3000–3600 mequiv) over 24 hours is sometimes given as a therapy for metabolic acidosis, but its unconsidered use can produce untoward effects.

Metabolic alkalosis

Alkalosis can occur in chronic laminitis as well as from exhaustion after long-distance cross-country work in hot weather; but a reduction in ventilation rate in the worst-affected horses may induce slight acidosis. During extended work a depletion of body potassium and chloride develops to varying degrees. H^+ ions are excreted as a substitute for the depleted potassium, and bicarbonate fills the anion gap after loss of chloride. Tetanic spasms in extremis may sometimes result from a decreased availability of ionized calcium brought on by alkalosis. Unfit horses may present signs of adrenal exhaustion. The consequent failure to secrete aldosterone precipitates an excessive urinary loss of sodium with potassium retention and a further decline in well-being.

Sodium bicarbonate therapy would severely aggravate alkalosis and the provision of a balanced electrolyte and glucose solution containing sodium, potassium and chloride (see also Ch. 11) is to be recommended. The want of potassium is not truly attested by its plasma level as a consequence of a shift from intracellular to extracellular space (see page 262). Electrolyte losses and hypocalcaemia, possibly related to a raised pH in exhausted endurance horses, are associated with a condition known as synchronous diaphragmatic flutter, in which heart and respiratory contractions coincide. In addition to electrolytes, calcium gluconate solutions are frequently given intravenously (Table 9.2).

Respiratory acidosis

This is the build-up of blood CO_2 (hypercapnia) that arises with the respiratory insufficiency of sprint work. It is obvious that the condition may also obtain with several diseased states of the lungs, including respiratory allergy, and the resulting decline in oxygen tension (hypoxia) in the blood may cause metabolic acidosis.

Respiratory alkalosis

Hard work in hot weather brings about overheating that directly affects the respiratory centre of the brain causing rapid breathing. Rapid ventilation rate flushes CO_2 from the blood. The pain of colic and laminitis also accelerates respiration rate with similar consequences. The effect is to induce a slight rise of blood plasma pH (Fig. 9.11), although an absence of change in endurance horses may reflect a concomitant fall in both CO_2 and bicarbonate.

Maintaining a balance of anions and cations

The normal arterial pH of horse blood is 7.5 and metabolism for all important functions occurs most efficiently when the pH approaches this value. Deviations are associated with losses of important ions through the kidneys and intestinal tract causing a burden on metabolism that has eventually to be rectified. A lowering of pH results from an excessive rate of acid production during exceptional work and from disease states of the intestinal tract and associated organs, kidneys and lungs in particular. Organic acids produced in metabolism have only short-term consequences as they should be ultimately disposed of by metabolism and respiratory compensation. Fixed acids and bases absorbed from digesta have a longer term influence and attention has been focused on the fixed cations and anions present in diet and their deducible effects on acid–base balance. Bone acts as a buffer to prolonged dietary imbalances of this kind. An acid diet, containing an excess of fixed anions, leads to bone resorption and osteoporosis so that renal excretion of those anions may proceed.

A quantitative measure of the acid–base status of a horse, as affected by metabolism, health and diet, is known as the base excess (BE) (Fig. 9.12). This is the base content of the venous blood measured by titration with a strong acid to a pH of 7.4 at normal CO_2 tension. Base deficit is the same as negative BE and its measurement requires titration with a strong base, again to a pH of 7.4. At that pH, BE is zero and plasma

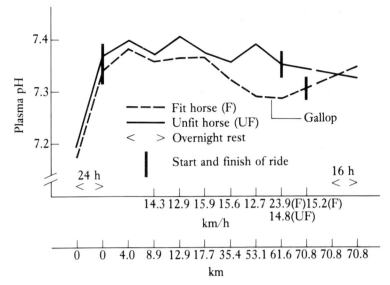

Fig. 9.11   The effect of an endurance ride on blood plasma pH in a fit Arab stallion and a less-fit pony gelding. (Data from Frape, Peace & Ellis 1979).

bicarbonate equals approximately 22–25 mequiv/litre. In the normal horse, bicarbonate should range from 24 to 27 mequiv/litre and the BE from 2 to 5 mequiv/litre.

The effect that diet and metabolic acid production have on BE is shown in Appendix D. Typical horse diets have a BE of 200–300 mequiv/kg estimated from the fixed-ion content. Thus, for a horse consuming 10 kg of feed per day, the excess base would amount to approximately 2500 mequiv, which is similar to that provided by 200 g of sodium bicarbonate – a quantity frequently given in therapy over a period of 24 hours. Horses with a serious deficit of bases may require, over that period, double the amount contained in a normal ration. The dietary optima assume that the protein content is probably higher than that provided for the average adult horse. For a diet containing 10 per cent crude protein, the BE could be of the order of 30 mequiv/kg less. The primary route of fixed-ion loss by working horses is sweat produced for the purpose of preventing an excessive rise in body temperature.

At high work rates heat production is forty to sixty times basal levels, and body temperature can rise appreciably. The mean muscle temperature of five Standardbred trotters increased from a normal of 37 to 41.5°C during a race of 2100 m at a mean rate of 708 m/min or about 26 mph (Lindholm & Saltin 1974). The principal mechanism for exhausting this waste heat from the body, when atmospheric humidity is not excessive, is the evaporation of sweat and of moisture from the surface of the lungs.

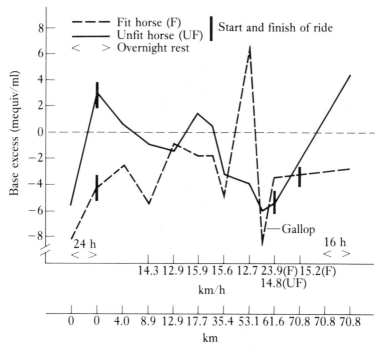

Fig. 9.12   Changes in plasma base excess during an endurance ride by a fit Arab stallion and a less-fit pony gelding. Base excess of the unfit horse declined from the start. Its respiration rate was consistently the higher, which could imply that a higher ventilation rate was removing more CO_2 from the blood and the carbonic acid reaction was moving to the left (last equation (4), Fig. 9.10b). Lap 8 was run at the lowest speed (Fig. 9.3) and, although the respiration rate of the unfit horse did not fall (Fig. 9.4), the low rate of the other horse immediately allowed the base excess to rise. This lap was followed by a rest period before the fit horse was galloped. (Data from Frape, Peace & Ellis 1979).

Carlson (1983b) calculated that if a horse could work at a moderate intensity for 1 hour with an oxygen consumption of 30–40 litres/min, the total waste heat would amount to 38 MJ. The dissipation of this quantity by evaporative processes alone would entail the loss of just over 15 litres of water. Although this is an oversimplification, it is a reasonable estimate of events in high environmental temperatures and low relative humidities. Sweat losses are quite modest in horses racing for distances of up to 3 km, but it seems that body water losses in sweat (and urine) and from the lungs during prolonged exercise can approach 10–12 litres/h and total as much as 40 litres. Typically, bodyweight falls 5–9 per cent, principally from evaporative losses, but the extent of loss depends on the level of fitness and the availability of water and electrolytes during the exercise. The evidence in Fig. 9.8 and Table 9.4 clearly indicates a greater rate of

Table 9.3 Composition (g/kg of dry matter) of mixtures for administration by stomach tube

Mix	Glycine	Sodium chloride	Monopotassium phosphate	Magnesium sulphate	Potassium chloride	Calcium carbonate	Calcium gluconate	Sucrose or glucose†
(1)*	470	270	190	13			57	
(2)*		325			325	175		175

Note: Mixes (1) and (2) may be squirted as a thick slurry to the back of the mouth, *after* water has been given to rectify a water deficit and thirst.

* In order to provide approximately isotonic solutions add 230 g of mix (1) to 6 litres of water and 120 g of mix (2) to 6 litres of water every 2–3 h for a 500-kg horse.

† May be replaced by molasses.

bodyweight loss and of sweat production (carrying the fixed ions of Na, K, Cl and Ca) by an unfit horse than by a fit one, although in both horses the molar sum of cations approximately equals that of the anions in the sweat.

Water losses are more aptly compared with total body water, which, in a horse of 450–500 kg bodyweight, would amount to 8–14 per cent of the total of about 300 litres in extended exercise. Of this 300 litres about 200 litres is in intracellular fluid (ICF) and 100 litres is in extracellular fluid (ECF), made up of the water in blood plasma, interstitial fluid, lymph and contents of the gastrointestinal tract. The 100 litres of ECF contain 14 000–15 000 mmol of readily exchangeable Na, representing nearly all the exchangeable Na of the body, the total Na content of which is about 1 kg. Of this total 40 per cent is located in the skeleton and during extended Na depletion this depôt appears to be partially mobilized. The 200 litres of ICF contain 20 000–30 000 mmol of readily exchangeable K, most of the body's reserve. The bulk of the exchangeable Cl is present in the ECF, where it is the major fixed anion, but its concentration is substantially lower than that of Na – probably of the order of 10 000–12 000 mmol/100 litres.

Dehydration through sweating entails a loss of both water and electrolytes with a contraction of the volume of body fluid. The electrolytes of horse sweat consist principally of Na and Cl, with lesser quantities of K (Table 9.4) and small proportions of Ca and Mg, amounting to about 10 mmol/litre of each. Changes in the composition of the blood plasma depend on the proportions lost of each of these constituents and of water and on the movement of ions into and out of ICF space. Extended exercise, with minimal water consumption leading to dehydration, will normally precipitate a substantial reduction in the concentration of plasma Cl; little change is frequently detected in the amounts of K and Na, although hypokalaemia is not rare. The explanation for the hypochloraemia is revealed by a comparison of the Na and Cl contents of sweat (Table 9.4)

Table 9.4 Electrolytes in evaporated sweat of horses of an endurance ride[*]

Horse	Cl	Na (mmol/litre)	K	pH of venous blood
Fit: mid ride[*]	910	710	215	—
Fit: finish[*]	1180	880	270	7.29
Unfit: finish[*]	3060	2120	780	7.36
Mean of other published data[†]	231	173	49	—

[*] Frape, Peace & Ellis 1979
[†] Meyer, Winkel et al. 1978; Carlson & Ocen 1979; Rose, Arnold et al. 1980; Snow Kerr, et al, 1982. (Exercise of various types)

with their respective concentrations in blood serum (Table 3.1), demonstrating that a much greater proportion of Cl than of Na in body fluid is lost. Extended work, in hot dry weather, by a 450–500-kg horse expressing the above plasma changes may yield losses of as much as 35 litres of water, 3500 mmol of Na, 1500 mmol of K and 4200 mmol of Cl (equivalent to 80 g Na, 59 g K and 149 g Cl). Supplements of electrolytes are given during endurance rides with the object of partially replenishing losses, it being recognized not only that natural feed given subsequently is likely to contain much more K than Na, but also that where electrolytes are given intravenously, excessive K produces toxic manifestations in cardiac muscle function (see Ch. 11). There should therefore be about twice as much Na as K and 1.2 times as much Cl as Na in such supplements. Organic anions make up the residue (Table 9.3). Small amounts of Ca and Mg may also be included.

The total concentration of electrolytes in sweat is higher than that in blood plasma, so a decline in the plasma concentration, despite considerable dehydration, is readily understandable. Thus most studies have revealed a decline in plasma electrolytes during endurance rides taking place in hot weather. For example, Carlson, Ocen & Harrold (1976) reported reductions of 4.2 mmol of Na per litre, 0.9 mmol of K per litre, 10.3 mmol of Cl per litre and no change in Ca. On the other hand the changes can be very variable depending on such factors as relative losses of water to electrolytes, plasma pH changes and time of collection after the ride. In horses losing a considerable amount of water, Rose, Purdue & Hensley (1977) detected increases of 6 mmol of Na per litre and 0.3 mmol of K per litre of plasma, no change in Ca and a decrease of 6.8 mmol of Cl per litre (sweat is rich in chloride and calcium). Thirst is in part controlled by the osmotic pressure of the blood and therefore it is frequently necessary to rectify electrolyte loss (Tables 9.2 and 9.3) before dehydrated horses will drink adequate quantities of water. Electrolyte mixtures are widely available commercially. If the horse is at the same time acidotic then plasma potassium may rise despite a potassium deficit, as intracellular potassium is exchanged for H^+ ions. Subsequently, renal losses of potassium may increase, ultimately causing severe potassium depletion. A further discussion of the means of assessing potassium status and the major causes of depletion and their therapy are given in Chapter 11.

Horses should be allowed to drink frequently during extended work and at least 2 minutes should be conceded on each occasion. If the weather is very hot then a drink at least every 2 hours is desirable. Hard work diverts much of the blood supply to the skeletal muscles away from the splanchnic bed of blood vessels serving the gastrointestinal tract. It is thought that this inhibits the efficient absorption of water so that the amounts consumed on any one occasion should be relatively small. The consumption of large amounts is also to be avoided because of the large difference between it and blood in osmotic pressure. After work, very dehydrated animals should be given about 4.5 litres (1 gallon) every 15 minutes, preferably containing 30 g (1 oz) of electrolytes. If the horse will not drink, administration by stomach tube is sometimes necessary. Dehydration is frequently accompanied by coldness and fatigue, muscular tremors, colic, thumps, lack of appetite and a low ratio pulse : respiration. Severely dehydrated animals are sometimes given a 5 per cent glucose-electrolyte solution intravenously, while heart rate is monitored, when their ability to absorb fluid from the gut is in doubt (Table 9.2).

Where there has been a contraction of blood volume the administration of electrolytes has only a transient effect on increasing that volume. Nevertheless, rectification of losses and of the acid–base balance must be considered. The total osmotic pressure of the blood depends to a large extent on the colloids it contains, the principal ones of which are the proteins, more especially albumin. However, protein loss during work will be minimal and reflect only its metabolism as an energy source apart from very slight losses through pulmonary haemorrhages. The rebuilding of energy reserves, particularly in respect of muscle glycogen, will take several days after very extended hard work.

In summary, the first requirement of exhausted dehydrated horses is water, followed closely by electrolytes. Bodily energy resources must then be rejuvenated and if the weather is cold the horse must be kept warm but not hot. After normal hard work a horse should be cooled by gentle exercise of the muscles through walking, but access to light grazing or hay should not be ruled out. After this relaxation of $1-1\frac{1}{2}$ hours, tepid water should be given before a light meal of concentrates.

Feed requirements for racing and eventing

Either underfeeding or overfeeding leads to inferior performance. A horse should not be fed in order to fortify it with reserves for future events, but rather the rates of feeding should be consistent with immediate needs. Overfeeding will create fatness, which causes a greater burden on the heart and the horse generally, and it interferes with the dissipation of heat. The synthesis of fat from excess carbohydrate, protein and dietary fat does not encourage those enzymes that participate in the breakdown of fat so necessary during work. Furthermore, overfeeding can lead to stocking up (oedema) of the legs, hives (bumps under the skin), forms of colic, founder, exertion myopathy and general overheating.

Concentrated feed should be given in a minimum of three meals per day with some hay, ample water and salt licks (a block of sodium chloride is satisfactory) available, in the morning, at noon and with ample hay in the evening. Many horses in training do not take sufficient salt from licks hampering progress and a feed source providing 60 g (2 oz) daily is recommended. Observations both in the United States and in the British

Isles show that horses racing on the flat from 2 years of age, and weighing 470–530 kg, consume daily between 13 and 18.5 kg of total feed, which amounts to between 2.7 and 3.7 per cent of bodyweight (Glade 1983a; Hintz & Meakim 1981; Mullen, Hopes & Sewell 1979; author unpublished data). Of this, concentrates with, for example, cereals, nuts, bran and linseed, amount to between 30 and 60 per cent of the ration. In one American study of 171 horses (Glade 1983a) the concentrates provided between 43 and 59 per cent of the digestible energy (DE) and between 39 and 64 per cent of the crude protein of the total ration. Moreover, the total ration provided 163 MJ of DE per 500 kg bodyweight and 1686 g of crude protein (270 per cent of the US National Research Council estimated minimum requirement). An analysis of the performance of these 3- and 4-year-old horses, under the supervision of seven trainers, in races over a distance of from 0.75 to 1.0625 miles (1207–1710 m), showed that there was a positive correlation between DE, or crude protein consumption, and time of finish. The time increased by 1–3 seconds for every 1000 g of crude protein consumed daily in excess of the NRC requirements. It is, however, not possible to tell whether the higher rates of consumption of energy and protein were the cause or a consequence of lower performance, or indeed whether the problem lay with energy or with protein, or both. Nevertheless, more is not necessarily better. In the United Kingdom, daily rates of protein intake among both flat and National Hunt horses amount to between 1000 and 1400 g/day (the author's measurements). These figures are well below the average of the American horses, largely as a result of the lower protein content of horse hays produced in Britain.

No evidence is available to show whether mineral, trace-element and vitamin requirements of racehorses are any different from those of other horses, although it has already been noted that differences in the demand at the cellular level for certain of the B vitamins may exist during hard work. Appendix B gives examples of dietary compositional errors encountered by the author in practice.

Horses participating in competitive long-distance rides are required to make effective use of body fat reserves as a source of energy and as a means of conserving glucose sources. A severe depression in blood glucose is a measure of fatigue. Feeding studies have shown, firstly, that dietary fat, in the form of corn oil, is well digested (Kane, Baker & Bull 1979), and that horses given a diet containing 8 per cent feed-grade animal fat required and ate 15 per cent less concentrate as a supplement to 2.7 kg timothy hay daily when being worked hard (24 km day [15 miles], 3 days/week at a rate of 201 m/min [7.5 mph] followed by 4 × 60-km [4 × 37-mile] rides at 2- and 3-week intervals) (Hintz, Ross et al. 1978a, b). Secondly, dietary fat lessens the decline in blood glucose during endurance rides (Hintz, Ross et al. 1978a) and seems to accelerate the rate of recovery of resting pulse and respiration rates during the first 10 minutes following such rides (Hintz, Ross et al. 1978b; White, Short et al. 1978), although no other beneficial physiological effects are evident.

The provision of diets containing 8–12 per cent of total fat may stimulate enzyme systems that participate in fat degradation, leading to an efficient use of free fatty acids, conserving blood glucose. This, however, is meagre evidence and much more is required before the widespread use of high levels of dietary fat for long-distance work can be vindicated. The practical problems of adding large amounts of fat to the diet also may have to be considered and overcome.

In preparation for endurance rides, the concentrate, or grain, allowance may be increased by up to 1 kg during each of the last 2 days before the event with an associated decrease in roughage intake of 1.5–2.0 kg. On the evening before the ride, an extra 0.5 kg of concentrates may be given, but generally speaking neither concentrated feeds nor hay should be given on the day of the race although access to water should continue.

In contrast, for short races, there is some justification for a light concentrate ration between 3 and 4 hours before the start; although timing is rather critical as the race should coincide with peak blood glucose and not peak blood insulin and many horses may refuse to eat at this time.

The build-up of feed and of training before races may take 8–12 weeks, and in the case of event horses a typical build-up may start with 5 kg of concentrated feed daily and finish with 8–8.5 kg 2 months later. The ration should be distributed amongst four daily feeds and the hay, one-third of which should be given in the morning feed and two-thirds in the evening, should be reduced to 3.5–4 kg/day for a horse of average size during the last part of training. The protracted and intensive training for dressage (Pl. 9.2) places emphasis on mental attitude and alertness, but it is equally important to achieve the right level of energy intake at each meal and overall.

In all cases feed intake should be increased as the work rate increases and the concentrated feeds should be severely restricted if work rate is reduced for any reason whether this be of short or longer duration. On rest days, a horse that would normally receive 8 kg of concentrates should then receive a maximum of 4 kg distributed in three feeds, but with a greater allowance of hay of up to 5–5.5 kg for the average-sized horse.

In Chapter 6 the phenomenon of waste heat generated during the digestion and metabolism of feeds was outlined and a mechanism for the 'hotting-up' effects of certain feeds was adduced earlier in the present chapter. Many trainers and horsemen and women are reluctant to use energy-rich cereals such as maize and barley because of the risks in this connection. However, the explanation here makes clear that where alternative feeds are rationed at rates that provide the same amounts of net energy, then energy-rich cereals will generate less rather than more waste heat, apart from any differences in metabolic rate caused by differences in the peak blood-glucose level achieved. This is controlled as much by the way feed is presented (see Ch. 5 and 6). The theory was recently put to the test at Cornell University in the United States (Wiltsie & Hintz in Hintz 1983). Polo ponies were maintained at a constant body weight, that is they

Plate 9.2 Wily Trout, a 16-year-old gelding, ridden by Chris Bartle and Pinocchio (15-year-old) ridden by Jane Bartle in 'Passage' during dressage training for the 1984 Olympic Games.

received approximately equal amounts of net energy from either maize and alfalfa hay, or oats and timothy hay. No significant differences in response, either before or after exercise, owing to diet (Fig. 9.13, p. 202) were noted. If anything, the maize-fed animals were less 'hotted-up'. Energy-rich feeds may possess other advantages for athletic horses – for example, causing less gut-fill, or non-functional weight. Where unnecessary problems of 'hotting-up' have arisen, it is undoubtedly the consequence of a lack of appreciation of the differences between cereal grains in their energy content and bulk density described in Ch. 5 and quantified in Table 5.2. The cooking of cereal starch may reduce 'hotting-up' by promoting digestion and thereby decreasing microbial fermentation.

Further reading

Snow D H, Persson S G B & Rose R J (eds) (1983) *Equine Exercise Physiology* (Proc. 1st Int. Conf., Oxford 1982). Granta Editions: Cambridge.

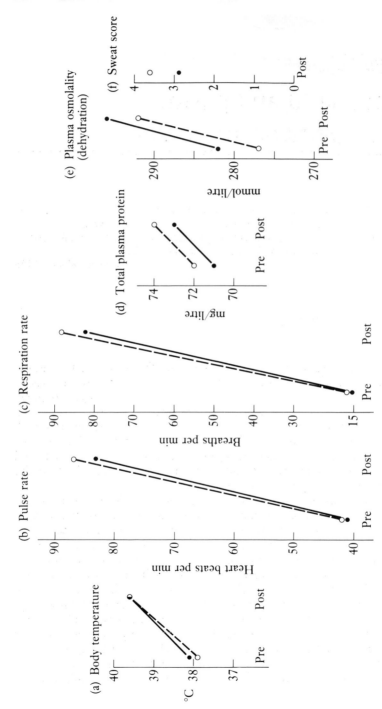

Fig. 9.13 Pre-exercise and post-exercise values for polo ponies given 6.8 kg alfalfa hay plus 3.2 kg maize (●) or 8.2 kg timothy hay plus 4.0 kg oats (○) continuously over 4-week periods of a reversal experiment to maintain constant bodyweight. (Data from Wiltsie & Hintz in Hintz 1983).

CHAPTER 10

Grassland and pasture management for horses

. . . but that grass which grows on wet grounds, or the winter-grass, abounds with little or no spirit, wherein a great deal of the true nourishment consists, and therefore it must needs beget a viscid and indigested chyle, which must also render those horses that are fed with it sluggish, dull and unactive.
William Gibson, 1726

Grassland types

In the humid temperate climate of much of the British Isles natural succession favours the replacement of grassland by scrub and then woodland and forest. To sustain high-quality 'permanent' pasture requires perserverance in land management, through the grazing of domesticated animals and through the treatment of grassland as a crop to be cultivated. These pastures constitute by far the greater proportion of grazing and they contrast with the uncultivated areas of mountain, moorland, heath and downland, where wild grazing and browsing animals contribute to the distribution of plant species that evolves.

The dominant species in most old pastures at the outbreak of the Second World War probably bore no relation to the seed mixtures originally sown, if the land was indeed sown down to grass. The acceptance of this fact led to the adoption of ley farming, in which seed mixtures are sown with the intention of reploughing the land as part of a rotation after a predetermined number of years, when the productive species will have ceased to dominate the sward. These grasslands provide annual yields of nutrients per hectare (ha) two to four times those obtained from average old pastures, and they approximate those of the best beef-fattening pastures. The latter have retained their productivity through the high natural fertility of the land and the best stock management, which has encouraged the persistence of grass and clover mixtures similar to those strived for in long-term leys. The very best of such land could, theoretically, support annually five, 500-kg barren or pregnant mares per ha from grazing and conserved forage.

Between 1940 and the last decade over half the old pastures in England and Wales were ploughed and although today there is less grassland than 45–50 years ago, a third of the permanent grass is less than 20 years old. Moreover, a quarter of all enclosed grassland can now be described as temporary leys (less than a 5-year life). This massive subjection of land to the plough, accompanied by a vast increase in the use of fertilizers, has spawned pastures relatively unknown to the horse of the 1930s, many of which are unsuited to its use except in skilled hands.

Permanent grassland (including land reseeded for an extended period) in England and Wales has been classified in three types according to botanical species. Class 1 grassland contains more than 40 per cent sown species with the most productive swards containing more than 30 per cent perennial ryegrass (*Lolium perenne*), a proportion of rough meadow grass (*Poa trivialis*), and the remainder of grasses consisting mainly of cocksfoot (*Dactylus glomerata*), timothy (*Phleum pratense*), other meadow grasses, Yorkshire fog (*Holcus lanatus*), species of bent grass (*Agrostis*) and fescue (*Festuca*). The proportion of white clover (*Trifolium repens*) depends very much on the use of nitrogenous fertilizers and the seasonal grazing pattern, but can amount to 25 per cent of the cover. Other broad-leaved plants vary in abundance according to management. Pastures in this class now constitute nearly a half the enclosed permanent grassland whereas in 1938 they formed only 4 per cent.

Class 2 grassland contains less ryegrass and 21–40 per cent of preferred sown species. However, the simple relationship between species composition and productivity is less clear than it was 50 years ago as fertilizers have increased the productivity of botanically poor pastures. This composition reflects rather more closely the extent of drainage. Where poor drainage has not been rectified, creeping bent (*Agrostis stolonifera*), Yorkshire fog, rough meadow grass and creeping buttercup (*Ranunculus repens*) thrive better than ryegrass does. Other less-productive grasses that invade swards of this class in significant numbers include meadow foxtail (*Alopecurus pratensis*), couch (*Agropyron repens*), crested dog's-tail (*Cynosurus cristatus*) and barley grass (*Hordeum murinum* and *H. pratense*), together with red clover (*Trifolium pratense*) and bird's-foot trefoils (*Lotus* spp.).

The decline in ground cover by sown species over the years occurs on all but the most fertile land regardless of the excellence of drainage. About 20 per cent of the cover by sown species is lost after 5–8 years and a further 10 per cent is lost during the next 4–12 years. Poorly drained soils provide a less suitable initial habitat for sown species and the proportion of them is less throughout the pasture's life. Class 2 grassland contains larger numbers of broad-leaved weeds, the most frequent of which are buttercups, daisy (*Bellis perennis*), plantain (*Plantago spp.*), nettle (*Urtica dioica*), creeping thistle (*Cirsium arvense*), ragwort (*Senecio jacobaea*), lesser yellow trefoil or suckling clover (*Trifolium dubium*) and yellow rattle

(*Rhinanthus minor*). White clover is still evident, although its frequency is less than in the 1930s probably as a consequence of increased fertilizer application.

During the past 30 years many old pastures, especially on dairy farms, have been improved by heavy treatment with fertilizers, drainage and intensive management. These swards are of open texture, subject to poaching in wet weather, but may contain a high proportion of ryegrass. Bare patches may be colonized by grasses such as soft brome (*Bromus mollis*) and annual meadow grass (*Poa annua*), and by mouse-ear chickweeds (*Cerastium* spp.). Lush pastures of this description with little bottom are unsuited for grazing by horses without great care and experience.

Class 3, the poorest of the three classes, contains less than 20 per cent of sown, or preferred species. It occupies a quarter of the permanent pasture area and contains a predominance of bent grasses, red fescue (*Festuca rubra*) and, in addition to the other poorer grasses, sweet vernal (*Anthoxanthum odoratum*) and a significant intrusion by rushes. In agricultural terms, most of these swards are only suitable for store cattle. They occur on a variety of both acidic and alkaline soils and contain less than 5–10 per cent ryegrass, but white clover should be present in those on alkaline soils. The diversity of habitats introduces a wide range of species in addition to those already mentioned.

In many river valleys, rushes and sedges appear in *Agrostis* pastures where drainage is impeded, or the land is otherwise neglected. The fertility may be potentially quite high, but in the more degenerate and derelict land, even in lowland areas, purple moorgrass or flying bent (*Molinia caerulea*), bracken (*Pteridium aquilinum*) and gorse (*Ulex europaeus*), quite useless for horses, may sometimes appear. On better-drained slopes of acid soils, between altitudes of 100 and 350 m (350–1100 ft) under annual rainfalls of 90–120 cm (35–45 in), fine-leaved fescues and bent grasses dominate pastures with scarcity of clovers in the latitudes of 50–57 °N of maritime climates. The specific distribution within these ranges depends on soil pH, latitude, aspect, soil drainage and grazing.

These areas merge into the uncultivated rough and hill grazings in which there is the invasion of bracken fern and gorse at lower altitudes along with the fine-leaved fescues. On moorland, matgrass (*Nardus stricta*) and purple moorgrass, rushes (*Juncus* spp.), heather (*Calluna vulgaris*) and bell heather (*Erica cinerea*) may occupy a larger proportion of the area. Excessive grazing by ponies of the better fine-leaved fescue and *Agrostis* areas may lead to their suppression and the encroachment of economically useless *Nardus*, bracken, etc.

Evolutionary changes in permanent pasture, leading either to improvement or degeneration in terms of productivity, depend mainly on skill and diligence in cultivation and management, grazing pressure and the extent of fertilizer application. Overgrazing and poor drainage threaten the existence of the better grass species and fertilizers generally will encourage them. The spread of shrubs and useless weeds may largely depend on drainage, the extent of cutting and the presence or absence of cattle.

The production by pasture of digestible nutrients for horses and ponies is clearly of economic importance. However, other issues sometimes play even a dominant part in the selection of pastures or their management for horse husbandry. For example, the thick cushion found in old pastures is better for exercise than is the open texture of leys and heavily fertilized swards, in which numerous stones, upturned during ploughing, contribute to leg injuries. Many horses needing rest and gentle exercise between periods of hard work, barren and pregnant mares and 1–3-year-old growing stock are turned out to subsist on pasture. In each of these cases, leys or highly fertilized permanent pastures would initiate rapid and unwanted fat deposition. This creates unnecessary problems in the early stages of subsequent work, in late pregnancy and early lactation, or it contributes to leg abnormalities in growing horses and to the incidence of laminitis and colic. Therefore, a high degree of skill is needed in evolving pastures for horses and ponies that provide useful grazing and also yield saleable and reliable stock. At one extreme the protein content in the dry matter of the pasture may be 30 per cent, which through the neglect of the principles of good husbandry can lead to digestive upsets, whereas at the other extreme rough grazing may contain little more than 4 per cent protein in the dry matter, insufficient even for maintaining horses (Owen, McCullagh et al. 1978). In the latter circumstance, the animals will seek out the more nutritious forage, which leads to its decline and to a precipitous regression in the grazing value of the land.

The thick, matted turf of well-drained old pasture resists poaching in wet weather and can, therefore, provide exercise and maintenance areas for out-wintered stock. However, not only is the total annual yield of digestible feed generally lower in these pastures, but, especially where drainage is poor, the season of herbage growth is shorter, a fact probably of much greater economic significance. The length of the grazing season is generally greater the higher the fertility of the land. Furthermore, permanent pastures contain a higher proportion of undesirable and poisonous weeds, particularly where husbandry is wanting.

Chemical composition of herbage

Amounts of crude protein, soluble sugars, nitrogen-free extractives (NFE) and ether extracts in the dry matter of herbage are highest during the period of rapid leaf growth in the spring, next highest during regrowth in the early autumn, lower during the period of flowering in mid-summer and normally poorest during the winter, when there is extensive dying back of the aerial parts of herbaceous plants. The months of the year when these phases occur in northern latitudes depend on the latitude, the lateness of the spring, rainfall, soil type and temperature. After the grazing of herbage, the first regrowth contains per unit of dry matter the highest protein, lowest crude fibre and highest NFE, or soluble carbohydrate and starch. These values change progressively as growth proceeds. For example, a study over 50 years ago (Fagan 1928) showed that in Italian ryegrass (*Lolium multiflorum*) from the second to the tenth week of growth, the crude protein composition of the aerial parts declines from 19 to 7 per cent, the crude fibre increases from 20 to 25 per cent and the NFE increases from 44 to 60 per cent. The changes can largely be explained by a rapid shift in the proportions of leaf to stem and leaf to flowering head (Tables 10.1 and 10.2). The horse digests fibre less easily than can domesticated ruminants so that shorter grass containing a higher proportion of leaf is a more valuable feed than herbage approaching maturity.

Table 10.3 gives some mineral values for herbage found by the author in stud paddocks at Newmarket. Although these values change with stage of growth their digestibility is less affected by maturation than is that of energy. The normal range does not necessarily imply adequacy, but simply what might be expected in lowland pastures in the southern parts of Great Britain.

Wild white clover constitutes a significant but variable contributor to the sward of permanent pastures. The chemical composition of clovers is rather similar, and the data in Table 10.2 attest to their being a rich source of calcium and protein. The ideal clover content of pastures for horses is hotly debated, but in highly productive leys clovers contribute annually the equivalent of 1000 kg of sulphate of ammonia per ha through virtue of the nitrogen-fixing nodules on their roots. Several management practices encourage the spread of clovers, such as liming, where the pH is 5 or less, the application of phosphate (P) and general restraint on the application of nitrogenous fertilizers, together with early season grazing as widely practised on sheep farms. On the other hand, clovers are suppressed by taking a hay crop, the use of high-nitrogen fertilizers in early spring without early grazing, and the use of selective weedkillers. Clovers tend to have deeper roots than do most grasses, especially the less-productive grass species. They draw moisture and minerals from lower horizons in well-drained soils, so that they remain green during summer drought when grasses have gone to seed, and they may rectify an imbalance of trace elements between the upper and lower horizons of the soil profile. Nevertheless, in the absence of conclusive evidence on their connection with metabolic upsets, their proportion in the herbage grazed in any one season should probably not be allowed to exceed one-third.

Herbs may be defined as broad-leaved plants with non-woody aerial parts and so could, of course, include clovers. Like clovers, many other herbs are rich in protein, minerals and trace elements relative to the common grasses; some are relished by horses and espoused by enthusiasts. The dry matter of nettle, for instance, contains nearly 6 per cent of lime, 5 per cent of potash and 2 per cent of phosphoric acid, but the fresh plant

Table 10.1 Effect of growth of timothy (*Phleum pratense*) on the ratio of leaf to stem and chemical composition (% of dry matter)

Sampling date	Leaf/stem	Crude protein		Ether extract		Ash		Crude fibre		N-free extractives	
		Leaf	Stem	Leaf	Stem	Leaf	Stem	Leaf	Stem	Leaf	Stem
May 20	2.57	21.7	14.1	3.8	2.9	7.1	9.9	19.1	23.5	48.3	49.6
June 2	1.30	17.2	11.4	4.7	2.5	6.5	7.9	23.8	29.7	47.8	48.5
June 16	0.39	18.5	7.6	4.1	2.6	8.0	6.6	26.1	32.6	43.3	50.6
June 30	0.35	12.3	4.4	3.3	1.7	8.8	5.0	26.9	31.7	48.7	57.2
July 14	0.20	11.1	3.4	3.2	1.3	9.0	5.0	30.6	32.4	46.1	57.9

(From Waite & Sastry 1949)

Table 10.2 Distribution of some mineral elements at early maturity (% of dry matter) and the comparative composition of the aerial parts of herbage plants

		Ca	P	K	Mg	N
Perennial ryegrass	Head	0.23	0.42	1.7	0.13	2.2
	Leaf	0.87	0.32	2.3	0.17	2.1
	Stem	0.30	0.27	1.7	0.09	0.8
Red clover	Head	1.1	0.41	2.1	0.28	3.7
	Leaf + petioles	2.1	0.29	1.7	0.34	4.5
	Stem	1.1	0.15	1.7	0.24	1.6
Yarrow, burnet, plantain and chicory (combined average)		1.4	0.36	3.2	0.84	3.2

(After Thomas, Thompson et al. 1952 and Fleming 1963, in Worden, Sellers & Tribe 1963)

Table 10.3 Mineral contents of grazed sward in stud paddocks at Newmarket, Suffolk (composition of dry matter)

		Found	Normal range in UK
Calcium	(%)	0.34–1.6	0.30–1.0
Phosphorus	(%)	0.20–0.54	0.15–0.45
Potassium	(%)	1.50–2.50	1.6–2.6
Magnesium	(%)	0.12–0.20	0.11–0.27
Sodium	(%)	0.03–0.34	0.1–0.6
Sulphur	(%)	0.22–0.43	0.15–0.45
Molybdenum	(mg/kg)	0.9–2.8	0.1–5.0
Copper	(mg/kg)	4.5–12.3	2–15
Selenium	(mg/kg)	0.025–0.049	0.02–0.15
Zinc	(mg/kg)	21–34	12–40
Manganese	(mg/kg)	44–220	30–115
Cobalt	(mg/kg)	2.9–7.6	0.03–2.0

(Data collected by author)

is usually not sought after by horses and ponies. A few relevant chemical values of herbs are given in Table 10.2. Herbs may be especially useful on marginal land, in which the upper layers are frequently leached of nutrients by excessive rainfall, and many tend to stay green in winter, so they provide a succulent winter bite although their regrowth is protracted. When herbs are present in abundance they depress total yield per ha of major nutrients, but this is less likely to be a critical issue in horse paddocks. In any case, their establishment in pasture as part of a normal grass and clover seed mixture is uncertain. Herb seeds are rather expensive but many of them inevitably become established in permanent pastures through natural agencies. In 1952, William Davies proposed that graziers determined to invoke the nutritional properties of herbs should sow the seeds in strips free from other very competitive plants in an otherwise normal grass meadow. Frequently the headland is the most suitable place. Table 10.4 gives a suggested seed mixture, which includes some relatively non-competitive grass species although these are not essential. Nevertheless, the economic worth of herb mixtures is unproven for horses in any general way.

Table 10.4   Suggested herb mixtures for sowing as a strip 8–10 m wide in horse paddocks

	Based on	
	Davies 1952	Archer 1978a*
	(kg/ha)	
Chicory (*Cichorium intybus*)	2	2.2
Ribwort plantain (*Plantago lanceolata*)	2	1.1
Burnet (*Sangiusorba minor*)	3	2.2
Yarrow (*Achillea millefolium*)	1	0.6
Cat's-ear (*Hypochoeris radicata*)	1	—
Dandelion (*Taraxacum officinale*)	—	0.3
Sheep's parsley, wild parsley (*Petroselinum crispum*)	1	0.6
Meadow fescue (*Festuca elatior*) or creeping red fescue (*F. rubra*)	10	(13)
Timothy (*Phleum pratense*) or smooth meadow grass (*Poa pratensis*)	5	—
	4	(7)
Crested dog's-tail (*Cynosurus cristatus*)	—	(7)
White clover (S100) (*Trifolium repens*)	2	—
Total	26 or 27	7.0 or 34.0

*   Without the inclusion of grass seed the mixture should be introduced into an existing paddock by direct seeding if ground is well harrowed and the sward cut short.

The ingestion of soil by animals while grazing can occur in significant amounts; it depends on the height of the herbage, the openness of the sward, the contamination of leaf by earth and on the species of animal. On rough terrain the soil intake by sheep is said to approach 20 per cent of their daily intake of dry matter and measurements have shown that a 500-

kg horse may ingest as much as 1–2 kg of soil daily while grazing. Apart from potassium (K), the mineral and trace-element contents of soil are generally higher than they are in herbage dry matter, although their availability, or digestibility, varies according to the element and soil type. Most common mineral elements (possibly apart from iodine and cobalt) are required by plants in their growth. However, the proportions taken up by roots differ considerably from one element to another. Plants do not absorb in significant quantities heavy metal poisons such as lead (cadmium may be absorbed in greater quantities), although the leaves can become contaminated by industrial fallout and the soil and subsoil can be polluted by industrial seepage. Soil consumption may then be a cause of, for example, lead or fluorine toxicity. High concentrations of toxic minerals in the soil may solely depend on their geological origin. The significance of seleniferous soils will be discussed later.

Productivity of pasture

The term pasture covers a heterogeneous bag of potential feed sources for horses, for which generalizations are bound to be incorrect in specific instances. Within any pasture the nutritional quality varies from area to area. Therefore the feeding value of the whole pasture will depend on the stocking density and the amount of the most attractive herbage at any one time. When lush herbage is available for several weeks, healthy, resting adult horses will become obese; if the energy digestibility of the edible dry matter is satisfactory, the protein requirements of adult non-lactating horses will normally be met. In temperate grassland areas – excluding acid areas with very high rainfall – the protein content of pasture tends to be directly correlated with rainfall and inversely with soil temperature during the growth period.

Horses restricted entirely to these pastures can experience periods of inadequate protein and energy intake (as indicated by blood measurements) (Owen, McCullagh et al. 1978) during warm dry weather. Pastures grazed by horses tend to produce the greatest yield of digestible energy and protein during May and June, after which there is a precipitous decline in productivity from July to August when grasses flower. Clovers and other legumes, when encouraged, prolong pasture growth and extend the summer grazing season. Where persistent leafy strains of grasses have been established on fertile soils, this mid-summer fall in productivity is much less noticeable. More fertile deep soils are less inclined to dry out and the leafy strains of grasses continue vegetative growth much later into the summer. By grazing these pastures, the formation of seed heads is delayed, or avoided, and tillering is encouraged so that their productivity is further enhanced. Regrowth of succulent leafy material occurs in early autumn, but work with sheep (Ribeiro, MacRae & Webster 1981) indicates that the metabolizable energy of autumn grass is utilized 40 per cent less efficiently than that of spring grass of the same crude chemical composition. This poorer value should be recognized when foals are weaned in the late summer on pasture without supplementary feeding (see Ch. 7).

In parched landscapes horses and ponies are deprived, first, of water (Table 7.3), energy and protein and, second, of phosphorus. There are normally large stores of vitamin A in the liver but after an extended drought there is a clinical deficiency as a result of protein and zinc deprivation coupled with a scarcity of green herbage. It is unlikely that a deficiency of any of the other fat-soluble vitamins D, E and K would occur among horses confined entirely to pasture. However, a few isolated pasture species not found in the British Isles (see vitamin D, Ch. 4) can cause vitamin D toxicity, bone demineralization and soft tissue calcification. A calcium deficiency is unlikely among grazing horses, even when the graze and browse are desiccated and parched. But stock can become deficient in calcium, phosphorus and magnesium if they are confined to wet acid soils covered by poor-quality, fine-leaved grasses. Many ponies coming off such hill land show signs of bone demineralization, such as big-head.

Horses seem to be less prone to grass tetany caused by magnesium (Mg) deficiency than are cattle, but a fall in serum Mg is possible when lactating mares are grazing on low-Mg soils and it has been suggested that part of the effect is through excessive amounts of K in lush herbage. Leafy material contains far more K than the horse requires in normal circumstances, and the needs for sodium and chloride are likely to be met in horses confined to pasture in temperate latitudes.

The trace-element status of grassland depends largely on the available supply in the underlying soil. For example, shale, mudstone and clay soils contain higher concentrations of selenium than chalk, limestone and sandstone soils (Thornton 1983) and many mountainous areas. Soils of low available selenium content are associated with low blood concentrations of selenium in horses. Although Brady, Ku & Ullrey (1978) were unable to induce a response in blood selenium concentration by supplementation of the diet over a 4-week period, the author has found that such supplementation should be continued for at least 2 months before measurable increases occur. Furthermore, work in three continents (Bergsten, Holmbäck & Lindberg 1970; Caple, Edwards et al. 1978; Basler & Holtan 1981) has demonstrated blood responses to supplementation of the diet.

Hypocupraemia in horses occurs less widely but it exists in several parts of the British Isles, particularly in southern Ireland. The effect may be seasonal and in the ruminant (Suttle 1983) the most frequent cause is an excessive uptake by plants of molybdenum and sulphate in the absence of generous amounts of available copper. A similar predisposing cause would seem to operate in the horse; although recent evidence suggests that the horse is much less susceptible to the effects of molybdenum and sulphate. Nevertheless depressed blood concentrations of both Cu and Se in grazing horses have been observed by the author, even in parts of southern England.

Seleniferous soils are notorious in various regions of the world where accumulator plants store toxic amounts of soil selenium. These accumulators leave residues that are apparently absorbed subsequently by other plants, causing alkali disease in grazing stock. Some inland continental areas, and even alkaline soils in central England, can induce symptoms of iodine deficiency in the young stock of grazing mares. When seaweed is used in excessive quantities as a source, signs of iodine toxicity similar to those of deficiency, have been observed. Deficiencies of iron, manganese, cobalt and some other more exotic trace elements, have not been recorded and are unlikely among grazing horses and ponies. Green leafy material is a rich source of folic acid, and comparisons made by the author between horses in training for flat racing, given a cereal-based diet supplemented with folic acid and vitmin B_{12}, and grazing in-foal and barren mares, foals and yearlings has provided evidence of a 23 per cent lower concentration of serum folate and a 33 per cent lower concentration of serum vitamin B_{12} in the horses in training. The nutritional significance of this has yet to be established.

Horses should be removed from a pasture as soon as they have eaten the available herbage if alternative land exists. They are more active than ruminants, and they can damage the soil's structure in wet weather and growing plants through prolonged trampling. In contrast also to cattle and sheep, which spend periods ruminating, horses may spend up to 60–70 per cent of the 24 hours searching for the most delectable foliage. Overgrazing, poaching of the soil, or stocking grassland during a heavy frost, damages plants and depresses the rate of regrowth, encouraging the spread of prostrate and opportunist annual weeds seen particularly in an arc around gateways, drinkers and feed troughs. One of the few advantages of old, matted, sod-bound pastures is that they may be less prone to damage in this way. However, the undergrazing of pastures tends to promote the nutritionally poorer grass species, partly through seeding, and parasite survival, as distinct from transmission, is extended.

The ideal stocking rate is no greater than that which will feed the horses in the growing season. It is preferable to stock few horses with the balance made up of cattle to clear the excess growth at the season's height. Mixed stocking, either on a rotational basis or together with horses, initially decreases the number of horses that can be maintained, but these few will receive a better diet and will imbibe fewer intestinal parasitic worm larvae. Moreover, the quality of the pasture can be maintained at a high level for many more years. By breaking up an area of land into paddocks, rotational grazing is facilitated and better parasite control is achieved, particularly where the grazing species are also rotated in each paddock. Occasionally lactating mares, or young stallions, will bully cattle, so careful judgement should be exercised and rotation rather than mixing of species thus carries certain advantages.

Wooden fencing is readily protected from chewing by running electric fencing wire along the top (Pl. 10.1). Furthermore, the paddock principle

Plate 10.1 The effective use of electric fencing installed on post and rail fencing to discourage wood chewing by horses.

of rotation allows a more complete recovery of the grass and prevents the excessive spread of flat weeds, such as plantain, daisy, buttercup and dandelion (*Taraxacum officinale*).

The theoretical carrying capacity of pasture land was earlier estimated by assuming no wastage, but for practical purposes much of the foliage is lost as a source of horse feed. Therefore, 1 ha of high-quality grassland should provide pasture and hay for three to four light horses of about 400 kg or four to six smaller ponies. Low-quality permanent pasture, however, may support only one horse per ha and, in the extreme, 25 ha of dry range may be required to supply the needs of a single horse throughout the year. Average-quality grassland can produce sufficient growth for two horses and when fertilized, for three horses per ha or, as summer pasture only, for double the number. Many Thoroughbred studs produce barely enough grass for one mare plus followers up to yearling sales per 1–1.5 ha and then provide none of their own hay. The latter should not in any event be produced from paddocks that have been grazed by horses during the previous year at the least, if worm control is to be practised assiduously.

Archer (1978a) found that only 10 per cent of the area of long-established horse pastures was grazed. After ploughing and reseeding following an arable rotation, the grazing area was extended to 20–30 per cent, most of which included the previously grazed areas. Horses will not graze near horse droppings and these areas are rejected if removal of the droppings is delayed for more than 24 hours after they have been voided. Horse urine does not engender a similar instinctive reaction. The inborn habit reduces the transmission of parasitic worm larvae and leads to the establishment of both grazing and camping areas with a consequent effect on productivity. Horses will, however, graze right up to cattle dung pats and graze evenly over areas after well-rotted cattle manure has been spread. Similarly, cattle will graze the longer grass around the horse dung pats. Ideally, therefore, horse droppings should be removed from pastures on a daily basis and not spread by harrowing. Although the latter practice will destroy more parasitic worms, it will enlarge the rejected area. The advantages of integrating ruminants with horse pasture management is obvious. Steers are better than milking cows as they remove less Ca, P and N from the soil and both are probably better than sheep for which, nevertheless, there are some staunch advocates. Pastures heavily fertilized for dairy cows are generally unsuitable for horses.

Even with the best will and resolute adoption of mixed grazing, some areas of rank growth of low feeding value will remain, swamping out any young basal growth. Such areas should be topped at least six times per season, with the toppings removed to avoid mounds of mouldy grass suppressing the underlying grass. The pursuance of this practice may prevent weeds from seeding, it will promote tillering and regrowth of young grass, and will destroy some of the infective larvae.

All grazing animals thrive better if they have the companionship of other stock and this may be particularly true with highly strung hotblooded horses. Even goats, sheep, chickens, ducks or ponies can fulfil a useful role. The old adage 'to get his goat' implied that a favourite racehorse could be nobbled by stealing his mascot before a race. Gilbert White in his *Natural History and Antiquities of Selborne* (1789) relates in a letter dated

15 August 1775 to Daines Barrington how a lone horse made an abiding companion of a solitary domestic hen, consoling and protecting it, in so far as it was possible, from the trials of avian life. In addition to companionship, all horses should have access to shelter from the sun at high noon and from cold, windy, wet weather. This shelter may be naturally formed by trees or be a simple three-sided covered structure. It should also go without saying that all horse paddocks, structures and fences must be free from protruding nails, barbed wire, loose wire, tins, broken bottles, large stones and anything that can ensnare horses, particularly where young stock are reared as they tend to be more curious and flighty. The tops of fence posts should be flat with no sharp edges and level with the horizontal poles nailed to their inner sides. A piped water supply should be clean and adequate (see Table 7.3) and where natural water courses are used, freedom from contamination for at least a few miles upstream should be confirmed. The opinion is held that some environmental factors contribute to bad behaviour, even cribbing, among horses at pasture. What can be considered a poor-quality environment as seen through horse's eyes is learnt by experience.

Grassland improvement

Permanent pasture should be well drained and where it is low lying, or the subsoil is heavy, consideration should be given to piped drainage and, in the appropriate areas, pumped low-level ditches. Improvement in drainage reduces poaching and allows the stocking density to be increased. By lowering the water table, plant roots are encouraged to permeate lower horizons of the soil so that they gain access to a greater variety of minerals and to moisture for longer rather than shorter periods during a drought. This extends the grazing season and encourages ryegrass, cocksfoot, timothy and clover, discouraging rushes, tufted hair-grass or tussock grass (*Deschampsia caespitosa*), *Agrostis*, couch, Yorkshire fog and water-tolerant meadow grasses. Fertilizers, therefore, can be more effectively used, and the increased potential growth of the resulting sward decreases the concentration of parasitic worms. It also decreases the risk of liver fluke infection. Although horses have a pronounced resistance to liver fluke, they can become infested and their economic worth is thus prejudiced. Where mixed grazing is practised, the sheep and cattle can, of course, be treated before they are introduced. The snail burden of the land may also be reduced by copper sulphate treatment or, for example, by grazing with ducks.

If piped drainage is impractical, the subsoiling of impervious heavy land will improve its structure and for a shorter period bring about the advantages of more permanent drainage. Sometimes such marginal land is infested with quite inedible species, including cottongrass (*Eriophorum* spp.), rushes and, in better drained acid areas, by thick mats of *Nardus* grass. Here, initial improvement can frequently be effected through burning and hard grazing with steers. Most poor-quality permanent pastures can be improved by occasionally discing or heavy harrowing to break up a surface mat. This allows aeration, encouragement of better grasses, and some control of moss (*Lycopodium* spp.), mouse-ear chickweed and buttercup. In dry areas, however, little penetration will occur unless there has been recent rain.

Although the cutting of pernicious weeds such as docks (*Rumex* spp.) and thistles before they seed is helpful, it must be done on a regular basis five or six times a season, and even then control of creeping thistle is doubtful. Thus, the use of selective weedkillers based upon methozone (MCPA) and/or 2,4-D to control broad-leaved weeds without damage to mature grass is recommended. Where small prostrate weeds, such as members of the buttercup family (Ranunculaceae), are widespread, spraying by tractor is most convenient; but where there are patches of creeping this-tles, docks and nettles, a knapsack sprayer can be used to advantage; this causes less damage to clovers and other valuable broad-leaved herbs. Spraying should always be carried out according to the manufacturer's directions: (1) when there is little wind; (2) at a time of rapid weed growth, but before flowering; (3) refraining in periods of drought but avoiding rain for a few hours after spraying; (4) not before newly sown pasture grasses have at least three leaves.

MCPA is effective against common nettle, daisy, dandelion, spear thistle (*Cirsium vulgare*), annual scentless mayweed (*Tripleurospermum inodorum*), black bindweed (*Polygonum convolvulus*), redshank (*P. persicaria*), cleavers (*Galium aparine*), common chickweed (*Stellaria media*), charlock (*Sinapis arvensis*), fumitory (*Fumaria officinalis*), poppy (*Papaver rhoeas*), fathen (*Chenopodium album*) and small nettle (*Urtica urens*). It will check broadleaved dock (*Rumex obtusifolius*), bulbous buttercup (*Ranunculus bulbosus*), creeping buttercup (*R. repens*), colt's-foot (*Tussilago farfara*), creeping thistle, curled dock (*Rumex crispus*), field horsetail (*Equisetum arvense*), perennial sowthistle (*Sonchus arvensis*) and soft rush (*Juncus effusus*). The most satisfactory control is achieved by spraying bulbous buttercup and chickweed in autumn and ragwort in early May. Bracken fern is very resistant to spraying, but the most effective time is August. Pastures should be left for 15 days before they are grazed in order to allow adequate time for the penetration of the herbicide and the dead residues of poisonous weeds should be removed from the pasture. MCPA and 2,4-D are of low toxicity to animals, and pastures properly treated with them have not caused fatalities regardless of how soon they have been grazed.

Horses and ponies do not normally graze growing poisonous weeds, but young stock or other stock freshly introduced to a pasture, following a period in stable, may eat avidly a variety of such plants. Furthermore, during times of drought some that are deep-rooted or are near watercourses and remain green entice the unwary horse. Weeds found in the British Isles that are poisonous to horses in both the growing and dried states, are listed in Table 10.5 and those reported worldwide to be toxic to horses are listed in Table 10.6. After cutting and drying or after being destroyed by sprays, weeds that retain their toxicity are more palatable than before and so should be removed and burnt.

Toxic weeds

In the plant world there is a vast array of substances known as alkaloids, which are organic basic compounds with a wide range of toxicity for the horse. They are rare in grasses, although poor growth of horses in the United States (Jordan & Marten 1975) grazed on reed canary grass (*Phalaris arundinacea*) during July and August has been attributed to a seasonal increase in the alkaloid content of the grass. Probably the commonest cause of poisoning among all grazing animals in the British Isles results from the consumption of common ragwort (*Senecio jacobaea*), which contains an alkaloid that causes permanent liver damage, termed seneciosis. The different resistances of grazing animals to this toxin reflect variation in the efficiency of urinary elimination of it (Holton, Garrett & Cheeke 1983). The horse is prone to intoxication; five out of twenty horses died in one outbreak (Giles 1983), apparently from the consumption of growing ragwort. Horses with ragwort-damaged livers should be given a well-balanced diet containing good-quality protein supplemented with B vitamins and trace elements, and should not be given access to groundsel (*Senecio vulgaris*), which contains the same toxin as ragwort but at lower concentrations. Other plants causing a similar liver condition are *Crotalaria* and heliotrope (*Heliotropium*).

The yew (*Taxus baccata*) is the most toxic plant in the British Isles, and little more than 100 g of it will kill a horse by cardiac arrest. The second most poisonous is laburnum (*Laburnum*), a member of the legume family that contains many plants known to cause damage to horses, among them broom (*Cytisus scoparius*) and lupins (*Lupinus*). The toxicity of the latter species is principally confined to the seeds, and the different strains differ in their potency. Sweet lupins (*L. luteus*) have a low alkaloid content and are grown on poor land as a source of fodder. If horses eat them death is rare and is caused by respiratory paralysis, not by liver damage. An accumulative poisoning associated with progressive liver damages (chronic lupinosis), occurs in sheep and horses in Australia. Here the causative agent is a fungus growing on the lupins that are eaten. Sweet clover or melilot (*Melilotus*) contains coumarin that is broken down to dicoumarol in hay made under bad harvesting conditions, or during moulding. Dicoumarol prolongs blood clotting time. Both white and red clovers may contain toxic factors. Both species contain appreciable amounts of oestrogens which are also found (in much higher concentrations) in subterranean clover (*Trifolium*

Table 10.5   Plants found in the British Isles poisonous to horses in fresh and dried states

Equisetum spp.	Horsetails
Pteridium aquilinum	Bracken
Taxus baccata	Yew
Helleborus spp.	Hellebores
Aquilegia spp.	Columbines
Aconitum napellus	Monkshood
Delphinium spp.	Larkspurs
Papaver spp.	Poppies
Chelidonium majus	Greater celandine
Hypericum spp.	St John's worts
Agrostemma githago	Corncockle
Saponaria officinalis	Soapwort
Arenaria spp.	Sandworts
Stellaria spp.	Chickweeds
Linum usitatissimum	Flax
Rhamnus catharticus	Buckthorn
Frangula alnus	Alder buckthorn
Lupinus spp.	Lupins
Laburnum anagyroides	Laburnum
Cicuta virosa	Cowbane
Oenanthe crocata	Hemlock water dropwort
Bryonia dioica	White bryony
Cannabis sativa	Hemp
Rhododendron spp.	Rhododendrons
Cyclamen hederifolium	Sowbread
Anagallis spp.	Pimpernels
Datura stramonium	Thornapple
Hyoscyamus niger	Henbane
Atropa bella-donna	Deadly nightshade
Solanum nigrum	Black nightshade
Solanum tuberosum	Potato
Digitalis purpurea	Foxglove
Senecio spp.	Ragworts
Convallaria majalis	Lily of the valley
Fritillaria meleagris	Fritillary
Colchicum autumnale	Meadow saffron
Paris quadrifolia	Herb paris
Iris spp.	Irises
Tamus communis	Black bryony
Lolium temulentum	Darnel

Table 10.6 Plants reported in the scientific literature to have caused poisoning in horses and ponies (for references see Hails & Crane 1982)

Species	Common name	Effects reported
Agrostemma githago	Corncockle	Hypersalivation, abnormal thirst, accelerated respiration
Allium validum	Wild onion	Haemolytic anaemia
Amsinckia intermedia		Walking disease, Haemolytic anaemia, liver necrosis, possibly nitrate toxicity
Aristolochia clematitis	Birthwort	Tachycardia, weak pulse, decreased appetite, constipation, pulyuria
Arum maculatum	Lords-and-Ladies, Cuckoopint	Purgative, irritation of gastrointestinal mucosa and kidneys, and fatty liver
Astragalus sp.	Locoweed	Eye lesions, locoism, and selenium accumulation in several species
Astragalus lentiginosus and A. mollissimus	Locoweed	Depression, anorexia, ataxia, hyperexcitability and especially violent responses to stimuli. Effects irreversible
Atalaya hemiglauca	Whitewood	Anorexia, dullness, irritability
Centaurea repens	Russian knapweed	Nigropallidal encephalomalacia; severe damage to nerve cells, muscular hypertonia
Centaurea solstitialis	Yellow star thistle	Nigropallidal encephalomalacia (chewing disease); hypertonia of muscles of muzzle
Cestrum diurnum		Calcinosis-hypercalcaemia. (Other *Cestrum* spp. cause severe gastritis and liver degeneration but not reported in horse)
Coriandrum sativum	Coriander	(No signs given in Hails & Crane 1982)
Crotalaria sp.		Hepatotoxicity, jaundice, dyspnoea, weak pulse and collapse
Crotalaria crispata		Kimberley horse disease, anorexia, dullness, staggering gait
Crotalaria dissitiflora var. *rugosa*		(No signs given in Hails & Crane 1982)

Crotalaria retusa		Kimberley horse disease; liver and central nervous system lesions, anorexia, dullness, staggering gait
Cuscuta spp.	Dodder	Enteritis, anorexia, nervous symptoms
Cuscuta breviflora	Peppery cuscus, dodder	Not described, but staggering, salivation, increased pulse and respiratory rates reported in cattle
Cuscuta campestris	Dodder	
Echium lycopsis (or *E. plantagineum*)	Purple viper's bugloss	Hepatotoxicity, blindness
Equisetum spp.	Horsetails, marestails	Thiamin deficiency – the horse very susceptible
Equisetum fluviatile	Water horsetail	Akin to bracken poisoning owing to presence of thiaminase
Equisetum hyemale	Rough horsetail or Dutch rush	Loss of condition, some diarrhoea, usually caused by eating hay containing horsetails – administer thiamin or dried yeast
Equisetum palustre	Marsh horsetail	
Equisetum variegatum	Variegated horsetail	
Eupatorium spp. (incl. *E. rugosum*)	Incl. white snakeroot and hemp agrimony	Trembles and lethargy
Eupatorium adenophorum	Croftonweed	Pulmonary oedema and possible nitrate toxicity
Galeopsis spp.	Hempnettles	Pulmonary oedema, enteritis, anorexia
Glyceria maxima	Reed sweet-grass	Cyanide poisoning
Helichrysum cylindricum	Everlasting flower	Blindness and encephalopathy
Heliotropium europaeum	European heliotrope	Hepatotoxicity
Hypericum crispum		Photosensitivity
Indigofera dominii and *I. enneaphylla*	Birdsville indigo	Birdsville disease, drowsiness, immobility, discharges from eyes and nose, dragging of hind feet, liver lesions
Juglans nigra	Black walnut	Acute laminitis
Jussiea peruviana		
Lathyrus nissolia	Grass vetchling	Incoordination and collapse during exercise
Lupinus spp.	Lupin	(1) Chronic: lupinosis-congenital defects, liver damage (due to mould growing on lupins). (2) Acute: respiratory paralysis due to lupin alkaloids when large amounts of the plant eaten

Contd

Table 10.6 Contd

Species	Common name	Effects reported
Medicago sativa	Lucerne	Photosensitization, possible oestrogen toxicity; certain strains cause anaemia and hepatotoxicity
Morinda reticulata		Selenium toxicity
Nerium oleander	Oleander	Highly toxic – convulsions, diarrhoea, colic, petechial haemorrhages
Oxytropis campestris and *O. sericea*	Locoweed	Nervous signs, anorexia, ataxia – damage permanent
Papaver nudicaule	Iceland poppy	Ataxia, convulsions
Perilla frutescens		Lung disease
Persea americana	Avocado	(No signs given in Hails & Crane 1982)
Phaseolus vulgaris	Kidney, haricot or navy bean	Central nervous system lesions, gastrointestinal disorder
Pimelea decora	Rice flower	Colic, diarrhoea, collapse, ulceration of mouth, tongue, oesophagus, gastritis (also St George disease in cattle)
Polygonum aviculare	Knotgrass, wireweed	Nitrite poisoning, gastrointestinal irritation
Prunus laurocerasus	Cherry laurel	Cyanide poisoning (also caused by other *Prunus* species, e.g. *P. serotina* – wild black cherry)
Pteridium aquilinum	Bracken	Thiamin deficiency
Quercus spp. (incl. *Q. rubra* var. *borealis*)	Oak	Acorn poisoning, hepatotoxicity, oak-leaf poisoning, dullness, diarrhoea, constipation, dark urine
Ricinus communis	Castor oil plant	Castorbean poisoning, dullness, incoordination, sweating, tetanic spasms, watery diarrhoea
Robinia pseudo-acacia	False acacia	Acacia bark poisoning, colic, diarrhoea, irregular pulse, hyperexcitability, paralysis
Senecio spp.		Hepatotoxicity

Senecio jacobaea	Common ragwort	Hepatotoxicity
Setaria sphacelata	Bristle grass	Oxalate poisoning, osteodystrophia fibrosa
Sinapis arvensis	Charlock	Cyanide poisoning, colic, gastroenteritis, diarrhoea
Solanum malacoxylon		Calcinosis – wasting, stiffness, hypercalcaemia, calcification of arteries
Sorghum almum	Sorghum	Cystitis, ataxia (growing grass)
Sorghum bicolor	Sudan grass	Cystitis, ataxia (growing grass)
Sorghum halepense	Johnson grass	Cyanide poisoning
Sorghum sudanense	Sudan grass	Possible lathyrism, equine cystitis and ataxia, ankylosis (growing grass)
Sphenosciadum capitellatum	Whiteheads	(No signs given in Hails & Crane 1982)
Stipa viridula	Sleepy grass	Hypnosis
Swainsona spp.	Darling pea	Depression, emaciation, ataxia, blindness
Tanaecium exitiosium		Weakness, staggering, collapse, frequent micturation, inflammation of stomach and heart
Taxus baccata	Yew	Heart failure – trembling, dyspnoea, collapse
Taxus cuspidata	Japanese yew	Heart failure – trembling, dyspnoea, collapse
Trichodesma incanum		Severe liver damage in horses and other farm stock
Viscum album	Mistletoe	Incoordination, dilated pupils, salivation, hypersensitivity

subterraneum) grown on light soils. Some reports indicate the presence of oestrogenic activity in moulds infecting the clover leaves. These hormone-like substances have been associated with infertility and increases in the teat length of sheep (references given in Saba, Symons & Drane 1974), and it is an open question as to whether any comparable problem occurs in grazing mares. Two genera of vetches in North America, milk vetches or locoweeds (*Astragalus*) and the related *Oxytropis* (see Table 10.6) are implicated in several abnormal conditions of horses. One caused by locoweeds is recognized as eliciting irreversible nervous signs. Although none of the poisonous members of these genera are found in the British Isles, certain pasture species of *Vicia* and *Lathyrus* found there are mildly toxic. In parts of the United States locoweeds begin their growth in late summer and remain green through the winter. They must be grazed for a period before poisoning is obvious. Some species are selenium accumulators and may contain up to 300 mg/kg; they are poisonous for this reason.

Several other widely recognized grazing ailments among horses have attracted attention. Horses grazing pastures that contain a high proportion of tall fescue (*Festuca arundinacea*) during hot summers in the United States may experience a rise in rectal temperature and respiratory rate and they lose their appetites. Lameness, diarrhoea, a rough hair coat and a disinclination to stand in the sun are typical signs of this so-called fescue toxicity. A water-soluble substance, associated with an endophytic fungus (Schmidt, Hoveland et al. 1983; Hemken, Jackson & Boling 1984) that infects the leaves and seeds of fescues, smut grass, broom sedge and *Panicum* species (unpublished observations of J. D. Robbins, C. W. Bacon, J. K. Porter and D. M. Bedell, Field Crops Laboratory, Agricultural Research Services, USDA, Athens, Georgia, USA), has been implicated and the only reasonable advice is to remove susceptible stock from the pastures in very hot weather.

Broodmares grazing pure stands of fescue in the mid-west of the United States may suffer agalactia, abortion, abnormally thick, tough placentas and prolonged gestations. Foaling difficulties have also been encountered. For this reason it is recommended in the USA that mares do not graze stands of pure fescue during the last 3 months of gestation. Fescue grown in parts of the mid-west contains low amounts of selenium, especially where little nitrogenous fertilizer is used, and this also can cause reduced fertility. If it is not possible to remove broodmares, then a good-quality legume hay should be introduced; this reduces the incidence of foaling difficulties, whereas grain does not. However, if the hay is not of the best, it will be rejected in favour of the grass and so will afford no benefit.

A non-contagious malady known as grass sickness (Gilmour & Jolly 1974) affects horses throughout Europe, including the British Isles. The principal lesion seems to be nerve damage and the onset may be sudden. It seems to occur mainly in horses between 2 and 8 years old that are being kept entirely out of doors with no supplementary feeding. It is also recognized among non-domestic equids and is associated with particular premises. The illness is fatal and refractory to treatment. The clinical symptoms are muscular tremors, patchy sweating, difficulty in swallowing, salivation, dilation of the bowels and stomach and chronic loss of weight. It is more common in horses that have been on the premises for less than 2 months. The incidence may be reduced by stabling the horses for part of the time and giving them hay and supplementary feed.

In bright sunny weather horses with unpigmented skin, and white or grey coats in particular, may be subject to photosensitization. This is a form of dermatitis that arises prominently in horses with damaged livers. Unwanted toxic compounds absorbed from the intestine are normally degraded by the liver and excreted. This process is obviously less efficient where there is a measure of liver dysfunction. Toxins contained in St John's worts (*Hypericum* spp.), buckwheat (*Fagopyrum esculentum*), bog asphodel (*Narthecium ossifragum*), species of *Vicia* and sometimes in lucerne (alfalfa) (*Medicago sativa*), alsike (*Trifolium hybridum*) and red clovers have been imputed as a cause. On other occasions therapeutic drugs may be incriminated.

In tropical and subtropical countries enzootic equine cystitis and ataxia occur in horses grazing fresh summer annual forages of the genus *Sorghum* (Johnson grass, Sudan grass and common sorghum). Arrow grass (*Triglochin*), wild black cherry (*Prunus serotina*), choke cherry (*P. virginiana*), pincherry (*P. pennsylvanica*), flax (*Linum*) and the sorghum grasses all contain cyanogenetic glycosides that are readily hydrolysed to hydrogen cyanide. They are most toxic during rapid growth immediately after a freeze. Heavy nitrogen fertilization, wilting, trampling and plant diseases may increase the hazard. Silage or haylage produced from the grasses is risky but hay properly cured and stored is safe. Cystitis, or inflammation of the urinary tract, and incontinence are more common in mares than in stallions or geldings, but posterior ataxia is manifested in all horses from which they seldom recover.

Oxalate toxicity will be briefly discussed under tropical grasslands (p. 236).

Fertilizers

A continuous supply of inorganic nutrients, including water, is required for the growth of pasture plants. In addition, roots require oxygen so that soil moisture must be present without waterlogging. For this reason, the structure, type and humus content of soil are critical in ensuring a continuous supply of moisture at all levels through which the roots of pasture plants permeate. The major nutrients for plants are water, nitrogen (N) in a fixed form, phosphorus (P), potassium (K) and calcium (Ca), which also has an important role in regulating soil acidity (pH). Other nutrients required include magnesium (Mg) and a large number of trace elements, prominently iron, copper, boron and manganese. It should be noted that some elements (e.g. boron) are required by plants but not by animals in measureable quantities and others are required by animals but not apparently by plants in detectable amounts (e.g. cobalt).

When first embarking on grassland management for horses and at intervals of, say, every 10 years thereafter, it is desirable to carry out chemical determinations on the soil in order to assess at the very least its pH, P, K and Mg status. Soil sampling must be carried out in a representative and sensible fashion so that distinctions can be drawn between clearly different types of land. Furthermore, the soil profile in old pastures can be such that the status is quite dissimilar in upper and lower layers of soil reached by plant roots. Many soils may show surface deficiencies of available P and Ca, accompanied by a lower pH, whereas lighter soils – especially where hay crops have been taken – are frequently K-deficient in the upper layers. A full response to N fertilizers should not be anticipated if these primary deficiencies have not first been rectified. A soil pH of approximately 6.5–6.8 should be maintained as below this range there will not be enough free Ca and much of the P will be fixed as ferric phosphate or aluminium phosphate. Above the range, several minerals, including iron, will be less available. The amounts of pit chalk or limestone required on acid soils, in order to raise the pH to within the range, will depend on the value determined and the texture of the soil; but amounts between 1.25 and 7.5 t/ha can be used as a top dressing. If the land is also deficient in P, then basic slag provides not only lime but also phosphates and it may be the most important fertilizer for horse pastures. The P is released gradually over a period of 3–5 years so that applications at those intervals of 600–650 kg/ha should be sufficient to maintain the P status of the soil at index 1–2 on the soil fertility scale of ADAS (the Agricultural Development and Advisory Service of the Ministry of Agriculture, Fisheries and Food of England and Wales). Basic slag will stimulate clover growth and where pastures are rich in legumes then in all probability the Ca and P contents of the soil are satisfactory. A visual survey of the frequency of clovers and scarcity of fine-leaved grasses can be used as a rule of thumb in predicting the well-being of the soil in these respects.

In using limestone, basic slag and, in particular, rock phosphates, it is critical that only finely ground material is purchased as this characteristic will influence the availability of the Ca and P to the roots. Rock phosphates are not generally recommended for studs and are really suitable only on soils of a pH below 5. Limestone contains variable amounts of Mg and manganese that may be useful. Although limestone soils are rarely deficient in Mg, acid soils frequently are and dolomitic limestone would provide a useful source of this element. Super phosphate is composed mainly of Ca $(H_2PO_4)_2 + CaSO_4.2H_2O$ and is a readily available source of P not requiring to be as finely ground as the other less-soluble forms. It therefore has an immediate effect on P-deficient pastures and is valuable for application to seed beds.

Table 10.7 gives the indices used by ADAS to classify grasslands in terms of their major nutrients. Where the value is 2 or over no fertilizer treatment for that particular nutrient is required at the time of measurement. Recommended rates of P and K treatment of grassland for grazing and for hay cropping are given in Table 10.8. Where fields are set aside for hay or silage a generous use of fertilizers is worthwhile.

Table 10.7 ADAS Soil Fertility Scale

Index no.	P	K	Mg
		(mg/litre of soil)	
0	0–9	0–60	0–25
1	10–15	61–120	26–50
2	16–25	121–240	51–100
3	26–45	241–400	101–175
4	46–70	401–600	176–250
5	71–100	601–900	251–350

Table 10.8 Recommended rates of phosphorus (P), potassium (K) and nitrogen (N) treatment of grassland

P or K index	For grazing (kg/ha) per year			For haymaking or silage–nutrients per cut (kg/ha)			
	P_2O_5	K_2O	N*	P_2O_5	K_2O	N High rainfall	Moderate[†] rainfall
0	60	60	20–50	60	80	60	80
1	30	30	20–50	40	60	60	80
2	20	20	20–50	30	40	60	80
>2	0	0	20–50	30	30	60	80

* The darker green the grass the less required. The larger quantities should be divided amongst three applications.
† 45–65 cm (19–25 in) per year.

A K fertilizer may be required only when a hay crop is taken, and it can be in the form of sulphate or 'muriate of potash' (potassium chloride), or as part of a compound fertilizer. The rates of N application for cutting are also given in Table 10.8; as horse pastures are frequently depleted of nitrogen, with yellowing of the grasses, the rates should be fairly high. Laminitis may occur after excessive N has been applied to grazing land. Even so, in order to maintain the quality of the herbage and to get an economic output from the pasture, moderate applications of N fertilizers should always be considered essential unless there is a reasonable proportion of well-distributed clover. A sward content of 25–35 per cent of red or white clover may yield the equivalent of 150–200 kg of slowly released N per ha annually through N fixation in the root nodules. N application rates of 20–25 kg/ha, when required, 3–6 weeks before the pasture is grazed are recommended.

The composition of some straight fertilizers is given in Table 10.9. Compound fertilizers contain two or more nutrients in a reliable form and the weight of a nutrient in a 50-kg bag of any fertilizer is given by dividing the percentage of the nutrient by two. Thus, a 50-kg bag of an N:P:K compound 20:10:10 will contain 10 kg N and 5 kg each of P_2O_5 and K_2O.

Table 10.9 Approximate composition of 'straight fertilizers' and farmyard manure

Straight fertilizers		Per cent
Ammonium nitrate	N	34
Ammonium sulphate	N	20.6
Nitrochalk	N	15.5
Ca cyanamide	N	20.6
Triple superphosphate	P_2O_5	45
Superphosphate	P_2O_5	18
Steamed bone flour	P_2O_5	29
Bone meal	P_2O_5	22
Guano	P_2O_5	13–27
Basic slag	P_2O_5	18–20 (45–50 lime)
'Muriate of potash' (KCl)	K_2O	60
Sulphate of potash with magnesia	K_2O	26 (5–6 Mg)
Kainit (Na, K, Mg)	K_2O	14
Kieserite ($MgSO_4$)	Mg	16
Calcined magnesite	Mg	60
Farmyard manure (kg/10 t)	N	15
	P_2O_5	20
	K_2O	40
	Mg	8

In all situations granulated, rather than powdered, inorganic fertilizers should be used and at least a week should elapse after treatment before horses and ponies are allowed into established pastures. This will give sufficient time for the granules to percolate down to soil level and so it avoids any significant amounts being consumed by grazing animals.

The principle organic fertilizer is farmyard manure, preferably excluding horse manure, and its very approximate composition is given in

Table 10.9. Its advantage over inorganic N sources is the slow release of N, but set against this is the expense of transport and distribution. It is probably more practical to distribute 50 t/ha on a few paddocks than to use half as much on a much larger area and there is some justification, particularly on light soils, for applications every 5–7 years. Fishmeal, containing 10–11 per cent N, is sometimes used in organic-based fertilizers and its advantages in terms of slow release are similar to those of farmyard manure.

Where land has been ploughed up and new permanent pastures, or leys, are being sown, it is essential that readily available nutrients are provided for the seedlings; recommended rates of seedbed application are given in Table 10.10.

Table 10.10 Recommended rates of nutrient application to seedbeds in establishing a pasture (kg/ha)

N, P or K index	Nitrogen						P_2O_5	K_2O
	Grass		Grass/clover		Grass/lucerne	Lucerne	(All mixtures)	
	Spring sown	Autumn sown	(a)	(b)				
0	125	50	40	75	50	25	100	125
1	100	50	0	50	25	0	75	75
2	75	0	0	0	0	0	50	30
>2	0	0	0	0	0	0	25	30

(a) If autumn sown or if it is intended to rely on clover as the main source of N for the sward.
(b) If little reliance is to be placed on clover as a source of N for the sward.

Reseeding paddocks

If there is little pressure on available grazing land and the appropriate equipment is available, it can make economic sense to reseed worn-out pastures after ploughing, as long as the land is potentially fertile. No permanent improvement would, however, result unless drainage, ditching, liming, fencing and hedging are first put in order. If there are large areas of pernicious perennial weeds in the old turf, they should first be destroyed with herbicides and 2–3 weeks should elapse before ploughing begins. For spring sowing, autumn ploughing of the old turf is desirable. Sometimes it may be necessary to use heavy discs or cultivators to break up the turf. Also, there is some justification for killing all the old turf by spraying with glyphosate, delaying ploughing for at least 2 weeks to ensure root kill. On very light soils it may be possible to plough during the late winter and sow in March or April, but it is essential to achieve consolidation of the ground and a fine tilth after ploughing.

A pasture seed mixture should be sown as soon as a good seedbed can be effected, before the surface soil is subjected to periods of desiccation. In Britain, sowing should be accomplished by March or April, otherwise a delay until July or August may be inevitable and undertaken then only if there are prospects of rain. It is also imperative that plant foods be provided; Table 10.10 gives suggested rates of application. In areas of low rainfall seeds should be sown 2.5 cm (1 in) deep; in areas of higher rainfall the sowing depth should be about 1.2 cm ($\frac{1}{2}$ in). Drilling deeper than 2.5 cm, which would be advisable in the height of summer, will result in seedlings of small seeds, such as poas and clovers, not reaching the surface; therefore for reasons of economy these should be omitted from the seed mixture. A fine firm tilth is then effected by harrowing and rolling in order to avoid a puffy, rapidly drying surface soil. Two seed mixtures are suggested in Table 10.11; these are composed of persistent strains that should not be given a cover crop and which suffer more than quicker

Table 10.11 Suggested seed mixtures and minimum seeding rates for permanent pastures when a good seedbed has been established (under adverse conditions the quantities may need to be increased up to double these amounts)

	kg/ha
Seed mixture 1	
Perennial ryegrass (Melle* or Contender)	10.00
Perennial ryegrass (Talbot, Parcour or Morenne)	3.50
Smooth meadow grass[†] (Arina, Dasas)	0.75
Creeping red fescue (Echo)	2.00
Rough meadow grass (VNS)	0.75
White clover (Blancho or NZ Huia)	0.75
Total	17.75
Seed mixture 2	
Perennial ryegrass (Trani or Springfield)	10.00
Perennial ryegrass (Melle)	5.00
Timothy (S48, S352, S4F)	2.00
White clover (Milkanova)	2.00
Total	19.00

* A mixture of strains similar to S23 and S24 could be used to give a range of heading dates.
† Kentucky bluegrass or smooth meadowgrass (*Poa pratensis*).

growing strains from drought in the seedling stage if they have been sown in late spring in poorly formed seedbeds.

In formulating one's own seed mixtures the smaller the seed the lower the weight required per ha. Thus, a lower weight of rough meadow grass, timothy and wild white clover, would be required than of perennial ryegrass or meadow fescue and less is required of persistent leafy tillering strains of grasses for permanent pastures than would be used of aggressive strains in short-term leys. The better late-flowering strains of ryegrass and timothy should not be mixed with aggressive early flowering strains in the establishment of a permanent pasture. On light soils smooth meadow grass or Kentucky bluegrass (*Poa pratensis*) may be a helpful constituent of the mixture by virtue of its underground stems, which knit the surface soil together. On heavier soils this species might be replaced with tall fescue.

For some time a new sward will be more susceptible to destruction by treading than would an established turf and so ideally it should not be grazed in the first year, but should nevertheless be topped to prevent the grasses and any annual weeds from flowering. If the latter are present in excessive quantities, a selective weedkiller could be used, although this would check the establishment of clover and in any event it should not be used before the grasses have developed two or three leaves. Another option might be grazing by sheep late in the first year, but even then not when the surface is wet. A hay crop should not be taken from a newly established permanent pasture during the first 2 years at the very least.

Where ploughing and reseeding are impractical, or it is essential to establish a new pasture quickly, the author has realized considerable success in horse paddocks by spraying the old turf with glyphosate once or twice during a period of rapid growth and leaving it for at least 14 days. Following this, there are two alternative procedures for reseeding: (1) The surface is thoroughly disced (Pl. 10.2), harrowed and rolled to effect a good tilth and seedbed before drilling the seed. This is delayed until the surface trash has decayed and if there are large amounts it can be burnt, or raked off. The essential point is that seedlings should be surrounded by mineral soil in a firm seedbed. Burning will destroy seeds of exotic plants that may have remained dormant in permanent pastures for many years and that can yield a fascinating but mildly bothersome crop in the young sward. Moreover, excessive vegetative matter, which may contain some herbicide residues, can suppress the growth of the seedlings, but the cultivations allow soil microorganisms to contact and rapidly destroy residual herbicide. In areas of low rainfall the cultivations will dry out the surface soil during the summer, which also suppresses seed germination. Therefore (2), after the vegetation has died, surface moisture is retained and time is saved by direct drilling the seed immediately into the soil with a special drill that has heavy disc coulters for cutting through the dead turf.

The same principles of top dressing and rolling at the appropriate times should be applied as for the more traditional husbandry described.

Another alternative is to graze intensively a worn-out paddock with

(a)

(b)

Plate 10.2 Preparation of soil for resowing permanent pasture at Upend Stud, Newmarket, Suffolk. Discing began 8 days after spraying with glyphosate, a satisfactory seedbed was achieved 21 days after spraying and seed was drilled after 24 days, followed by rolling. Parts (a) and (b) show stages in seedbed preparation and (c) shows pasture one year after sowing. Previously this paddock was rife with docks and thistles.

(c)

steers, then disc and harrow heavily at a time when the surface soil is sufficiently moist to allow penetration of the implements. After a reasonable seedbed has been attained, sow, harrow, roll and fertilize as before. Heavy regrowth of the old turf can be controlled by regular topping until the new seedlings are well established. This procedure is the least expensive but unlikely to give complete satisfaction, especially where the previous sward was dense or its regrowth too rapid during the seedling stage.

Trace-element status

Australian evidence (Langlands & Cohen 1978) suggests that general pasture improvement increases the uptake by grazing animals of Cu, Zn, Mn, P, Ca and Mg. Improvement in the drainage of waterlogged soils tends to increase Se and Zn availability but may reduce the availability of Fe, Mn, Co and Mo, and an excessive use of N fertilizers may decrease the concentration of several trace elements in the sward. However, the relationships are complex (Burridge, Reith & Berrow 1983). The effect on Mo availability may be advantageous as peaty, poorly drained soils found in parts of Somerset and southern Ireland precipitate Cu-deficiency problems through low availability of Cu and high availability of Mo in the soil, particularly where the soil pH is also high. These soils (pH in excess of 7.6–7.7) also tend to be deficient in available Mn and Co. Some of the soils contain more than 20 mg of Mo per kg and an increase of Mo by 4 mg/kg depresses Cu availability to grazing ruminants by 50 per cent (for the horse see p. 212). Soils subject to a high rainfall, waterlogging and a low soil pH are prone to Se-deficient herbage, as may occur in hill areas, but deficiency also occurs on sands and gravels, in, for example, Newmarket. By contrast, seleniferous soils containing very high levels of Se associated with toxic signs in grazing animals, exist on glacial lake deposits in southern Ireland.

The correction of trace-element deficiencies by applying minerals to the soil is unsatisfactory for some elements as the uptake is scant and repeated treatment is necessary. Better absorption is generally achieved with foliar sprays, but these are expensive and translocation is slight so that frequent treatment is unavoidable. Injections of Se have proved successful in grazing horses, but these are relatively expensive and repeated treatment at intervals is again necessary. When horses are held for extended periods on grazing lands, supplementary feeding with relatively concentrated sources of trace elements seems at present to be the most practical solution.

Silage and haylage

Horses and ponies will acquire a taste for good-quality silage, which can fulfil a useful role as a substitute for hay, or as a means of overcoming the problem of dust associated with hay. Silage should be made with safe raw materials free from weeds and contaminating pollutants. It should have undergone an adequate acetic acid fermentation and possess a high content of dry matter without any evidence of moulding. This can occur in the space at the tops of silos where the conditions are hot and humid. Good-quality haylage is quite satisfactory and material compressed in plastic sacks with a mild fermentation and with about 50 per cent of dry matter is normally reasonably safe, but is usually on the expensive side. To prepare haylage, stands of pure grass (see p. 64) are cut at the bud stage and wilted for 18–24 hours before baling tightly. Careful harvesting should minimize contamination of the herbage with earth. The baled product has a relatively high pH (about 5.5) and should not be fed to horses for at least a week after baling. Haylage from an unfamiliar source should be given in small quantities initially even where the stock have previously received silage or haylage. Owing to its moisture content its feeding value is about half that of high-quality nuts. However, its high content of soluble dry matter and ease of mastication facilitate a rapid intake of available energy; some horses seem more prone to gas colic if given inordinate amounts.

Some clostridial spores will inevitably be present on the ensiled crop. Current evidence in the United Kingdom (Ricketts, Greet et al. 1984) strongly suggests that horses and ponies, possibly through the absence of a rumen, are much more susceptible than are ruminants to pathogenic clostridial bacteria. The toxins of *Clostridium perfringens* invoke severe enteritis and enterotoxaemia; the neurotoxins of *Cl. botulinum* are particularly dangerous because they cause a descending paralysis commencing with tongue and pharyngeal paresis, with dysphagia, progressive paralysis, recumbency and eventual death. These botulinum bacteria multiply in the silage clamp or bale during abnormal fermentation.

It is particularly important, therefore, that clostridial growth is entirely prevented in ensiled forage intended for horses. The fermentation of sufficient water-soluble carbohydrate in this material brings about a fall in pH, which will inhibit the growth of clostridia if the content of dry matter (DM) is at least 25 per cent. Where the DM content is only 15 per cent, a pH of 4.5 will not inhibit the proliferation of clostridia and secondary fermentation of the silage occurs; lactic acid is degraded to acetic and butyric acids, carbon dioxide is evolved and there is extensive deamination and decarboxylation of amino acids, yielding amines and ammonia. These changes are accompanied by a characteristic smell and loss of the attractive vinegary aroma.

The likelihood of clostridial proliferation is hampered by wilting the fresh material for 36–48 hours before ensiling. In big-bale silage and haylage DM contents of 30–60 per cent are thus achieved. An adequate decline in pH is facilitated in normal silage by harvesting forage containing not less than 2.4–2.5 per cent of water-soluble carbohydrate in the fresh material. To ensure that this level is attained, it may be necessary to add molasses. However, viscid liquids of this type are difficult to apply uniformly so that additives such as formic acid, at a rate of 2.3 litres/t, are often used to ensure that the necessary pH of 4.2–4.6 is attained. Subsequent fermentation will reduce the pH further to a level that makes clostridial fermentation unlikely. If, however, a stable pH is not reached in silage with low DM content, then saccharolytic clostridia, which are present in the original crop as spores, will multiply and will initiate the secondary fermentation referred to above. The resulting rise in pH activates the less acid-tolerant proteolytic clostridia. The ammonia they release brings about a further increase in pH.

Big-bale silage, widely used in farming, and other plastic-wrapped silages have high DM contents, minimal fermentation, no additives and a pH between 5 and 6. They depend entirely on high osmotic pressure (low water activity) to inhibit clostridial proliferation and on the absence of tears in the plastic to prevent mould growth. There have been several deaths in the United Kingdom among horses consuming big-bale silage associated with the presence of clostridial organisms and/or toxins, without any contemporary effects on cattle. Great care should thus be exercised in selecting silage with the appropriate aroma, a high DM content and a pH of 4.0–4.5. These are essential prerequisites of silage or haylage used in horse feeding if health risks are to be minimized.

Hydroponics

The practice of germinating barley seeds in lighted trays under humid conditions produces a high-quality feed when given to horses as rapidly growing young plants. However, the cost per unit of dry matter in particular, owing to the inclusion of a realistic rate of depreciation for capital equipment and the high labour commitment, make the practice difficult to justify on economic grounds. The barley grains should not have been through a grain drier, which would severely damage their capacity for germination, nor should they have been treated with mercurial seed dressings. They should be bright and free from broken grains. The time interval from germination to consumption should be minimised by establishing optimum conditions of 20 h light per 24 h and a temperature of 19–20°C. Slow growth increases the likelihood of moulding and its attendant risks. A build up of mould spores in the room must be avoided by routine hygiene. About three-quarters of the product is moisture and therefore the DE content is only 2.5 MJ/kg despite its high digestibility.

Tropical grassland and forages

A characteristic of most tropical forages is a high yield of dry matter, which is relatively rich in fibre but impoverished of N and P and from which the rewards of animal production are relatively meagre. There are, nevertheless, large differences between the products of wet and dry seasons. In one study (Kozak & Bickel 1981) the crude protein content of grasses decreased from a range of 6–10 per cent in the rainy season to 4–5 per cent in the dry season in Tanzania without any change in the crude fibre content, yet digestibilities of the crude protein were 34–58 per cent and 16–25 per cent respectively. Kozak & Bickel also noted that the digestibility of the dry matter of pasture forage decreased from 47–63 per cent at heading to 30–53 per cent at flowering. It has been asserted that the yield per ha of digestible nutrients is at best only half what can be achieved from temperate grassland. The low digestibility of tropical grass is apparently an effect of its higher lignin content than that of temperate grass and analyses reveal that as the environmental temperature rises there is a fall in the cellulose content of tropical grass but a proportionate rise in the hemicellulose and lignin contents. Malaysian experience has shown that horses fed as much cut tropical grass as they can eat, plus oats or other concentrates, lose weight. Their general health and their performance can be improved by limiting their access to grassland, or its products, and confining them for greater periods of time to the stable which affords shade, cool ventilation and more satisfactory control of insect-borne disease. Furthermore, where the land is light, significant amounts of sand may be consumed by grazing stock and this may interfere with bowel function.

In the author's experience working horses introduced to the tropics and required to subsist on indigenous forages and cereal grains frequently exhibit decreased performance, lameness – changing from one leg to another and involving both creaking hip and shoulder joints – arching of the back, swelling of the facial bones and muscular wasting of croup and rump. Similar observations and the poisoning of grazing cattle have been recorded in various parts of South East Asia, the Philippines, Brunei and North Australia (Blaney, Gartner & McKenzie 1981a,b; Seawright, Groenendyk & Silva 1970). The poor nutritional quality of tropical grass has frequently been overcome by the introduction of high-quality clovers, trefoils, medics (*Medicago* spp.) including lucerne, selecting those varieties low in tannins, as increased tannin concentrations are prevalent when legumes are grown on tropical soils deficient in P.

The signs described above are those of osteodystrophia fibrosa, big-head, found by the author and others (Blaney, Gartner & McKenzie 1981a) to be caused by large amounts of oxalate and small amounts of Ca and P in many tropical grasses. Most published reports have concerned species of the genus *Setaria* (Blaney et al. 1981b; Seawright, Groenendyke & Silva 1970), which may contain 30–70 g of oxalate per kg of dry matter. A simple method for the quantitative estimation of oxalate in tropical grasses has been proposed by Roughan & Slack (1973). Lesser, but harmful quantities of oxalates have been found in the widely used Napier (*Pennisetum purpureum*) and Signal (*Brachiaria* spp.) grasses and Paragrasses. For several reasons Napier is an inferior grass for horses; some chemical characteristics of each of these tropical grasses, for comparison with temperate grasses, are given in Table 10.12. The samples were shown to be deficient also in selenium. Ponies given a diet containing 10 g/kg of oxalic acid plus 4.5 g/kg of Ca exhibit decreased urinary Mg and are in negative Ca balance through reduced Ca absorption (Swartzman, Hintz & Schryver 1978). It has been concluded that problems in horses may arise where the dietary dry matter contains more than 5 g/kg of total oxalates with a Ca:oxalate ratio of less than 0.5 (Blaney et al. 1981a). Horses presenting typical signs, and which were in negative balances of both Ca and P, supplemented once per week with limestone, rock phosphate or dicalcium phosphate and 50–60 per cent molasses, consumed up to 200 g of Ca and 50 g of P per hour and changed to positive Ca and P balances (Gartner, Blaney & McKenzie 1981). Success has also been achieved by supplementing horses with 125 g of limestone per day (providing 45 g Ca) until the initial abnormality has subsided and then reducing the amount to 100 g daily. If rock phosphate is introduced to the diet, the fluorine intake should not be allowed to exceed 50 mg/kg of the total diet.

Table 10.12 Dry matter composition of four tropical grasses fed to horses compared with a temperate (Newmarket, UK) grass sample (author's unpublished data)

	Young Napier grass	Napier hay	Mature Napier hay	Signal grass	Signal grass hay	Paragrass hay	Guatemala grass	Newmarket grass
Crude protein (%)	5.7–10.0	5–6	2.7–2.9	11.0–11.5	5–6	5.4–9.8	7.4–7.5	12.7
MAD fibre (%)	43–51	54–55	54–55	41–42	51–52	42–44	46–47	27.5
Ash (%)	3.2–4.5	3.6–3.8	2.1–2.3	7.3–7.8	3.0–3.2	3.4–8.0	6.0–7.0	9.95
Ca (%)	0.046–0.16	0.08–0.10	0.10–0.12	0.09–0.13	0.09–0.10	0.11–0.20	0.05–0.06	0.93
P (%)	0.06–0.16	0.11–0.13	0.10–0.12	0.17–0.18	0.14–0.15	0.15–0.23	0.170–0.175	0.31
Mg (%)	0.21–0.32	0.20–0.22	0.22–0.25	0.26–0.27	0.11–0.12	0.18–0.36	0.087–0.088	0.16
K (%)	0.9–1.3	0.8–0.9	0.38–0.40	2.8–3.3	0.95–1.05	0.66–2.60	2.1–2.2	1.9
Na (mg/kg)	170–690	200–300	170–190	510–650	170–180	1280–11300	150–170	1800
Zn (mg/kg)	21–38	34–36	20–22	27–29	ND	35–43	ND	38
Se (mg/kg)	0.055	ND	ND	ND	0.03	0.03	0.04	0.1
F (mg/kg)	8.4	ND	ND	ND	ND	ND	ND	ND

Abbreviations: MAD, modified acid-detergent; ND, not determined.

Further reading

Davies W (1952) *The grass crop, its development, use and maintenance.* E. & F. N. Spon: London.

Ministry of Agriculture, Fisheries and Food (1984) *Poisonous plants in Britain and their effects on animals and man* (Reference Book 161). HMSO: London.

Forbes T J, Dibb C, Green J O, Hopkins A & Peel S (1980) *Factors affecting productivity of permanent grassland.* A national farm survey. Grassland Research Institute: Hurley, Maidenhead, UK.

Green J O (1982) *A sample survey of grassland in England and Wales 1970–1972.* Grassland Research Institute: Hurley, Maidenhead, UK.

Hails M R & Crane T D (1983) *Plant poisoning in animals. A bibliography from the world literature, 1960–1979.* Commonwealth Agricultural Bureaux: Slough, UK.

CHAPTER 11

Pests and ailments of horses and ponies related to diet, to stable design and to grazing areas

For a surfeited horse. Take a handful of pennyroyal, half a handful of hyssop, an handful of sage, an handful of elder leaves or buds, an handful of nettle tops, fix large sprigs of rue, and handful of celendine, cut small and boiled in three pints of stale beer, which must be boiled to a quart.

Sir Paulet st John, 1780

Arthropod parasites

Lice

There are two species of horse lice; *Haematopinus asini* is a blood sucker and *Damalinia equi* lives on skin scales. The females lay eggs on the hair and a greater problem of scratching or rubbing is frequently observed in the winter than in the summer so that many are lost when the winter coat is shed. Control is achieved by dipping, spraying or dusting with insecticides but a second treatment should be undertaken to kill those that will hatch from eggs already laid.

Ticks

Grazing horses, particularly those sharing ground with wild grazing and browsing animals, are prone to infestation by ticks, which can transmit diseases. However, the cattle and sheep tick in the British Isles does not usually cause symptoms in horses. In the United States, the soft tick (*Otobius megnini*) lives deep in the ears, but the adults, which do not feed, live in cracks in stables, fences and under troughs where they also lay eggs. Larvae of hard ticks (*Dermacentor andersoni; Amblyomma americanum*) are found in various places on the horse, and the adults, after mating, fall to the ground where eggs are laid in secluded locations. The tropical horse tick (*Dermacentor nitens*), whose primary host is the horse, transmits equine piroplasmosis, a protozoan blood disease. Insecticide should be applied to all parts of the skin where the ticks may be attached, including the ears, and as the parasites spend long periods off the host, the grass and other areas around the stable should also be treated. With slight infestations, the tick can be detached by application of chloroform to release their mouthparts.

Mites

Mites (*Psoroptes equi, P. cuniculi, Chorioptes bovis*) cause itch or scabs and may be controlled by dipping or spraying with insecticide. As a general rule, high-pressure sprays frighten horses so that low-pressure hand-pumped sprayers are preferable and the horse should be confined to a chute during treatment.

Biting midges

An intensely itching dermatitis called sweetitch, which occurs during the summer months and is quite common in Ireland, is probably caused by species of *Culicoides* (Baker & Quinn 1978), a blood-sucking midge, whose saliva induces an immediate-type hypersensitivity. Stock should not be grazed over wet areas where the midges are found and they should be stabled before dusk. Some control is achieved with antihistamines.

Flies

Several species of fly are more of a nuisance than a direct cause of trouble. The warble fly (*Hypoderma lineatum*) can cause some damage, mainly in young horses when the larva penetrates the skin of the legs and wanders under the skin to the back. When it is nearly ready to emerge it should be poulticed. The screw worm fly (*Callitroga hominivorax*) does not occur in Western Europe and has probably been eliminated from the United States. It causes wounds in the skin in which it lays eggs from which the larvae hatch. Direct treatment of the wounds with insecticide is appropriate. Several species of botfly (*Gastrophilus*) are widespread. The adult lays eggs on the breast and around the mouth and gums of the horse. The larvae are swallowed and attach by hooks to the stomach or small intestine, detaching when fully grown and pupating in the manure. Some control can be achieved by sponging areas of the skin on which eggs are attached to hair, using water at a minimum of 49°C (120°F) and where necessary ivermectin or other insecticidal anthelmintic can be given orally.

General stable hygiene is a major factor in the control of all flies, including the immediate removal of dung, contaminated feed and bedding. Routine grooming may assist not only by removing potential trouble but ensuring that there is a regular scrutiny of the horse's coat.

Blister beetles

These insects (*Epicauta* spp. and *Macrobasis* spp.) are lethal when ingested by livestock. Various species are distributed throughout Canada and the United States. They range from 0.8 to 2.7 cm in length and they may be black, black with grey hairs, black with red or yellow contrasting stripes, yellow with black stripes, metallic green or purple. They travel in swarms and feed on flowering plants, such as alfalfa or clover. When alfalfa hay is harvested, the insects can be crushed and incorporated in the bale. Upon ingestion cantharidin, an extremely stable toxin, is released for which there is no known remedy. It is claimed that 6 g of the beetle are sufficient to kill a horse. Cantharidin causes severe inflammation of the oesophagus, stomach and intestines and during urinary excretion causes severe irritation to the urinary tract. The horse develops colic and dies within 48 hours.

Hay baled in July and August is more likely to be infested than that cut earlier. If the insects are detected, they may be present in considerable numbers as a consequence of their swarming nature. By knocking biscuits of hay before feeding at least one or two may fall out when infestation occurs. The existence of the risk is, however, not a justification for the exclusion of alfalfa hay from the diet.

Worm infestations

In temperate countries helminth infection (helminthiasis) in horses is limited to gastrointestinal nematodes, including lung worm, and to liver fluke (trematodes). In tropical countries, however, horses suffer spirurid and filarial infections as well. Foals may be heavily infected with migrating large strongyle nematode larvae with a prepatent period of 6–12 months and therefore injurious infections can be present for many months before eggs are detected. Adult horses may become severely parasitized by migrating larvae, even if wormed, when sharing pastures with horses that are not wormed.

The determination of the severity of worm infestation is no simple matter. Faecal egg counts simply reflect the presence of egg-laying worms. The only reliable means of establishing the degree of parasitism by gastrointestinal nematodes is to analyse serum proteins. Alpha and beta globulins peak in concentration 6 months after infection and thereafter the latter of these proteins declines. There is a coincidental depression in serum albumin and eventually in haemoglobin. Worm egg counts are, however, of use in assessing anthelmintic efficacy in control schemes. Raised eosinophil counts reflect only migrating larvae so that these counts may not differ between treated and untreated animals. The counts tend to be highest in July and August in the Northern Hemisphere and so are not diagnostic. Both small and large infections with strongyles cause an elevation in the immunoglobulin IgG (T) concomitantly with depressed serum albumin before a patent infection occurs.

Liver fluke

Reference to the infectivity of liver flukes (*Fasciola hepatica*) in horses was made in Chapter 10. Their presence may be detected by faecal egg counts and their influence by liver-function tests, indicating liver damage. Untreated cattle and sheep encourage their spread and snails are an essential intermediate host.

Lung worm

Infestation with lung worm (*Dictyocaulus arnfieldi*) is very common among donkeys and not uncommon in horses. Most horses possess some resistance and do not develop patent infections. However, where they do, the prepatent period before larvae may be found in the faeces is 3 months. The examination of faeces for larvae is useful in detecting carrier animals responsible for spreading the infection. The larvae may remain in a state of retarded development and can elicit coughing for periods exceeding 1 year, during which time resort to veterinary lavage of the trachea for the detection of larvae in the washings is a rational move in seeking proof of the cause, even though the larvae are not readily demonstrated. Although lung worm evoke eosinophilia there is no detectable change in serum proteins.

It should be re-emphasized that the important managemental aspect of lung-worm infection is to locate the carrier which, for example, may be a donkey or unhealthy mare shedding faecal larvae without showing signs. This animal should then be removed and treated with effective anthelmintics. Fluke and lung worm are not nearly as economically significant among horses in the British Isles as gastrointestinal nematodes.

Gastrointestinal worm control

The essence of control is to reduce the number of infective larvae on pastures grazed by susceptible stock, particularly those under 3 years old. There is some evidence that a tolerance is developed to both strongyle and ascarid nematodes so that stock should probably not be kept entirely isolated from infective sources. Faecal egg counts reflect only the activity of adult worms in the intestines and may not give a good indication of the seriousness of a strongyle infection. Horses should be treated orally with anthelmintics on arrival at a stable or as directed by the veterinary surgeon; but an initial dose, much larger than normal, of an effective wormer may be prudent in cases of severe infection with strongyles as such doses of thiabendazole, or fenbendazole, can be larvicidal. Veterinary guidance is essential. Recently, moderate but closely defined doses of oxfendazole (Duncan & Reid 1978) or ivermectin (Dunsmore 1985) have been shown to possess efficacy against adult ascarids and small and large strongyles at all stages from eggs to adults, including migrating larvae. Ivermectin also controls botfly larvae.

Young stock should always have access to the cleanest pasture until they have developed some tolerance to worms. Mares must therefore be properly treated so that they do not pass on any severe infection to their offspring; droppings can be removed expeditiously by the use of vacuum cleaner attachments to tractors. Stabled horses should also be treated routinely, particularly where they have been given access to pasture, even for short periods, in the summer. (Table 11.1 gives a simple routine of treatment.) The principal gastrointestinal worms are strongyles (or strongylids) and ascarids, specifically *Strongylus vulgaris* and *Parascaris equorum* (Fig. 11.1). *Strongyloides westeri* (threadworm) larvae are found in the colostrum and milk of mares, presenting a source of infection to their foals.

Strongyles (redworm)

Routine treatment should entail alternating between wormers of the thiabendazole group and pyranthel embonate monitored on studs by faecal egg counts both in the spring and mid-summer. Srongyle eggs develop into infective larvae only in the period between March and October – especially in warm weather. Infective larvae can, however, survive the winter in the British Isles, but in the spring there is a rapid disappearance of these larvae from pasture with increasing ambient temperature. Overwintered larvae die out by June. Nevertheless, in the early grazing season this source augments that from eggs passed by other horses throughout early grazing. The high level of infectivity accumulating during the summer on pasture can be contained by regular anthelmintic dosing at 4–6-week intervals, which complements the management procedures given in Chapter 10. For stron-

Table 11.1 Minimum routine worming programme for grazing and stud horses and ponies in northern latitudes

---

Late March
Parasitological faeces examination of all horses
Anthelmintic treatment of all horses

Early June
Parasitological faeces examination of early foals
Anthelmintic treatment of all horses

August
Parasitological faeces examination of late foals
Anthelmintic treatment of all horses

October
Parasitological examination of all horses
Anthelmintic treatment of all horses

December
Anthelmintic treatment of all horses

---

gyle control there is little point in dosing foals less than 2 months old as the prepatent period of small strongyles is 8–10 weeks and the developmental stages are not susceptible to most anthelmintics in the normal dose range. Badly infested pastures may need ploughing and reseeding, or at least should be rested till June by which time overwintered larvae will have largely gone. However, young stock should not be given access to them until July or August before which grazing should be restricted to cattle or sheep.

Ascarids (roundworms)

Ascarid infection is common in horses under 3 years old, by which time a considerable measure of resistance will have developed. Foals are especially susceptible and it is thought that nearly all become infected without necessarily developing signs, owing to anthelmintic control measures and increasing immunity.

The migrating larvae damage successively the liver and lungs within 14 days of infection. Eggs occur in the faeces from 80 days of age. Eggs acquired by the young foal through coprophagia of the dam's droppings are normally immature and pass passively through the foal's intestines.

Clinical signs of severe infection include pyrexia, coughing, nasal discharge, nervousness, colic and unthriftiness. To preclude this foals should be treated at 4-week intervals from 1 month of age for the control

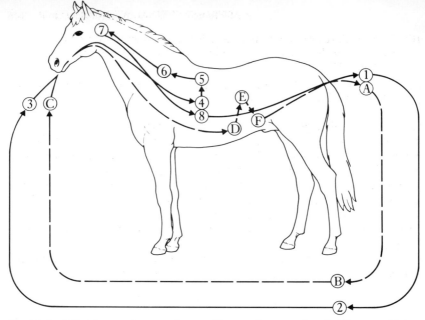

Fig. 11.1 Life history of a roundworm (*Parascaris equorum*) and red worm (*Strongylus vulgaris*).

Ascarids

Parascaris equorum
(in young stock)

(1) Immature eggs appear in droppings from 12–15 weeks after infection
(2) Eggs may become infective in pasture or stable within 2 weeks, or remain dormant for up to several years
(3) Mature eggs ingested by susceptible foal or yearling
(4) Second stage larva hatches from egg and penetrates intestinal wall
(5) Larva reaches liver from 7 days after ingestion of infective egg
(6) Larva reaches lungs from 7–14 days after ingestion
(7) Larva coughed up and swallowed
(8) From fourth to approximately thirteenth week larva grows and matures in small intestine and begins shedding eggs 12–15 weeks after infection

Strongyles

Strongylus vulgaris
(in young and adult stock)

(A) Non-infective eggs shed on pasture
(B) Larva matures within egg case on the pasture during the grazing season
(C) Third-stage infective larva ingested
(D) Larva penetrates submucosa of intestine within a few days
(E) By 14 days after ingestion larva has reached the anterior mesenteric artery where it develops over a 4-month period

(F) Young adult returns to intestine (especially caecum or colon) where it becomes sexually mature by 6 months after ingestion when egg shedding commences

(Strongylus edentatus is similar but larvae migrate only as far as the colonic wall before returning to the lumen)

of the intestinal stages. A mixture of thiabendazole and piperazine citrate is effective and when this is used in rotation with pyrantel embonate, fenbendazole, or mebendazole, a broad-spectrum worm control is achieved where hygiene measures and pasture management are good. Eggs can remain viable on pasture over winter and in suitable conditions some may persist in the environment for many years. The pasture management advised for strongyles is also applicable to the control of ascarids. Table 11.2 lists effective wormers for horses and Table 11.3 gives dose rates and species of internal parasite controlled.

For a more detailed discussion of parasitic worm control in the stud the reader is referred to Rossdale & Ricketts (1980).

Table 11.2   Effective anthelmintics for horses and ponies

Registered trade name	Drug	Manufacturer
Benzimidazole derivatives for nematodes		
Fenbendazole	Panacur	Hoechst
Mebendazole	Telmin	Janssen Pharmacential
Parbendazole	Helmatac	Smith Kline & French
Cambendazole	Equiben	Merck Sharp & Dohme
Thiabendazole	Equizole	Merck Sharp & Dohme
Oxibendazole	Equitac	Smith Kline
Oxfendazole	Systamex	Wellcome
Organic phosphorus compounds for nematodes		
Dichlorvos (pellets)	Equigard	Shell Chemicals
Haloxon	Equivurm	Crown Chemicals
Other compounds for nematodes		
Piperazine	Various	—
Pyrantel	Strongid	Pfizer
Ivermectin	Eqvalan	Merck Sharp & Dohme
Organic phosphorus compounds for botfly larvae		
Dichlorvos (gel)	Equigel	Shell Chemicals
Dichlorvos (pellets)	Equigard	Shell Chemicals
Trichlorfon	Neguvon	Bayer

Table 11.3 Dosage and efficacy of worming agents in horses and ponies based on published evidence (veterinary and manufacturer's specific directions should be followed in the use of these drugs)

Drug	Dosage (single treatment unless otherwise stated) (mg per kg bodyweight)	Nematode species or bot fly larvae controlled; percentage kill given in parentheses. (Where there is yet insufficient quantitative evidence term 'effective' is used)
Dichlorvos	33–43 (single dose rather than divided dose preferred	Larval forms: *Gasterophilus* (bot) (90+) Adult forms: *Parascaris equorum* (90+); *Strongylus vulgaris* (90+); *Strongylus edentatus* (65–75); *Oxyuris equi* (90+); *Probstmayria vivipara* (90+)
Fenbendazole	5–7.5 30 60	Adult forms: *S. vulgaris* (100); *S. edentatus* (99); small strongyles* (92); *O. equi* (100); *P. equorum* (80) Mucosal stages: *Trichonema* spp. (93) Migrating larvae: *S. vulgaris* (83); *S. edentatus* (89)
Ivermectin	0.2	Effective against: Adult forms: *S. vulgaris*; *S. edentatus*; *Strongylus equinus*, *Triodontophorus* spp., small strongyles*, *O. equi*; *P. equorum*; *Trichostrongylus axei*; *Habronema muscae* Immature forms: *O. equi*, small strongyles*, *Gasterophilus* spp. Migrating larvae: *S. vulgaris*, *S. edentatus* Cutaneous: microfilariae of *Onchocerca* sp.
Mebendazole	5–10	Adult forms: *P. equorum* (100); *Trichonema* spp. (98); *Triodontophorus* spp. (99+); *Oesophagodontus* (99+); *Trichostrongylus axei* (99+); *S. equinus* (99+); *S. edentatus* (99+); *S. vulgaris* (99+); *Poteriostomum* (99+); *O. equi* (effective); *P. vivipara* (effective)

Drug	Dose	Efficacy
Oxfendazole	10	Adult forms: T. axei (99+); P. equorum (99+); O. equi (99+); S. vulgaris (99+); Trichonema spp. (99) Developing forms: Trichonema spp. (97) Migrating larvae: S. edentatus (effective); S. vulgaris (50)
	50	Developing forms: Trichonema spp. (99+)
Parbendazole	2.5	Adult forms: O. equi (99); P. vivipara (99+); small strongyles* (96–99); S. vulgaris (98); S. edentatus (98) Developing forms: O. equi (99) Adult forms: P. equorum (95+)
	10	Developing forms: P. equorum (95+) Migrating forms: S. vulgaris (effective)
Piperazine	200 (divided into 2–3 doses in feed)	Adult forms: S. vulgaris (95+); P. equorum (95+)
Thiabendazole	100 (divided into 2–3 doses in feed)	
Pyrantel pamoate	6–7	Adult forms: S. vulgaris (98); S. edentatus (69); S. equinus; (93); Trichonema spp. (90); Triodontophorus (99); O. equi (81); P. equorum (94)
Pyrantel embonate	20	
		Larval forms: Gasterophilus (effective)
Trichlorfon (do not fast before treatment; causes mild colic and diarrhoea)	40	Larval forms: Gasterophilus (97)

* Small strongyles include Triodontophorus spp., Oesophagodontus robustus, Trichonema spp.

Ailments related to diet

Several common ailments in horses are associated with acidosis, some physiological aspects of which were discussed in Chapter 9. Diet is a major cause of acidosis and the effects are typically expressed as laminitis or colic. Therefore, further discussion of the predisposing events and consequences of this diet-related disturbance is appropriate.

An unusually large meal of cereal grains or other feeds rich in starch and protein will overtax the normal digestive powers of the horse so that the caecum and ventral colon are flooded with an unaccustomed proportion of undigested material. Starch-fermenting organisms grow much more rapidly than do those that ferment cellulose, and the starch fermentors proliferate at their expense. The rapid production of volatile fatty acids (VFA) brings about a fall in the pH of the contents of the caecum and ventral colon, killing those organisms that cannot withstand the acidity and encouraging those organisms that can. Prominent among the latter are certain anaerobic lactobacilli and streptococci, the populations of which increase vastly. In addition to VFA, they produce lactic acid, which is more acid (has a lower pK value) than the VFA so that the intracaecal pH may decline from 7 to 4 within 12–24 hours – the lowest value commonly detected after a 'grain overload'.

There are several consequences of this unaccustomed rate of lactic acid production: (1) Organisms that use lactic acid as a source of energy are not present in sufficient numbers to cope with the surge and their numbers may decline as some are unable to exist at the very low pH values attained. (2) The ciliate protozoa, which are much larger and slower growing than bacteria, normally engulf starch – fermenting it at a relatively slow rate. They then act as a starch reservoir preventing an excessive rate of bacterial starch fermentation, but are likewise killed by the acid environment and so no longer function as a buffer. As these normal homeostatic mechanisms are destroyed, acid production proceeds at an accelerating rate. (3) This medium kills significant numbers of bacteria of the Enterobacteriacae. These Gram-negative organisms are disrupted, yielding large quantities of endotoxin particles that previously formed part of their cell wall.

These underlying processes indicate that feeding methods should be imposed to foster dietary health. By slowly increasing the concentrate ration, the bacteria that ferment lactic acid and the protozoa that engulf starch are encouraged to multiply (Table 11.4). These increased numbers of organisms act as a substantial buffer against a decline in pH of the large intestine. The concentrate portion of the ration should never be increased by more than 200 g day for a 550-kg horse. This means that 40 days should elapse in raising the concentrate portion of the diet from 0 to 8 kg.

Lactic acid is, of course, normally produced in the muscles during

Table 11.4 The effect of the proportion of concentrate cubes (C) in the diet and the level of feeding (low, L; high, H) on the mean number of ciliate protozoa (no. per g faeces) × 10^{-4} in the faeces of horses and ponies on the twenty-second day of each feeding period

Expt 2		Expt 3			Expt 4		
			L	H		L	H
C	Protozoa	C	Protozoa		C	Protozoa	
0.24	5.2	0.24	6.9	6.3	0.23	5.0	6.7
0.38	8.8	0.67	5.9	5.9	0.57	14.0	3.3
0.52	7.5						
0.67	9.6						
0.82	10.8						
c.v. of means	7.1%*		154%			27%†	

(From Frape, Tuck et al. 1982)
* Highly significant linear increase in faecal protozoa with increasing cube proportion ($P < 0.01$).
† Significant interaction between feeding level and cube proportion ($P < 0.05$).

anaerobic work as described in Chatper 9. Why then is the lactic acid produced in the intestine potentially more lethal? First, anaerobic work can be sustained for only a few minutes, after which aerobic conditions lead to the complete metabolism of the lactic acid. In contrast, lactic acid fermentation may persist for 24–36 hours. Second, the lactic acid produced by bacteria is a racemic mixture of D- and L-lactic acid, whereas that produced in the muscles is solely of the latter type. Although D-lactate seems to be metabolized quite efficiently, its rates of absorption and of metabolism by the liver are slower than those of the other type. Its protracted existence leads to greater tissue damage. Horses need a fully functional liver to catabolize large amounts of lactic acid and those that have suffered liver damage through disease and infestation, and from bad feed management, are less able to cope with rapid increases in the dietary energy allowance.

The horse is very sensitive to endotoxins released and absorbed during the disruption of bacteria for some 6–12 hours after carbohydrate overload and a series of events is initiated: the endotoxins increase the level of circulating vasoactive prostaglandins, which cause severe circulatory disturbances, including decreased blood perfusion of vital organs and increased perfusion of the gastrointestinal tract. There is a decrease in capillary blood flow and slow refilling of the capillary bed, so the horse develops cold extremities. Incomplete perfusion of the lungs through capillary shunting causes incomplete oxygenation and lowered oxygen tension of the blood (hypoxaemia) and this added to the restricted blood flow through other tissues and organs, leads to anaerobic glycolysis and further production of lactic acid. A restricted hepatic blood flow reduces the removal and metabolism of lactic acid; restricted extraction of lactate by the kidneys aggravates the situation. Increased capillary permeability and diarrhoea cause dehydration and haemoconcentration, worsening the circulatory situation and causing increased heart rate. Within 5 minutes of the release of significant amounts of endotoxin there is arterial hypoxaemia, despite intense hyperventilation, because of pulmonary arterial hypertension. Within 30 minutes severe abdominal pain (colic), diarrhoea and metabolic acidosis are observed. If horses survive the acidotic phase, respiratory alkalosis, anorexia and severe lameness develop.

It is clear that diet-induced endotoxaemia is a major factor in augmenting the effects of bacterial lactic acid production, which can yield grave symptoms of cardiovascular collapse. Endotoxaemia and lactic acidosis are implicated in obstructive bowel disease, equine colic precipitated by abrupt increases in the intake of starch and protein, and in laminitis.

As endotoxins are lipopolysaccharides they are not neutralized by antibody and so treatment for endotoxaemia is directed against the mediators of the condition. It involves expulsion of the ingesta and the use of antiinflammatory and other vasoactive drugs, fluid therapy and glucose injections. If a horse has overindulged in rich feed it is vital that action is taken before signs of illness are presented, otherwise there is considerable risk of a fatal outcome. Moore, Garner et al. (1979) detected an increase in caecal fluid endotoxin 3, 6, 12 and 18 hours after starch overload. Caecal lactate was raised 6, 12 and 18 hours after the meal and caecal fluid pH approached 5 at 6 hours and was less than 5 at 12 hours. Although starch fermentation in the large intestine yields lactate, tissue lactate is produced directly from the arterial hypoxaemia caused by the action of endotoxin (Moore, Garner et al. 1981). Plasma lactate may be raised for 48 hours and remain above normal concentrations for even longer in marked contrast to the effects of galloping.

Garner, Hutcheson et al. (1977) reported that the greatest increase in blood lactate normally precipitates circulatory collapse and death with or without symptoms of laminitis. Lesser increases in blood lactate are more frequently associated with laminitis and the lowest increases with neither effect. The development of laminitis is the most frequent outcome of lactic acidosis. In this study 70 per cent of the cases developed this condition, whereas only 15 per cent suffered fatal cardiovascular collapse. Garner, Hutcheson et al. (1977) concluded that the rate of increase in blood lactate, as determined by blood measurements at 8 and 16 hours, gives a fair indication of whether the horse will die or will contract laminitis, and so it provides the basis for appropriate therapy. Other work has shown that where arterial plasma lactate exceeds 8 mmol litre, death is inevitable, whereas survival is probable with maximum values below 3 mmol litre (Coffman 1979c).

A comparable but less-frequent problem may also result from the excessive consumption of concentrates. In this particular condition there is a rapid proliferation of *Clostridium perfringens*, a normal inhabitant of the gut that secretes an enterotoxin. In small amounts this enterotoxin is apparently harmless, but when there are large numbers of the bacterium excessive gas is produced and the toxin causes damage to the intestinal mucosa and precipitates diarrhoea. The toxin is neutralized by antibody, but immediate therapy involves the replacement of depleted tissue water and electrolytes and the relief of gastrointestinal tympany (ballooning).

Laminitis

In this ailment the hoof heats up through painful inflammation of the laminae of the foot between the rigid structures of the bone and the hoof. Pressure and decreased blood flow to the toe cause pain and damage to the laminae. Often, a strong digital pulse can be detected at the fetlock, sometimes petechial haemorrhages are observed in the buccal cavity and oedema is seen along the underline of the belly.

Laminitis may be caused by overwork, concussion of the feet in horses unaccustomed to hard ground, infections, the consumption of several gallons of cold water before cooling down after hard exercise, abortion, high fever, drug-induced complications and the consumption of certain toxins, especially where liver function is abnormal. But by far the commonest cause (compounded by inadequate exercise) is the overconsumption of concentrates or of grass by animals unaccustomed to them. After a grain overload, horses seem to be more likely to survive severe cardiovascular stress than are ponies. Ponies seem to have a greater risk of laminitis than do other equine animals and entire males seem to be at greater risk than are geldings (Dorn, Garner et al, 1975). Geldings may be more prone than are mares according to a survey in England (J. Ridgeway, personal communication). This same survey recorded that 70 per cent of the cases occurred among horses on pasture, particularly in the months of more rapid growth (April, May, June and October), and that in over half of them the horses were grey. The cell-mediated immune mechanisms of photosensitization, alluded to in Chapter 10, might here be imputed, although a purely chance effect may have been observed.

Once laminitis has occurred a horse is more susceptible to its recurrence. Where diet is the cause, both grain and excessively lush grass or legume pasture are often at fault and if overconsumption is suspected the situation requires immediate veterinary care. One should certainly not wait until signs develop as diet causes the most severe type of laminitis. Signs of anxiety and trembling may herald a more critical disorder. The front feet are frequently the most severely affected, when the animal adopts a straddled stance with the front feet pushed forwards and lameness in them occurs in about 60 per cent of the cases.

Treatment of acute laminitis includes the use of ice packs on the legs and feet, standing the horse in cold water or alternating cold and warm water, or even use of cold-water bandages on the legs. The expulsion of the digesta which is yielding lactic acid and causing endotoxin release should be invoked immediately by dosing with mineral oil. Veterinary use of antibiotics to kill the lactic acid-producing bacteria and treatment with drugs to reduce pain and hypertension may be required.

The objective of further treatment is to hamper the appearance of the severest forms of chronic deformation. Chronic inflammation initiates a more rapid growth of the hoof wall than would otherwise occur so that the toe of the hoof extends and curls up, heavy rings developing on the wall in response to inflammation of the coronary band. The hoof may separate from the foot, and without care infections can focus on cracks developing between the hoof wall and the sole. By standing the horse in sand to increase the pressure on the sole, the downward rotation of the third phalanx bone, which assumes a position no longer parallel to the wall of the hoof, may be partly suppressed. Where the angle between the hoof wall and the surface of this phalanx, measured on X-ray photographs, exceeds 60° the condition is considered to be hopeless. The malformation can be partly overcome by rasping and corrective shoeing and any overweight condition should be rectified.

A poor appetite is a normal sequel but dry feed should be withheld for several days and then a maximum of 1 kg of hard hay, or straw, per 100 kg bodyweight can be given until optimum bodyweight is achieved. Fresh clean water must be available at all times.

Although veterinary attention is appropriate if there is any suggestion that laminitis may be in the offing, prevention is always better than cure and an increased work rate or energy demand should be accompanied by no more than a 200-g daily increase in the concentrate allowance for a 500-kg horse and proportionately less for smaller horses. For similar reasons, very restricted access, particularly initially, should be allowed to lush pastures. Where the dietary Ca:P ratio is low, the addition of limestone to the diet has both therapeutic and prophylactic properties for laminitis (C. M. Colles, personal communication). Large routine intakes of bran associated with an elevated phosphate clearance are apparently often antecedent events.

Colic

The word colic means abdominal pain, which may wax and wane in concert with intestinal smooth muscle contractions and which is present in several abnormal conditions. As this implies no diagnosis it is apposite to discuss the various types and causes of colic and the management favouring a healthy prognosis. Probably all equine animals experience colic several times in their life, so that in various degrees of severity it is very common and in its most severe forms is associated with disorders which are the commonest causes of death. Records show that 80 per cent of cases recover spontaneously within 1–2 hours, but in the remaining 20 per cent, unless immediate action is taken, a disturbance that may initially be mild can become fatal.

Colic usually accompanies a rise in blood lactate, accompanied by the signs previously discussed, and the severity and outcome are closely correlated with this increased value. Lactate concentrations in the peritoneal fluid are also typically higher than in the blood, except for cases of impaction.

Most colics are characterized by some of the following postures and reactions in various forms and intensities: tail twitching, pawing the ground and restlessness in which the horse may get up and down frequently, playing with its food and water, submersing the nostrils and blowing bubbles, and generally losing appetite, the head is frequently turned towards the flanks and, in the extreme, the horse rolls and thrashes about risking further damage. However, one might enter the box to find the horse cast, with no intestinal sounds, no droppings, or a very few small ones, and a much distended abdomen. The horse may attempt frequent staling (urination) in an endeavour to relieve the pressure on its bladder. The rapidity of heart beat and respiration rate and the extent of sweating and fever will depend on the severity of the disorder. Normal heart rate lies generally between 38 and 40 beats per minute but the rate may rise to 68–92 per minute in moderate colic and to over 100 per minute in severe pain. Similarly respiration rate, normally 12–24 per minute, may exceed 72 per minute and the normal body temperature of $37.7 \pm 0.3°C$ ($100 \pm 0.6°F$) will be elevated. Other signs may include diarrhoea and undigested cereals in the faeces, foul-smelling breath and ingesta in the nostrils, frequent stretching, and on some occasions skin changes in the form of a nettle rash. Capillary perfusion time is increased as measured by thumb pressure on the gum, after which the white patch regains its colour over a longer period than the normal 1–2 seconds. Dehydration is also expressed as a delay in the return of the skin to its normal posture after being pinched.

Unless one is very familiar with the sequence of events in a particular horse, veterinary help should be sought immediately signs are initiated. Feed should be removed but clean water provided. If the animal shows signs of injuring itself through violent actions the horse may be walked, otherwise leave it alone in a box free from projections or structures that might endanger it. In all colic cases the horse should be kept warm in cold weather and during recovery a warm bran mash is helpful.

Many colics involve the presence in the stomach, or intestines, of a thick sticky mass of fermenting feed or a compacted mass of roughage. The general use of a mineral oil or a kaolin–pectin paste is safe and will accelerate the expulsion of the offending masses. General treatment also has the objective of preventing rupture of some part of the gastrointestinal tract, or displacement of its parts by control of pain and tympany, by evacuation of the bowels, by arresting rapid bacterial fermentation and by reestablishment of normal peristalsis. Where the derangement resides in the stomach, a stomach tube is used to allow the expulsion of gases and to permit the administration of antifermentatives such as chloral hydrate or turpentine (an oil obtained from various species of *Pinus*) in raw linseed oil. Liquid paraffin may be given by a nasogastric tube for impactions at rates of 2–6 litres per 500-kg horse once or twice per day for several days, and 1–2 pints (0.5–1 litre) of raw linseed oil (acting as an emollient cathartic) may be used in obstinate cases together with warm salt water to stimulate thirst. Again 30–60 ml of turpentine may be added to the oil for flatulent colic.

Where the trouble is in the large bowel normal movement is encouraged by cold-water enemas and massage of any impaction at the pelvic flexure via the rectum. The general use of tranquillizers and pain-killing drugs removes the need for continued forced exercise to prevent the animal from damaging itself and they reduce the risk of a simple impaction becoming a volvulus (twisting of the intestine on its mesenteric axis). Walking may only be necessary to distract the horse's attention where drugs are unavailable or ineffectual. Quick action in mild colic can forestall a more serious derangement precipitating endotoxin shock and death.

Impaction colic

About 30 per cent of all colics are of this form and of these most impactions occur in the large intestine. As briefly mentioned in the first chapter, these are located typically at points where there is either a change in the diameter of the colon or at flexures where it turns acutely. More frequent sites are the pelvic flexure and where the right dorsal colon empties into the small colon, but occasionally the sternal and diaphragmatic flexures may be involved. Impaction may also occur at the ileocaecal valve. The closer to the ileum the large intestinal blockage occurs, the more dangerous it is as it will severely restrict water resorption in the caecum and ventral colon, which can lead to dehydration and hypovolaemic shock. Sometimes impaction of the small intestine, which is typically of the ileum, may result from excessive cereal intake, in which case ingesta may be present in the nostrils. But where the colon is the source, the horse will frequently look at its flank, emit no intestinal sounds, and void small mucus-covered droppings; palpation of the impaction is possible. An enema of 2–3 gallons (9–13.5 litres) of warm soapy water is frequently given in addition to the lubricants already mentioned.

Impactions are fairly common in old horses with bad teeth restricted to poor-quality hay with little water after experiencing lush grass. Recent evidence suggests that the production of butyric acid in larger than normal quantities within the intestine tends to inhibit its motor activity leading to a cessation of intestinal sounds. Impaction colic in the small colon may also occur in foals between 2 and 6 months old when roughage feeding is initiated. Here also an enema of mild soapy water is very helpful.

Gas or flatulent colic

This disorder may be secondary to a blockage or an impaction and is extremely painful. Distension of the small bowel is rarely noticeable and if the abdomen is unusually large, tympany of the ventral colon may well be present. Sometimes on postmortem examination several regions of the gastrointestinal tract appear to be involved. An impaction and lack of movement of the intestines inhibit expulsion and minimize absorption of gas into the blood. The latter route of removal is more important than may be realized as about 150 litres of carbon dioxide and methane can be absorbed daily from the intestinal tract.

If gastric tympany occurs, the condition becomes evident within 4–6 hours of eating. Intubation with a nasogastric tube is essential to relieve the pressure, which can otherwise cause rupture of the stomach. A warm 4 per cent salt solution, which decreases the viscosity of the fermenting mass and encourages water consumption, is sometimes administered in small quantities at intervals by allowing it to drain into the stomach through the tube. The horse may have adopted a typical dog-sitting attitude, or it may stand without moving the feet, especially where a rupture or a serious intestinal infarct exists. After stomach rupture, in particular, ingesta are sometimes observed in the nostrils. With different intensities of gas colic the horse may feel cold, despite experiencing a fever, it may exhibit congested mucous membranes of the eyes and have a sour and vinegary breath. Powerful anodynes prevent violent rolling and self-inflicted trauma. Where there is risk of violent action the horse should, if possible, be kept on its feet, but quiet horses are best left alone.

Gas colic is frequently the sequel to a rich diet of cereals, or lush legumes, or even to the inadvertent consumption of a pile of grass cuttings. Quick action is essential as again violent reactions on the part of the horse may lead to ruptures or to twisting of the gut with a poor prognosis.

Sand colic

This is probably the most common cause of colic in areas of very sandy soils. Some horses contract the bad habit of consuming large quantities of sand and soil. The problem therefore can frequently recur and is associated with periods of inappetence, diarrhoea, anxious pacing up and down with groans on lying down, pawing the ground or a crouched stance with a turned head. Sand may be present in the droppings. Treatment again includes repeated large doses of liquid paraffin. Occasionally, however, enteroliths (large stones) are formed, apparently on a nidus of ammonium magnesium phosphate. This gradually enlarges and its removal requires surgery.

Twists and rotations

Torsion (rotation on its own axis), volvulus (twisting on the mesenteric axis) and intussusceptions (infolding), usually of the terminal ileum into the caecum, all require immediate surgery with rather poor prospects. The membranes of the eye and lips are typically dry and pain induces a rapid rise in pulse and respiratory rates so that the loss of carbon dioxide causes alkalosis despite a raised blood lactate.

Parasitic worms

Strongyle larvae cause damage to the lining of blood vessels, particularly that of the anterior mesentric artery and its branches, and this can lead to various degrees of occlusion and inhibition of blood flow (ischaemia). A thromboembolism can be a major contributory cause of the complete loss of blood flow to, and death of, a portion of the intestinal tract leading to obstructive colic. Where the blockage is incomplete, recurrent colic will be experienced. In acute cases surgery is needed but thorough worming at 30–60-day intervals will help to contain the situation and a defined dose of specific wormers (Tables 11.2 and 11.3) under veterinary guidance will have some beneficial impact on the larval stages. (These stages cause a rise in intestinal alkaline phosphatase activity of peritoneal fluid – see Ch. 12.) In young horses, impaction of the small intestine with ascarid worms can occur where management is bad.

Spasmodic colic

There is an increase in bowel movement in this type of colic, which may be precipitated by a sudden change of feed, work and chilling. Spasms may last for a few minutes or up to half an hour, and may occur repeatedly over a period of hours, the signs being typically those already discussed. Recovery occurs without treatment but relief of pain and the use of spasmolytic drugs is helpful in amelioration.

Foal colic

This is very common in the first 48 hours after birth and is caused by the meconium blocking the large intestine at various levels. Lubrication of the impacted mass with orally administered liquid paraffin (200 ml) or glycerol, the use of enemata of soap and water, and the relief of discomfort are normally sufficient remedies. If no response is registered within a few days, volvulus or intussusception may be suspected. Abdominal pain can also occur at this age through rupturing of the bladder, which is effectively repaired by surgery, and in older foals discomfort may coincide with eruption of permanent teeth. Umbilical hernias can cause colic in young foals, but these usually correct themselves by 6–8 months of age. Inguinal hernias, especially in colts, also have similar effects.

Other colics

Abdominal pain as indicated above is not the reserve of the intestinal tract and may result from bladder and kidney stones, urinary infections, pericardial effusions and sometimes liver disease.

Predisposing factors in colic

1. Overheating – sudden access to large quantities of cereals or stands of green clover and lush grass.

2. Stress caused by changes in routine, changing stables, and mares and foals in new surroundings.

3. Irregular work, horses standing idle on full feed or changes in the timing of feeds.

4. Working a horse on a full stomach. Even during protracted slow work large feeds should not be given and bulk feeds should be excluded. Bulky feeds should be given in the evening, waiting until the digestive powers have been restored by rest.

5. Work itself can precipitate colic, especially towards the end of an exhausting day – both feed and work should be regular. Extended work should be interrupted by short rests every 2–3 hours when a few mouthfuls of concentrated feed and water are provided.

6. The consumption of excessive amounts of cold water after severe and hot work before the horse has cooled down and/or providing heavy feed at this time.

7. Insufficient good-quality roughage or mouldy corn and mouldy silage.

8. Large quantities of cut green feed.

9. Unfit horses changed abruptly to an increased work rate and a concentrate-rich diet.

10. Failure to provide fresh clean water at all times.

11. Lack of teeth care. 'Quidding' may be noticed in which small balls of partly chewed feed are dropped into the manger. This is usually associated with cheek teeth that require rasping.

12. Greedy feeders that bolt feed, or greedy bullies in group-fed herds.

13. Inadequate work control.

Prevention of colic

1. Each horse has its idiosyncrasies so that before a new horse is placed on a rich working diet its habits and particular requirements should be studied and understood.

2. Hard-worked animals should receive small feeds of concentrates at frequent and regular intervals, and the timing of feeds should be regular even at weekends.

3. Increased demands for energy should be met by an increased feeding rate for concentrates of no more than 200 g/day in a 500-kg horse and a proportionately lesser rate of increase in smaller horses.

4. A sensible and regular exercise programme should be instituted.

5. Where a horse is changed from one stable to another, its old routine should go with it and if necessary changed gradually.

6. All animals should be checked last thing at night.

7. Good-quality roughage free from contaminating weeds and, in the United States, also free from blister beetles is required.

8. Stores of horse feeds and dangerous chemicals should always be held in rooms, the doors of which cannot be opened by horses.

9. No horse should be overworked and, following strenuous work, no substantial feed or water should be given until the animal is cool and rested and then should only be given in moderate amounts.

10. Cribbers or wind suckers (the swallowing of air into the stomach) (Pl. 11.1) should be fitted with cribbing straps, or, if cribbing is known to precipitate colic in the individual, surgery may be necessary.

Plate 11.1 A 'wind-sucker' mare (cribber) – a vice in which the incisor teeth grip a solid object, the horse pulls down and swallows gulps of air. Sometimes a leather strap is fastened snugly around the neck just behind the jaw to deter the horse from the practice.

11. Teeth should be inspected at regular intervals. These may require the filing of the upper and lower molars and premolars or removal of teeth with infected roots – decaying teeth may be inferred from excessive salivation.

12. Individuals that bolt their concentrate ration should have it mixed with chaff or dry bran. The placement of stones the size of tennis balls, in the manger may retard the rate of feed consumption.

13. A proper worming programme is essential in all horses and a pasture rotation system should be instituted where horses have access to grass.

Overfeeding

In addition to the consequences already discussed, rank overfeeding can have a number of other deleterious effects in horses. Some of the more obvious are listed below:

1. Obesity. This is said to reduce fertility and to present difficulties at foaling in mares, to affect the work rate of horses and to accelerate the onset of fatigue.

2. Obese horses that suddenly experience food deprivation, as in a drought, may be subject to anorexia secondary to colic, which sometimes causes hyperlipidaemia (high concentrations of blood lipids). Pony mares in late pregnancy or in peak lactation and subject to pasture changes are prone to this problem. Treatment can be problematic and certainly requires veterinary advice. Anorexia may also be a sequel to acidosis.

3. The overfeeding of young horses, in particular, colts, especially when the food is given in separate and discrete meals and where the ration is unbalanced in respect of its mineral content, may cause bone disorders. The principle one is epiphysitis – typically of the distal epiphyses of the radius, metacarpus, tibia and metatarsus – and recognized by bony enlargements and lipping of the physes.

4. Contracted tendons, as previously discussed in Chapter 7, can be associated with overnutrition of protein and energy in the foal and yearling. There is no evidence that they have a contributory hereditary cause.

5. Enterotoxaemia in young horses fed in groups is occasionally precipitated in the largest and most aggressive individual. A flatulent colic occurs in which the intestines are loaded with rich feed and gas. Symptoms of dyspnea and subcutaneous oedema may be presented and the cause is apparently a rapid proliferation of the bacterium *Clostridium perfringens*.

Chronic diarrhoea

This is normally of large intestinal origin resulting from some upset to the normal balance of the intestinal flora. It can follow stress such as the prophylactic use of oxytetracycline or some other antibiotics. Where the diarrhoea is of small intestinal origin, it may be connected with a want of certain digestive enzymes detected by xylose and other tolerance tests (Ch. 12). It should also be remembered that the adult horse loses the ability to digest lactose when about 3 years old so that large intakes of milk sugar after this time may induce diarrhoea. Chronic diarrhoea can also be prompted by parasitism, mesenteric abscesses or some disorder of vital organs.

Acute diarrhoea

Severe diarrhoea is frequently caused by salmonella infection precipitated by stresses of transport and particularly by strongyle worm infection. Antibiotic treatment to eliminate salmonella is of questionable value and in fact oxytetracycline may trigger the onset of the infection. Salmonellosis is seen in the United Kingdom in closely stocked groups after mild winters and is associated with heavy worm burdens. It forms a major risk to foals and may be transmitted by adults which are asymptomatic carriers. Suspected cases should be isolated in a box with a very strict programme to ensure that contaminated faeces are not transmitted to other stock. The bacteria are not continuously excreted in the faeces, but careful veterinary examination may be necessary to detect any carriers that are shedding the organisms. Feed and rodents may also be suspect reservoirs of potential infection.

All cases of acute diarrhoea are associated with a critical loss of fluid, potassium, sodium and chloride, and unless replacement therapy is quickly instituted the consequences are rapidly fatal in young stock. Appropriate fluids are listed in Table 9.2. Fluid losses of 20–50 litres, sodium deficits of 2000–6000 mmol, potassium deficits of 700–3000 mmol and bicarbonate deficits of 1000–2000 mmol may exist and should be made good in adults and young. Foals may experience absolute losses amounting to 15–20 per cent of that of the adult. In acute diarrhoea where there is hypotonicity and dehydration, the use of hypertonic solutions yields an immediate response. Where the animal is dehydrated and hypertonic, hypotonic solutions are given. The normal plasma values for sodium, potassium, chloride and bicarbonate in horses are respectively 139, 3.6, 99 and 26 mmol/litre.

A mild foal diarrhoea typically occurs at the time of the first postpartum oestrus from a digestive upset said to be caused by a change in the composition of the milk at the time and associated with a change in electrolyte excretion.

Dehydration and potassium status

Although the carcass of a 500-kg horse may contain by weight only 15 per cent as much sodium (Na), or potassium (K), as of calcium (Ca), on average the 1100–1200 g of K are subject to a much greater flux than is the Ca owing to its higher solubility in tissue fluids. The volume and water content of muscle cells and of red cells are modulated primarily by the control of their Na^+ and K^+ contents. The cell membrane is relatively impermeable to small cations – that is they diffuse slowly, whereas small anions diffuse freely – but haemoglobin acting as a large anion remains as an intracellular entity. However, the equilibrium distribution of charged particles between red cells and plasma differs from that predicted by normal diffusion processes. The slow passive movements of Na^+ and K^+ are balanced by an active outward movement of Na^+ and an inward transport of K^+ in each cell, mediated by several hundred discrete pumps fuelled with ATP. If glycolytic mechanisms yielding ATP break down, or if the cell membrane is damaged such that diffusion leakage increases, then there is a decline in resting membrane potential and the pump mechanism is incapable of maintaining a physiological cellular environment. That an intact glycolytic pathway in red cells is necessary for the maintenance of a physiological cation distribution between cells and plasma has been amply confirmed in numerous experiments.

It is clear that the horse must maintain cellular potassium within strict concentration limits for normal health to prevail. The measurement of plasma or serum K^+ concentrations, as a guide to body K status, although frequently done, is misleading. Measurements of large numbers of horses have failed to detect any correlation between serum and cellular concentrations of K^+ (Frape 1984b; Muylle, Nuytten et al. 1984) and serum contains on average only 3.7–4.3 mmol of K^+ per litre that is 3.8–4.4 per cent of the mean concentration in red cells. In fact, the extracellular fluid of the body in total contains only 1.3–1.4 per cent of the total body K. In maximal anaerobic exercise, serum K^+ tends to increase, whereas it has a tendency to decrease in endurance work without comparable changes in the cells.

In severe diarrhoea, the bodily loss of K by a 500-kg horse may approach 4500 mmol (175 g), associated with a fall in red cell K^+ from 97.5 to 75 mmol/litre, but without significant change in plasma K^+ (Muylle, Nuytten et al. 1984). Protracted work in hot weather apparently leads to losses of K^+ and Na^+ in a horse of this size of 1500–1800 mmol and 4000–5000 mmol, respectively. Thus the measurement of red-cell K^+, the concentration of which appears to be well correlated with that of muscle-cell K^+ (Carlson 1983b), is recognized as a more reliable means of assessing K^+ status and possibly of understanding the underlying processes obtaining in setfast (tying-up) and azoturia. Husbandry and dietary measures may then be devised to limit the frequency of their occurrence, and the situations, or individuals, in which the problems are more likely to arise may be more easily predicted.

A fall in red-cell K^+ concentration below 81 mmol litre is associated with weakness of skeletal and smooth muscles, tremors and, in severe depletion, with recumbency, cyanosis and eventually respiratory and heart failure. The K^+ ions that are released from muscle cells during hard

exercise act as potent arteriolar vasodilators and they stimulate cardiores-
piratory reflex activity. Thus there is a close correlation between the extracellular increase in K^+ and the increase in both muscle blood flow and O_2
consumption and therefore in performance. The blood flow is insufficient
in K^+ depletion precipitating hypoxia, anaerobic glycolysis and metabolic
acidosis. This turn of events is pathological, and the ensuing damage to
muscle-cell membranes leads to further cellular K^+ loss, which cannot be
restored by a Na^+/K^+ pump deficient in readily available energy. Recent
investigations in Belgium (Muylle, van den Hende et al. 1984) have
revealed an anomalous situation in which about 10 per cent of 436 horses
examined possessed a normal red-cell K^+ concentration distributed inde-
pendently of the remainder. Their mean red-blood cell K^+ concentration
was 83.8 mmol/litre, some 13.7 mmol less than the normal for the other 90
per cent despite similarities in management and diet. Other studies by the
same group (Muylle, van den Hende et al. 1983) indicated that in a smaller
sample of 43 horses, 11 were in this low red-cell K^+ range and, of these, 9
were performing unsatisfactorily on the racetrack and they expressed a more
nervous temperament.

A rank dietary K deficiency is unlikely to be more than an occasional
cause of K^+ depletion, although several investigations implicate dietary
magnesium deficiency as a cause. A 500-kg horse given a diet composed
of a 50:50 mixture of grain and hay may absorb daily 90–100 g of K^+, well
above the maintenance level of requirement. Excess dietary K is rapidly
excreted so that a generous dietary content is of no avail in the acute K^+
depletion of diarrhoea, or of abnormally high sweat loss, when in any event
appetite is depressed. Dosing with an appropriate solution is the only
reasonable approach in overcoming the worst of the deficit. Potassium in
excess is a moderately potent toxin to heart muscle so that only limited quantities may be given intravenously while cardiac action is monitored. Intravenous infusion rates of 11.5–13.7 mmol of KCl per min bring about plasma
K^+ concentrations exceeding 8.0 mmol/litre and consequential cardiac
arrhythmias and abnormal electrocardiograms, through a transient but
excessive alteration to the gradient of the transmembrane K^+ (Epstein
1984). Thus the bulk of such dosage must be given orally or by nasogastric
intubation. By this route Muylle and colleagues (1984) administered a
solution containing 50 g/litre glucose, a commercial amino-acid mixture
(0.05 litres/litre), KCl (5 mmol/litre), $CaCl_2$ (3 mmol/litre) made isotonic with
NaCl, which was partially replaced by Na acetate according to the
acid:base balance of the horse; the quantities given daily were proportional
to the body deficit calculated from red-cell K^+ values (see Ch. 12 for
assessment of red-cell potassium).

Choke and grass sickness

Choke is caused by a foreign object, or feed, lodging in the oesophagus.

Normally the obstruction will clear after a while and it may be common in individuals possessing gullets with an abnormal structure. A similar event may be present in grass sickness, but stomach distension is frequently involved, as failure of the normal function of the stomach and of the oesophagus seems to result from neural impairment of the muscles controlling the contraction of these organs. Nasal return of ingesta indicates such oesophageal impairment. Neural toxins in herbage are suspected as a cause.

It has been calculated that the intestinal tract of the average horse holds about 100 litres of fluid, which in water deprivation – as may be induced by choke – is drawn on to maintain homeostasis. If choke prevents feeding for more than 6–7 days, dehydration precipitates prerenal azotaemia (raised blood urea). Once the obstruction has been removed the horse should be watered and fed several times per day with small quantities of wetted nuts or other feed, and a stone should be placed in the feed box. Some scarring of the oesophagus will have occurred, which may cause repeated trouble.

Wasting diseases

Some of the ailments already described cause a loss of body condition and weight. If a horse is presenting these signs, then one of the following causes might be suspected:

1. Jaw and dental abnormalities, including sharp points on the upper and lower molars and premolars or abscesses below the teeth.

2. Roundworm infestation damaging the gut, which has interrupted the uptake of nutrients, and larval stages injuring blood vessels that supply the intestines and other organs and causing anaemia.

3. Diarrhoea.

4. Tuberculosis which very occasionally involves the digestive system.

5. Liver malfunction resulting from ragwort damage or the presence of some chronic septic focus in the body.

6. Windsucking.

7. Shy horses, low in the social order, subjected to group feeding.

8. Heart abnormalities and anaemia.

9. Arthritis and other causes of chronic low-grade pain.

10. Cancer, especially in older horses. Grey horses are prone to internal melanomas.

Muscular ailments

There are several metabolic abnormalities in this group, for which many of the signs are shared and attached to which the nomenclature is frequently confused.

Tying-up myositis or setfast

This is a muscular inflammation not to be confused with the more severe degenerative changes presented in azoturia. It is associated with the depletion of muscle energy that occurs after several hours of work. Onset may happen within a few strides when the horse falters and stops, generally refusing to move. The muscles are not palpably abnormal, but are stiff and sore. The horse may adopt a cramped stance and if willing to move it does so reluctantly and slowly. Lesser attacks may occur following cooling off. The muscular stiffness, which passes within 1–3 days, is probably caused by the muscular accumulation of lactate. The activity of serum CK returns to normal after 4–6 days, but that of AAT may take 4–5 weeks to do so. In addition to a depletion of muscle glycogen (principally in the fast-twitch fibres), there is also a depletion of ATP and CP. However, muscle glucose as well as lactate concentration is raised, suggesting local hypoxia (insufficient oxygen reaching the muscle cells from the blood). A leakage of the muscle enzymes AAT, CK and lactic dehydrogenase is the cause of their raised blood concentration.

The horse should be kept warm; effective treatment normally includes the intravenous administration of calcium gluconate, magnesium and phosphate ions and vitamin D. Prevention requires training to increase condition, more frequent rest periods, and the administration of electrolyte solutions during and after physical activity. There is, moreover, some evidence that tying-up is associated with elevated phosphate clearance and that feeding calcium carbonate is helpful (C. M. Colles, personal communication). The condition is also occasionally noticed in fillies coming into season (oestrogens increase the activity of l-hydroxycholecalciferase, which further implicates vitamin D).

Azoturia (exertional rhabdomyolysis)

This is a much more severe derangement than tying-up and is associated with degenerative changes in the muscles. It may indicate poor management of training and commonly occurs within the first hour of exercise, following a period of inactivity on full rations. Azoturia may be initiated by muscular spasms causing a sudden restriction in the blood flow to the muscles and a consequent rapid build up of lactic acid. It is unusual for the ailment to follow a rest period of 1 day or of one as long as 14 days and commonly occurs following 2 days' rest on full rations. Large stores of muscle glycogen may be present, and during its anaerobic breakdown lactic acid accumulates in the muscles at a rate exceeding the rate of its removal in the blood. The fall in pH appears to cause a coagulation of muscle protein and liberation of myoglobin, which escapes in the urine. However, the diagnostic methods used to distinguish this protein from any urinary haemoglobin contain inaccuracies. Azoturia occurs in both race and draught horses on heavy cereal rations, whereas tying-up is more commonly restricted to light horses. It is also seen in horses kept on lush pastures during the week and ridden only at weekends.

Muscle masses rich in glycogen develop hard painful swellings. The muscle enzyme CK attains a peak concentration in the blood 6–12 hours after a single severe episode of muscle damage and returns to normal frequently within a week if no further damage occurs. AST reaches a peak about 24 hours after the episode, but normal values are not returned to for 2–4 weeks. Very high plasma concentrations of CK and raised AST are also observed following strenuous exercise. This may not represent muscle damage and normal values are achieved after 24 hours. Muscle damage during azoturia also causes the release of myoglobin that colours the urine dark red-brown and this in turn can lead to nephrosis and uraemia. The horse sweats profusely, there is a reluctance to move and pulse and respiration rates are rapid. In fact it is vital that the horse is not allowed to move at all as recovery requires that complete rest is immediately instituted. It may remain on its feet or become completely recumbent and severe pain and distress are often accompanied by repeated attempts to rise. The horse should be removed to its stable as quickly as possible in a low-loading trailer, where every effort should be made to keep it standing by slinging or other means, as this may prevent the development of uraemia.

If exercise and movement have been stopped immediately the horse may recover in 2–4 days by treatment with narcotic drugs and corticosteroids administered intravenously to control the swelling and to stimulate energy metabolism, and by intravenous or oral administration of electrolytes in large quantities to maintain a high rate of urine flow at an alkaline pH so that myoglobin precipitation in the renal tubules is prevented. Although acidosis is a normal feature of azoturia this should not be assumed without analysis of plasma acid-base and electrolyte status. Some horses have been found to be alkalotic, when, of course, treatment with sodium bicarbonate would be harmful. Intramuscular injections of 0.5 g thiamin repeated daily seems to be warranted, and the inclusion of pantothenic acid and riboflavin, also involved in oxidative energy metabolism, may be justified. Injections of α-tocopherol have been recommended, although there is no evidence that vitamin E and selenium are involved. If the painful swelling of the muscles is not reduced, pressure on the sciatic nerve can induce a secondary degeneration of other muscles, but the maintenance of proper kidney function is vital if health is to be restored.

The risk of azoturia is reduced if horses are always warmed-up slowly, their concentrate allowance halved at weekends, or at other times when they are not worked, and their work is reinstated gradually with a gradual increase in the consumption of starchy and high-protein feeds. Recent observations suggest that the crisis may entail an abnormality in calcium metabolism.

After hard work it is an advantage to trot or canter slowly as this will stimulate the transport of lactic acid from the muscles to the liver in healthy horses so that muscle and blood pH return to normal more rapidly. Moreover, this light exercise stimulates the flow of oxygen to the muscles accelerating the conversion of lactic acid back to glycogen within the muscles themselves. Accordingly the static changes that accompany tying-up and azoturia are avoided.

Stress tetany (hypocalcaemic tetany)

Extended exertion, particularly in hot weather, leads to dehydration, loss of electrolytes and energy depletion, and the signs of fatigue presented may reflect a combination of these losses even though the diet is quite satisfactory. Calcium losses in the sweat can amount to 350–500 mg of Ca per hour, and continued hyperventilation is associated with alkalosis (discussed in Ch. 9). An acute life threat may be posed by hypocalcaemia in which plasma levels can fall to 1.5 mmol/litre (6 mg/dl), whereupon muscular twitching and cramps are manifested. Occasionally hypomagnesaemia may be present. As normal muscle function, including that of cardiac muscle, requires the concentration of calcium in the blood to remain within strict limits, the most immediate need is for careful intravenous administration of a calcium gluconate solution (Tables 9.2 and 9.3) while heart function is monitored, as an excessive rate of administration can be fatal. Where the clinical signs of hypocalcaemia are clear, a greater risk is entailed in awaiting confirmation by laboratory determination so that immediate veterinary treatment of this kind is indicated.

A fall in plasma calcium and of magnesium also occasionally occurs in lactating mares, although clinical signs are quite rare. Where tetany does occur this is usually precipitated by additional stresses of exertion, transport, weather or disease. The same treatment is instituted. In the dairy cow hypocalcaemia causes paresis rather than tetany.

Electrolyte losses in extended exercise

Several days rest are required for the regeneration of muscle cell glycogen reserves after extended work. The exhaustion of this and loss of cellular potassium in sweat contribute to a sense of fatigue and, as suggested in Chapter 9, recovery during the next few days may be accelerated by providing potassium chloride as well as common salt in the feed (10 g sodium chloride plus 5 g potassium chloride per kg total feed is recommended). Isotonic dehydration depresses thirst so that rehydration is also stimulated by the provision of these electrolytes. Salt licks are supplied in most horseboxes, but many horses will not consume sufficient in this form and they normally contain sodium chloride only. Thus a powdered feed supplement available from feed merchants containing both salts yields a more satisfactory outcome. In fact, during summer weather generally many horses in training for flat races do not consume enough salt from licks and therefore present a higher packed cell volume and plasma viscosity than should be the case. By encouraging water consumption, electrolytes can have the effect of improving performance. Similarly during long endurance events those horses that do not drink fatigue more easily and are less likely to finish. Thus, by satisfying a need for electrolytes at restpoints, a thirst response is induced, water consumption is increased and dehydration is deferred.

Losses of the electrolytes calcium, chloride and potassium throughout long rides may cause 'thumps' or synchronous diaphragmatic flutter during or after the exercise. Losses of these electrolytes and 'thumps' may also be brought on by severe diarrhoea. A decrease in the plasma concentration of calcium, chloride and potassium is thought to change nerve irritability initiating a contraction of the diaphragm muscles in unison with that of the heart beat. This is seen as sudden movements of the horse's flanks. Treatment consists of replacing lost electrolytes, which are always beneficial and never detrimental as long as water is available. The quantities to be administered are given in Chapter 9. The exhausted endurance horse is usually alkalotic, contributing to the hypocalcaemia, so that Ringer's solution, which is slightly acid, is preferred for immediate use. Sodium bicarbonate should not be used in these circumstances unless metabolic acidosis has been demonstrated.

Lameness

In one survey of 314 Thoroughbreds, 53 per cent suffered lameness at some time and in 20 per cent of the cases lameness prevented subsequent racing (Jeffcott, Rossdale et al. 1982). Undoubtedly, the condition represents a considerable embarrassment to the industry and is a problem in horses and ponies of all types. Lameness can be defined as a disturbance of gait, that reduces the weight on the affected limb. Although there are a multitude of causes, one study carried out by the author on horses in the Far East revealed that faulty mineral nutrition, as estimated by phosphate clearance, all too frequently was associated with vertebral fractures, and probably with fractures of other kinds. Unsoundness of joints may result from sprains, strains and jarring forces causing inflammation, which may also result from osteoarthritis. Although this and many bone disorders – for example, splints, ring bone, osselets, bone spavin, curb, capped hocks and thoroughpin – are unlikely to have a major dietary involvement in their causes, the severity of response may well have dietary implications. Abnormalities in mineral, trace-element and vitamin nutrition are frequently associated with various types of lameness but little research work has been undertaken to achieve any objective assessment of the scale of that involvement. Other reasons for lameness include bruised feet, bowed tendons, navicular disease and spinal lesions, which may in part implicate poor hoof care; inspection of hooves daily frequently avoids long-term problems.

Epiphysitis has already been discussed in Chapter 7 and some reference was made to contracted tendons (Pl. 7.1). The latter usually occurs in foals that are doing well with mares having ample milk on good grass. The speed of onset in a foal is surprising in that within 24–48 hours the heel will rise and a slight concavity develops on the front wall of the hoof. Wear at the heel is decreased and increased tension occurs in the extensor tendons. These do not contract, but apparently fail to develop at a rate commensurate with bone growth, and the fetlock joint also tends to enlarge. It is essential to spot the abberation in the early stages so that it can be counteracted by weekly rasping of the heels, exercise, the removal of concentrates and reduction in milk intake. Ultimately, surgical remedy may be the only means of rectifying severe angular deformities of leg joints (Campbell 1977). Exercise is important in the prevention of leg-growth abnormalities (O'Moore 1972), which may be more prevalent where mares and foals spend long periods in their boxes during adverse weather without any cut in feed intake (Owen 1975). It may be preferable to allow mares and foals to remain out all the time in the summer regardless of weather, so long as some form of shelter is available.

Navicular disease

Lameness and damage to the navicular bone of the hoof are caused by thrombosis of the navicular arteries (Østblom, Lund & Melsen 1982). Work at the Equine Research Station in Newmarket has led to the successful treatment of this condition with warfarin (dicoumarol) (Vogel 1984). This drug interrupts blood coagulation by extending prothrombin time. The dose has to be carefully titrated to extend clotting time by 2–4 seconds from the standard of 14 seconds. An excessive dose will lead to bleeding. If colic advertently occurs during treatment, an extended prothrombin time resulting from depressed liver function necessitates a removal of warfarin and the institution of vitamin-K treatment.

It is essential that horses treated with warfarin are fed consistently, particularly in respect of the amount of green feed, which is rich in vitamin K, and as the level of work affects clotting time any change in activity must be imposed gradually. If practical, regular work should be instituted and blood samples taken at intervals of at least one per month during rest periods, but immediately after exercise. Sampling should also occur 5–7 days after any change in work, or feed, routine. Veterinary treatment with warfarin must always err on the side of caution as excess can be fatal. The drug seems to act by reducing the viscosity and increasing the flow of blood, which improves the nutrition of the navicular bone; this is probably more relevant than the prevention of thrombus formation. More recently some other treatments have been successfully deployed in an endeavour to avoid the risk of haemorrhage encountered with warfarin. These include the drug isoxsuprine hydrochloride (Rose, Allen et al. 1983) and the egg-bar shoe (Østblom, Lund & Melsen 1984) that has indicated the reversibility of the primary disease.

Malabsorption of fat-soluble vitamins

Occasionally the efficiency of absorption of vitamins A, D, E and K is reduced and the most frequent cause may be an interruption to biliary flow by obstruction of the bile duct. The immediate effect is a failure in the blood-clotting mechanism but this can be overcome by injections of vitamin K. Vitamin K administration is also successful in counteracting the bleeding syndrome of dicoumarol poisoning. Plant sources of this poison are referred to in Chapter 10.

Chronic obstructive pulmonary disease (COPD)

Coughing is a sign, and the art of its long-lasting treatment is first to determine the cause rather than to alleviate the symptoms. The cause must be ascertained, not only at the level of the horse but also in the context of its environment. Respiratory diseases directly impair the usefulness of a horse and they vary in intensity among individuals and between environments from slightly incapacitating coughs to very acute illnesses. One such disease entity is COPD, which affects individual horses with varying degrees of severity and if left untreated increases with time. Slight illnesses can incapacitate competition animals. One investigation conducted with horses diagnosed as having the condition, but capable of being ridden at moderate work rates, demonstrated that at the same velocity as healthy horses, their heart rates on average were 20 beats per minute faster (Littlejohn, Kruger & Bowles 1977).

Horses suffering from COPD rarely have a purulent discharge from the nostrils, but if exposed to sensitizing substances such as mouldy hay they cough more frequently and their respiratory rate increases. If the source of aggravation in their environment is not removed, the condition becomes progressively worse and leads to degeneration of parts of the lung. Later, breathing is characterized by a 'double lift', seen in the abdomen on expiration, when the normal abdominal contraction is followed by a second lift as the horse endeavours to expel more air. The disorder is known as 'broken wind' or 'heaves' in which parts of the lung tissue lose their elasticity. Individuals between 6 and 10 years old are most frequently affected, but the problem is more common among ponies than among Thoroughbreds owing apparently to a more frequent exposure of ponies to poor-quality fodder and bedding in poor housing.

COPD is not as irreversible as originally thought – the major lesion in the lungs being bronchiolitis with emphysema only occurring in small areas. Recently, the prophylactic use of sodium cromoglycate has been found effective in sensitive individuals. However, prevention is better and less expensive than a partial cure.

Research during the last decade (Lawson, McPherson et al. 1979; McPherson, Lawson et al. 1979a) has shown that the major environmental agents associated with the condition are two moulds (*Micropolyspora faeni* and *Aspergillus fumigatus*) that cause respiratory hypersensitivity. Some horses are sensitive to other organisms and even to ryegrass pollen (McPherson, Lawson et al. 1979a). When particles of these moulds are inhaled they more frequently cause coughing in horses suffering from COPD. Furthermore, precipitins to these mould antigens are found more frequently in sera of COPD horses than in those not suffering from the ailment. The aetiology is further complicated by a number of horses, without precipitins in the blood, responding to inhalation challenge with the antigen and some horses that did not suffer from COPD had serum precipitins to the causal organisms (Lawson, McPherson et al. 1979). Some researchers therefore hold the view that COPD horses may demonstrate precipitins to these fungi as a *consequence* of impairment to pulmonary function.

COPD occurs in individual horses and is not infectious but there are other respiratory diseases that spread from horse to horse and cause impairment and damage to pulmonary function. The principal transmissible diseases are caused by bacteria, parasites and viruses. Viral infections seem to be of increasing incidence among horses and ponies and include equine influenza, equine herpes viruses I and II, rhinoviruses I and II and adenovirus, particularly in foals. It is thought that the widespread occurrence of viral infections may be associated with the trend to have larger numbers of horses in close proximity in totally enclosed buildings, greater national and international traffic in horses and, it must be said, a greater and increasing awareness and understanding of viruses. Lung damage caused by these infections may leave horses more subject to the allergic responses of COPD; certainly one aggravates the other and both are influenced by building design.

It has been clearly demonstrated that poor ventilation increases the chances of horses becoming affected with COPD (McPherson, Lawson et al. 1979b) and undoubtedly it aggravates the spread of secondary infections if not of the primary viral agents of transmissible diseases. Where horses are suffering from COPD, the major source of the two mould antigens is hay and straw bedding (McPherson, Lawson et al. 1979b). All hay contains moulds, but some is visibly mouldy or musty and therefore presents a greater hazard than hard, stemmy, clean and shiny hay. Thus there is reasoned justification for the use of such safe, but nutritionally poor, hay for horses

– the lack of nutrients and energy being made up with a high-quality compounded nut, or other concentrate.

Where horses are affected they should be turned out to grass. If this is impracticable, hay should be soaked, rather than merely dampened prior to feeding, or it should be replaced by silage, haylage or high-fibre compounded nuts. Straw bedding should be replaced by peat, shredded paper or by softwood shavings and the ventilation of the box should be improved. Recently, scientists at the University of Edinburgh (Thomson & McPherson 1984) concluded that the pathophysiological changes occurring in equine COPD are reversible. They were led to this conclusion by observing symptomatic horses to become asymptomatic in 4–24 days (mean 8.4 ± 4.8 days) when these horses were bedded on shredded paper and routinely fed a complete cubed diet. The pulmonary function values of asymptomatic horses did not differ significantly from those of normal horses. The ingestion of mould spores rather than their inhalation, as would occur with the feeding of soaked hay or nuts, is *not* a cause of the problem as it depends on a direct reaction between the inhaled particle and the surface lining of the lung alveoli. Meal feeding should also be avoided as some evidence indicates that allergy may be caused by oat dust and some other sources of feed dust.

Ventilation in horse buildings

Horse buildings should be designed so that the atmosphere carries a minimal number of dust particles and transmissible disease agents. Low temperature and high humidity favour the viability and inhalation of pathogens and so are the worst housing conditions. Unfortunately few stables are properly insulated and any restriction of ventilation to maintain a reasonable temperature can cause condensation of moisture. Natural systems of air flow should have a controllable inlet area of up to 0.3 m^2 (1 ft^2) per horse (Sainsbury 1981). Horses kept in close quarters with one another require substantially more ventilation than do individuals or those separated into small groups because of cross-infections and the build up of pathogens and harmful irritants, including pollen, fungal hyphae and spores. If building design impedes natural air flow, fan assistance should be provided at the base of the outlet chimney. This will function as an aid to natural ventilation and is a more desirable solution than the installation of a pressurized system for which the costs are greater and the numbers of suspended air particles increased by greater air turbulence. There should, for similar reasons, be no recirculation of air. Abrupt changes in air flow as a result of 'on-off' regulators are to be avoided. Bottom-hinged air inlets are recommended as shown in Figs 11.2 and 11.3, which deflect cold air up in cold weather and may be fully open in hot for cross-draughts. Their height above the floor should be such as not to interfere with the

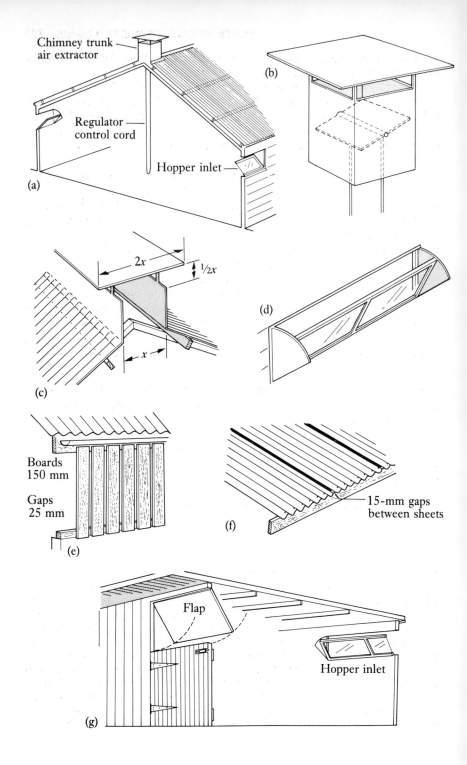

(a) Chimney trunk air extractor
Regulator control cord
Hopper inlet

(b)

(c) $2x$ $\frac{1}{2}x$ x

(d)

(e) Boards 150 mm
Gaps 25 mm

(f) 15-mm gaps between sheets

(g) Flap
Hopper inlet

stock and they must, of course, be of safe construction. Additional inlets will probably be necessary for very hot weather.

Where horses are grouped in covered yards a copious air flow is essential at all times. This will be facilitated by a wide open ridge, 0.3–0.6 m wide with a covering flap. Space boards at the top half of the wall are successful (150 mm wide boarding with 20 mm gaps, Fig. 11.2), or on exposed sites, narrow sliding boards in which one set slides over the other.

Probably the most successful housing in the United Kingdom consists of various forms of monopitch lean-to buildings (Figs 11.3 and 11.4), which contain single open-fronted boxes facing the warmest wind and the greatest sunlight. These buildings have a low back and high front with hopper-flap air inlets such that the air flows from back to front. An extension to the roof at the front gives cover and cross-partitions act as load bearers. These divisions should be solid as the partly open ones allow crossflow of air, the incidence of wall kicking increases and where the division contains iron bars, even at quite high levels, shod horses can become ensnared.

In summary, ventilation systems should have a copious and continuous air flow, affected to a minimal extent by the direction and intensity of the wind. The simplest systems involve the exhaustion from the apex of rising warm and stale air so that excess moisture, dust particles and disease

Table 11.5 Range of climatic environmental requirements of stabled horses

Ambient temperature	0–30°C
Relative humidity	30–70%
Air movement	0.15–0.5 m/s
Ventilation rate	0.2–2.0 m³/h per kg bodyweight
Outlet ventilation area	0.1 m² per horse
Inlet ventilation area	0.3 m² per horse

(From Sainsbury 1981)

Fig. 11.2 Arrangements for the natural ventilation of stables. (a) Ventilation suitable for all stables using extractor chimney trunks and hopper inlets for fresh air. (b) Detail of the extractor chimney trunk, which may have a regulator or electric fan placed in the base. (c) Simple open ridge suitable for extraction ventilation of covered yards; *x* is normally about 300 mm in yards up to a width of 13 m and 600 mm in yards over 13 m and up to 25 m wide. (d) Hopper window suitable as a fresh-air inlet. (e) Spaced boards, giving draught-free ventilation: normally 25-mm gaps between 150-mm wide boarding. (f) 'Breathing roof' – corrugated roof sheets fixed with 15-mm gaps between for extractor ventilation. (g) Mono-pitch house showing hopper flap at back and ventilating flap at the front. Note overhang on roof to protect the horses from rain, sun and wind. (From Sainsbury 1981).

(a)

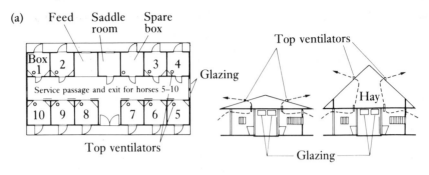

(b)

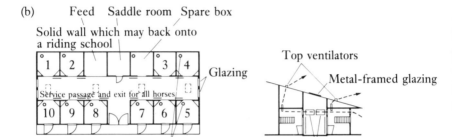

(c)

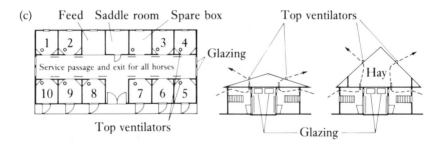

(d)

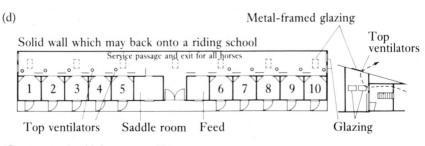

(Caption to fig. 11.3 on page 278).

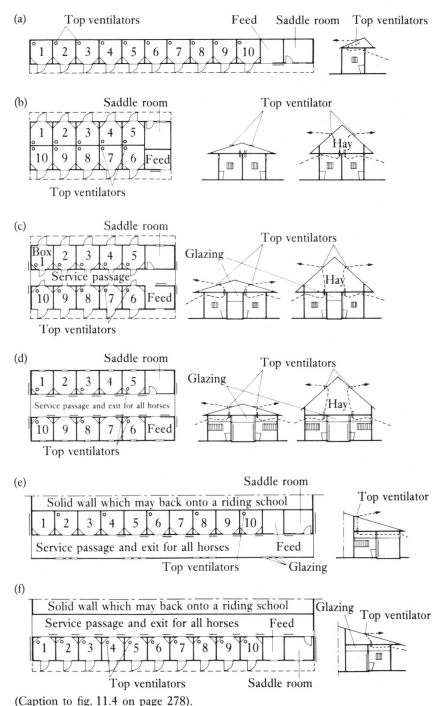

(Caption to fig. 11.4 on page 278).

Fig. 11.3 Four layouts of ten horse boxes, which combine the advantages of indoor and outdoor boxes without draughts. (a) All the boxes open to the outside: boxes 1–4 have large exterior service doors; boxes 5–10 have interior sliding doors and shutters opening to the outside. (b) Stable with a solid back wall and interior service; boxes 5–10 have shutters opening to the outside. (c) Stable with a side exposed to bad weather and with interior service; boxes 5–10 have shutters opening to the outside. (d) Stable with a solid back wall and with interior service; all boxes have shutters opening to the outside. (From Ministère de l'Agriculture, Service des Haras et de l'Equitation, Institut du Cheval, Le Lion d'Angèrs (1980) *Amènagement et Equipment des Centres Equestres, Section Technique des Equipment Hippiques*, Fiche N° CE.E.5).

Fig. 11.4 Simple layouts for ten horse boxes. (a) A row of outside boxes with external services. (b) Back-to-back boxes with external services. (c) Boxes with external horse access and internal service; solid doors and walls in the passage keep draughts to the minimum. (d) Boxes with internal (central) servicing without access for the horses' heads to the outside (a less satisfactory arrangement). (e) Boxes with internal (lateral) servicing, without access for the horses' heads to the outside (a less satisfactory arrangement). (f) Boxes with internal (lateral) servicing and with access for the horses' heads to the outside; solid doors and walls onto the passage keep draughts to the minimum. (From Ministère de l'Agriculture, Service des Haras et de l'Equitation, Institute du Cheval, Le Lion d'Angèrs (1980) Amènagement *et Equipment des Centres Equestres, Section Technique des Equipments Hippiques*, Fiche N° CE.E.4).

agents produced by the horse, its feed and bedding are eliminated. Proper air flow is encouraged by good insulation. Individual horses should be in sight of each other without being contaminated by a neighbour's air. Table 11.5 summarizes the climatic requirements of stabled horses.

Further reading

Bogan J A, Lees P & Yoxall A T (1983) eds *Pharmacological basis of large animal medicine*. Blackwell Scientific: Oxford.

Ministère de l'Agriculture (1980) *Aménagement et équipment des centres équestres Section technique des équipments hippiques*. Le Lion-d'Angèrs.

Lewis L D (1982) *Feeding and care of the horse*. Lea & Febiger: Philadelphia.

Rossdale P D & Ricketts S W (1980) *Equine stud farm medicine*. Cassell (Baillière Tindall): London.

CHAPTER 12

Laboratory methods for assessing nutritional status

If the urine of a horse be somewhat high coloured, bright and cleare like lamber,
and not like amber, or like a cup of strong March beere, then it sheweth the horse
hath inflammation in his blood . . . Now for the smell of his dung, you must
understand, that the more provender you give, the greater will be the smel, and the
lesse provender the lesse the smell.

Markhams Maister-Peece, 1636

Many methods of nutritional assessment are only appropriate for large
stables where normal variation in values is overcome by determinations
carried out in a number of horses fed in a similar manner. Nonetheless,
whether one is dealing with individuals or groups, it is essential to establish
normal values for the individuals or groups concerned. The normal value
of a particular parameter in an individual may differ from the population
means, as age, sex, time of year, system of management, stage of training,
diet and breed all influence the norm. Thus, by establishing some routine
practice it is possible to assess quickly whether a particular value has
shifted from its norm, which enables one to uncover a disturbance in its
early stages. Furthermore, a range of tests is undoubtedly necessary truly
to understand a situation.

Metabolic profiles have been adopted as measures of the nutritional
and physiological health of dairy herds and some progress has been made
towards similar techniques in stables. Nevertheless, in assessing nutritional
status, it is essential to have a full knowledge of the ingredient and chemical
composition of all the feeds and the actual weights of each type fed. This
will facilitate an objective diet evaluation, which, with the monitoring of
management and disease, will indicate the most appropriate laboratory measurements to be undertaken. Such a procedure will save unnecessary
expenditure and will yield quick, direct and appropriate results.

Many metabolic tests are now available for gauging nutritional status
in the horse and the following are but a summary of some of the more
commonly used measures. It should, however, be emphasized that single

measurements, let alone single determinations, are practically valueless. A salient reason for this is that few, if any, of these methods are specific in determining the status of a particular nutrient.

Metabolic tests

1. Serum albumin. Blood albumin is a reserve protein and a decline in serum concentration may indicate trauma, severe liver disease, protein-losing enteropathy (loss of protein into the gut), parasitic-worm infestation, inadequate dietary protein and possibly excessive starch intake in the presence of adequate protein. Insufficient dietary protein is more likely to depress blood haemoglobin than are low amounts of dietary iron where normal dietary ranges are concerned.

2. Plasma concentrations of nutrients and their products are used to assess the status of those nutrients. These include magnesium, potassium (but see p. 262), phosphorus, zinc, selenium, copper and T_3 (a thyroxine metabolite reflecting iodine status). Whereas a deficiency of selenium is characterized by a depression in serum concentration of the element, the situation with zinc is unclear. In growing children, dietary deprivation of zinc causes a depression in the serum concentration of the element only where there is sufficient dietary protein to promote a normal growth rate. Where other deficiencies are at hand some success has been achieved in detecting zinc inadequacy by measuring the more labile zinc content of leucocytes. In other species, at least the activity of serum alkaline phosphatase is sensitive to zinc status, as zinc is an important cofactor. The relationship in the horse has not been studied in detail, but as the activity per se is not specific for zinc status, the activity coefficient (i.e. the activity ratio with and without added zinc) may be a more appropriate response parameter (see page 281). Serum ferritin provides a good index of hepatic and splenic iron and can be used to evaluate iron storage in horses. Among the fat-soluble vitamins the measurement of plasma α-tocopherol in relation to plasma lipid and plasma retinol (vitamin A) concentrations are informative. However, the concentration of vitamin A in the liver varies more closely with the dietary intake of vitamin A. The homeostatic control of many blood components and the interaction of many nutrients imply that variations in blood concentrations require careful interpretation. In so far as vitamin K status is concerned, the determination of blood prothrombin time, discussed in Chapter 11, can be used. Among the B vitamins, quick and routine methods are now available for assessing whole blood vitamin B_{12} (cyanocobalamin) and serum folic acid, which clearly reflect diet, in the experience of the author. Among other B vitamins, methods have been successfully developed for assessing thiamin, riboflavin, pyridoxine and biotin adequacy through measuring the activity coefficient of the appropriate enzymes.

Enzymes that the author has found convenient are, respectively, erythrocyte transketolase, glutathione reductase, several transaminases and several carboxylases. The principle of these methods is that the B vitamin, or a metabolite, acts as a coenzyme essential in the functioning of the enzyme. When the enzyme is not fully saturated with its cofactor, then the addition *in vitro* of the appropriate B vitamin or its metabolite increases the activity of the enzyme *in vitro*. A large difference between the activities with and without the addition indicates that the nutritional status in respect of that vitamin is low. Recently, the activity *per se* (not the coefficient) of transaminase in red blood cells of sow pigs has been proposed as a good indicator of pyridoxine status. The serum activity of AAT and AST, routinely measured in horses is greatly influenced by the extent of leakage from liver and muscle cells and therefore the activity of horse red cells may reflect pyridoxine status.

3. A deficiency of vitamin E and/or selenium can cause muscle and liver damage that non-specifically leads to an increase in the activity of certain plasma enzymes, particularly AST and CK, owing to leakage from the tissue cells concerned. More specific measures of adequacy are, however, the determination of red-cell fragility under the influence of hydrogen peroxide for vitamin E and the activity of glutathione peroxidase ($GSH-P_x$) in respect of selenium.

4. Blood-plasma concentrations of inorganic phosphate reflect the dietary intake of phosphorus and in the deficient state there is also an increase in the activity of plasma alkaline phosphatase. Plasma measurements of inorganic phosphate and calcium are necessary, together with the urinary measurement of phosphorus, in the clearance test for calcium adequacy to be discussed below. Creatinine clearance tests are also used in the assessment of sodium, potassium and chloride adequacy.

5. Alkaline phosphatase is produced by several tissues. During ischaemic colic of the bowel, intestinal alkaline phosphatase is released into body fluids. A measure of this aids both diagnosis and a decision over immediate action in acute cases. Intestinal alkaline phosphatase is not bound by 1-phenylalanine, whereas other sources are, thus providing an easy means of differentiation and detection of ischaemic colic when this source is shown to be elevated in peritoneal fluid (Davies, Gerring et al. 1984).

6. The chemical analysis of hair samples has been suggested as a means of measuring intakes of proteins, minerals, trace elements and toxic heavy metals. The author has found increased amounts of lead in hair in lead toxicosis. However, the mineral composition is influenced by hair colour (Hintz 1980b) and the intake of both minerals and trace elements may be more reliably quantified by other means. The bulb diameter of hair among horses under range conditions has been used successfully in the assessment of protein consumption (Godbee, Slade & Lawrence 1979).

7. Glucose and xylose absorption tests have been used to measure the gross function of the small intestine. The recommended procedure is to give 0.5 g D(+)-xylose per kg bodyweight as a 10 per cent solution by stomach tube. By measuring the peak plasma-xylose concentration after 90 minutes it is possible to discriminate between normal and abnormal absorption. A normal control animal should also be measured under the same conditions.

8. Feed allergen tests. These have been used successfully by the author in horses for determining whether feed entities are causal in certain oedematous skin and respiratory irregularities. The horse seems to be prone to such reactions. Blood and skin tests are appropriate in the detection of dietary and mould allergens.

9. Haematology. The measurement of numbers of red and white blood cells has been used in folic acid, vitamin B_{12} and protein assessments. For consistency, samples should be drawn immediately after strenuous exercise; lower erythrocyte parameters should be expected for ponies and cold-blooded horses. Normal values are also influenced by sex, age, stable and season, and, among racehorses, by training level.

10. The measurement of $[K^+]$ in red blood cells should be undertaken within 2 hours of removal of the blood from the horse. The metabolic energy required for the active transport of K^+ is not generated in blood samples stored in the refrigerator and consequently K^+ leaks by diffusion from the cells and Na^+ penetrates their membranes. This movement can be reversed by incubating the blood with glucose, or in fact by reinjecting the cells into the circulation.

Creatinine clearance tests

Electrolyte status is a function of intestinal absorption, renal tubule reabsorption, tissue deposition, sweat loss and renal excretion. The kidney is the principle organ modulating the supply of electrolytes to the horse. Thus the collection of urinary losses over a period of several days would indicate the quantities surplus to requirement. Such a urinary collection is in practice problematic, and the measurement of urinary concentration of electrolytes in samples from a single collection may merely reflect water intake that can vary enormously between individuals. However, by relating the concentration of the electrolyte to the concentration of creatinine, which is a relatively constant function of the muscle mass of a healthy horse, the urinary losses can be estimated. The creatinine clearance test is used in assessing sodium, potassium, chloride and calcium adequacy.

A single urine sample and a serum sample collected at the same time, are required. This should be achieved without resort to the use of diuretics because they affect sodium and, to a lesser extent, potassium and chloride losses. The clearance of an electrolyte equals the concentration in the urine times the urine volume divided by the concentration in the serum. The clearance ratio is the clearance of the electrolyte divided by the clearance of creatinine (Table 12.1). In this ratio, the urine volume cancels out and so need not be measured. The equation for which the determinants are required is given below:

$$\text{Per cent electrolyte clearance (\% Cr)} = \frac{[E]_u}{[E]_s} \times \frac{[Cr]_s}{[Cr]_u} \times 100$$

Where $[E]_u$ is the concentration of electrolyte in urine, $[E]_s$ is the concentration of electrolyte in serum, $[Cr]_s$ is the concentration of creatinine in serum and $[Cr]_u$ is the concentration of creatinine in urine.

Table 12.1 Normal creatinine clearance values of some electrolytes in horses

Na	0.02–1.0%
K	15–65%
Cl	0.04–1.6%
PO$_4$	0–0.5%

High-concentrate rations tend to give a raised phosphate clearance and a depressed potassium clearance – the reverse being the case for high-roughage rations. These effects are quite normal. Potassium clearance, together with measurements of blood and urinary pH, is useful in evaluating the type of acidosis and in assessing potassium depletion in exhausted horses, as the urinary excretion of potassium and hydrogen ions tends to display a reciprocal relationship. Potassium clearance is depressed in chronic laminitis. A raised sodium clearance may indicate excessive sodium intake in the form of common salt, Addison's disease, dehydration or tubular malfunction.

Epiphysitis in young growing stock and chronic non-specific lamenses in adults are frequently expressions of secondary nutritional hyperparathyroidism, by far the most common cause of which is faulty calcium and phosphorus nutrition and an improper balance between these two minerals. Thus a simple means of evaluating calcium adequacy is required. Calcium concentrations in serum vary to a small degree in relation to calcium intake and where they are measured in a number of horses, significant differences can be detected between deficient and normal groups (Fig. 3.2). However, the method is insufficiently sensitive to be of practical use. Greater sensitivity is achieved with creatinine clearance ratios. The horse seems to regulate serum calcium more by renal excretion than by controlling intestinal absorption. Intact proximal tubules are required for phosphorus reabsorption and calcium excretion and in renal failure serum calcium is increased and phosphorus is depressed. The fully functioning kidney, on the other hand, excretes excess calcium in the urine and therefore an inadequate intake might be revealed by a reduced urinary excretion, if it were not for the fact that calcium sediments in horse urine owing to its alkaline nature. Repeatable estimates of the calcium content of urine cannot therefore be readily attained.

Phosphate clearance may fall to zero, where the diet is marginally adequate in phosphorus and the clearance will be increased when the intake is greatly in excess of need, or when the horse is suffering from secondary nutritional hyperparathyroidism. Thus a means of assessing calcium adequacy without measuring urinary calcium is at hand. An inadequate dietary intake of calcium stimulates the secretion of parathormone, which increases the reabsorption of calcium by the renal tubules, mobilizes bone calcium, and increases the loss of phosphate through the tubules by decreasing tubular reabsorption of phosphate. The net effect of this is to stabilize the ionized concentration of serum calcium and to depress serum phosphate. Concentrations in serum of calcium tend to be below average, but still within the normal range. An increased phosphate clearance with a normal or slightly depressed serum calcium is therefore frequently indicative of insufficient dietary calcium when other causes have been eliminated from consideration.

A range of simple laboratory techniques appropriate to the assessment of many aspects of nutritional status has been discussed. It should, however, be re-emphasized that they may be deployed with assurance only when there has been a rigorous and quantitative inspection of the diet used over, at least, the previous 2 months.

Conclusion

The feeding of horses and ponies during the second half of this century has, if anything, become less enlightened rather than more so. The productivity and reproductivity of horses and ponies have not generally improved during an era in which the economies of keeping other domestic stock have been enhanced and made more 'cost-effective' out of all recognition compared with earlier achievements. The stabling and management of horses and ponies are expensive and therefore ripe for 'streamlining'. What is required is a wider appreciation of the benefits of observing, measuring and recording important activities and changes, a wider acceptance and adaptation of information accumulated on other domestic species and their feeds, and intelligent application of recent knowledge acquired in equine studies worldwide.

A feasible and challenging goal for a newborn foal should be the achievement of a healthy working life of 20 years. Although this entails a degree of luck, the probability of its realization requires both the diligent application of acquired experience and the intelligent use of an increasing fund of scientific evidence. It is trusted that each chapter of this book has contributed in a distilled form some of that evidence for its wider adoption.

Appendix A

Calculation of the dietary composition (of 88% dry matter) required for a mare of 400 kg bodyweight in the fourth month of lactation

Example

1. To calculate the proportions of hay and concentrate in the total diet, divide the required daily energy intake (MJ/day), by the desirable daily total feed intake (88% dry matter) giving the average energy density of the total diet. Then form a simple equation containing the energy densities of the roughage and of the cereal available (88% dry matter).

Required daily energy	84.5 MJ of DE per day
Desirable total feed intake	9.0 kg
Oats	12.1 MJ of DE per kg
Hay	7.3 MJ of DE per kg

Let x be the proportion of cereal in the diet and $1 - x$ be the proportion of hay.

$$12.1x + 7.3 (1 - x) = 84.5/9.0$$
$$12.1x + 7.3 - 7.3x = 9.39 \text{ MJ of DE per kg}$$
$$4.8x = (9.38 - 7.3) \text{ MJ of DE per kg}$$
$$x = 2.09/4.8 = 0.435 \text{ or } 43.5\% \text{ oats (435 g/kg)}$$

If 43.5% of the diet is oats then 56.5% is hay.

2. Now calculate the protein, calcium (Ca) and phosphorus (P) contents of this simple mix from the information given in Appendix C (p. 295).

Oats (12.1 MJ of DE) contains (per kg):

95 g   crude protein
0.8 g   Ca
3.3 g   P

Hay (7.3 MJ of DE) contains (per kg):

55 g crude protein
3.5 g Ca
1.7 g P

Initial dietary composition (g)

	Crude protein	Ca	P
Contributed by oats:			
$\dfrac{43.5}{100}$ × each oat value above	41.3	0.35	1.44
Contributed by hay:			
$\dfrac{56.5}{100}$ × each hay value above	31.1	1.98	0.96
Total initial composition (g/kg)	72.4	2.33	2.40
Requirement (g/kg)	120.0	5.00	3.00
Deficit of minerals (g/kg)		*2.67*	*0.60*

Now calculate how much soyabean meal is required to make good the protein deficit (other protein concentrates could be used in amounts that are inversely proportional to the lysine contents of the protein source under consideration and that of soya). About 2.0–3.0 g more protein than is required should be allowed for because minerals will also be added to the diet. Soyabean meal from Appendix C contains 440 g of crude protein per kg. As before:

$72.4x + 440 (1 - x) = 120.0 + 3.0$ g of crude protein per kg
$72.4x + 440 - 440x = 123$ g/kg
$-367.6x = -317$ g/kg

Change signs on both sides. $x = 0.862$ or 86.2% oats plus hay, and therefore $(100 - 86.2)$ soya forms the remainder, i.e. 13.8% soyabean meal.

The soya contains more energy than oats and hay but this will be approximately compensated for by the complete absence of energy in the mineral and vitamin supplement.

3. Now 0.6 g of P is still required per kg, being the deficit shown above. Dicalcium phosphate contains 188 g of P per 1000 g from Appendix C. Therefore an addition of:

$\dfrac{1000}{188}$ × 0.6 g = 3.19 g of dicalcium phosphate ($CaHPO_4$) per kg total feed will provide the necessary P.

Now dicalcium phosphate from Appendix C also contains 237 g of Ca per 1000 g. Therefore 3.19 g provides

$\dfrac{237}{1000}$ × 3.19 g = 0.76 g Ca

The original deficit was 2.67 g of C and it is now

$$2.67 - 0.76 \text{ g} = 1.91 \text{ g Ca}$$

This can be made up with limestone flour ($CaCO_3$) containing 360 g Ca per 1000 g. Therefore

$$\frac{1000}{360} \times 1.91 = 5.31 \text{ g } CaCO_3 \text{ per kg total diet}$$

4. The diet should also contain 5 g of common salt (NaCl) per kg and a proprietary vitamin/trace-mineral mixture suitable for horses.

5. The complete diet is now as follows:

	g/kg	or	Per cent
Oats	283		28.3 (43.5–13.8–0.32–0.53–0.5)
Soyabean meal	138		13.8
Dicalcium phosphate	3.2		0.32
Calcium carbonate	5.3		0.53
Salt	5		0.5
Vitamins/trace minerals	+		+
Hay	565		56.5
Total			100.

The composition of the concentrate portion of the ration is as follows:

		Per cent
Oats	$\frac{28.3}{43.5} \times 100$	65.1
Soyabean meal	$\frac{13.8}{43.5} \times 100$	31.7
Dicalcium phosphate	$\frac{0.32}{43.5} \times 100$	0.74
Calcium carbonate	$\frac{0.53}{43.5} \times 100$	1.22
Salt	$\frac{0.5}{43.5} \times 100$	1.15
Vitamins/trace minerals		0.1
Total		100.0

6. The total daily ration was to be 9.0 kg and of this the above concentrate would form $43.5/100 \times 9.0 = 3.9$ kg daily. The remaining 5.1 kg of hay could be given in excess as some will be wasted and horses would naturally consume their ration of concentrate before filling up on hay unless they are under racing stress.

The concentrate ration should be divided into a minimum of two feeds per day and introduced gradually in increasing amounts until the full ration is provided. Small adjustments of the quantity can be made to allow for differences in condition between individuals. Where growing horses are being fed, particularly of faster growing breeds, the overriding concern must be the condition of the legs and if there is any tendency towards contracted flexor tendons, epiphysitis or crooked legs the concentrate allowance should be reduced until the condition subsides (See Ch. 8 for details).

Appendix B

Dietary errors of common occurrence in studs (a) and training stables (b)

(a) The range of dietary composition for foals and yearlings in ten studs where home mixes have been prepared and when restricted access to pasture of moderate quality is provided (% of total diet, air-dry basis; dashes imply none used)

	Dietary range		Typical poor-quality diets at specific studs			
	Weaned foals	Yearlings	Foals (1)	Foals (2)	Yearlings (1)	Yearlings (2)
Oats	37–70	11–58	48	—	57	9.5
Boiled barley	0–1.6	0–26	1.6	—	—	20
Flaked maize	0	0–6	—	—	—	2.5
Bran (wheat)	0–12	5–28	5	—	28	16
Coarse mix (sweetfeed) of low quality	0–43	0–43	—	42	—	—
Barley chaff	0–7	0–7	—	7	—	—
Cubes of moderate quality	0–12	0–60	12	—	—	23
Soyabean meal	0–4	0	—	—	—	—
Linseed (boiled)	0–7	0–9	1.6	—	5	7
Palm kernel cake	0	0–3	—	—	—	—
Milk pellets	0–13	0	—	—	—	—
Locust beans	0	0–3	—	—	—	—
Egg	0	0–+	—	—	—	—
Molasses	0	0–2	—	—	—	—
Carrots	0	0–3	—	—	—	—
Grass hay	6–18	14–29	16	18	10	11
Lucerne hay	0–9.6	0	—	9.5	—	—

	Dietary range		Typical poor-quality diets at specific studs			
	Weaned foals	Yearlings	Foals (1)	Foals (2)	Yearlings (1)	Yearlings (2)
Limestone flour	0–2.2	0–2.2	—	2.2	—	—
Dicalcium phosphate	0–0.5	0	—	—	—	—
Vitamins, trace elements and minerals	+	0–+	—	—	—	—
Pasture	12–36	14–40	16	22	—	11
Some chemical characteristics						
Crude protein (%)	10–16	10–17	11.1	10.0	11.8	11.7
Total lysine (%)	0.4–0.6	0.4–0.7	0.48	0.40	0.45	0.45
Calcium (%)	0.32–0.9	0.22–1.2	0.33	0.65	0.23	0.44
Phosphorus (%)	0.2–0.46	0.2–0.65	0.42	0.19	0.65	0.57

General comments

— Widespread use of poor-quality hay for young stock
— Failure to rectify this by adjustment of concentrate composition
— Insufficient checks on rates of growth
— Failure to compensate for inadequacies of pastures

Common errors

(i) Very variable protein and lysine intakes exacerbated by variable pasture quality and availability
(ii) Suboptimum protein: energy ratios for weaned foals
(iii) Extreme variation in calcium intake
(iv) Suboptimum Ca:P ratios
(v) Excessive intakes of vitamins A and D but possible inadequacies of several other vitamins
(vi) Lack of control of growth curve leading to poor conformation, epiphysitis and abnormal alignment of legs
(vii) Use of poorly formulated micronutrient supplements
(viii) Use of more than one micronutrient supplement lacking complementary effects
(ix) Pasture trace-element and other deficiencies, which contribute to metabolic and conformational problems

Faults in diets specified above

Foals	(1)	Insufficient protein, lysine and calcium Excessive use of vitamins A and D_3 Deficiency of selenium and marginal zinc status
	(2)	Insufficient protein and lysine Excessively wide Ca:P ratio Selenium depletion, zinc inadequacy and suspected induced manganese deficiency Poor conformation
Yearlings	(1)	Marginal protein and lysine inadequacy Rank calcium deficiency with adverse Ca:P ratio Epiphysitis evident Several vitamin inadequacies and suspect trace-element status
	(2)	Lack of objectivity in ration formulation and unnecessary complexity Adverse Ca:P ratio

(b)   The range of dietary composition for Thoroughbreds in eight racing stables where home mixes are prepared (% of total diet, air-dry basis; dashes imply none used, or analytical value unavailable)

	Dietary range	Typical poor-quality diets at specific stables	
		(1)	(2)
Oats	32–59	49	51
Bran (wheat)	0–16	—	1.1
Coarse mix (sweetfeed)	0–20	—	—
Chaff	0–3.7	—	—
Cubes	0–24	—	2.0
Soyabean meal	0–2.2	0.8	0.7
Linseed (boiled)	0–1.5	0.8	1.5
Molassed peat	0–0.5	—	0.4
Carrots	0–2	—	—
Grass pellets	0–6	—	—
Molasses	0–1.5	—	1.4
Grass hay	28–53	49	40.6
Limestone flour	0–1.1	—	1.1

	Dietary range	Typical poor-quality diets at specific stables	
		(1)	(2)
Salt (sodium chloride)	0–0.1	—	0.07
Corn oil	0–0.57	0.11	—
Vitamins, trace elements and minerals	0–4	0.2	0.1

Some chemical characteristics

Crude protein (%)	7.2–11.5	7.3	7.4
Total lysine (%)	0.35–0.5	0.36	0.40
Ca (%)	0.15–1.38	0.15	0.68
P (%)	0.24–0.43	0.25	0.28
K (%)	—	1.5	—
Na (%)	—	0.16	—
Mg (%)	—	0.18	—
Zn (mg/kg)	—	—	24
Mn (mg/kg)	—	—	46

Common errors

  (i) Variable protein and lysine intakes
  (ii) Suboptimum protein: energy ratios
 (iii) Extreme variability in calcium intake
 (iv) Suboptimum Ca:P ratios
  (v) Excesses of vitamins A and D_3
 (vi) Inadequate allowances of folic acid and possibly of other water-soluble vitamins and of salt
(vii) Incorrect quantities of vitamins, trace elements and minerals provided by most supplements

Faults in diets specified above

Stable (1)   Lameness, metabolic upsets (e.g. azoturia, setfast)
              Dietary faults include excesses of vitamins A and D_3 and excess iodine
              Insufficient Ca and adverse Ca:P ratio
              Insufficient sodium and folic acid, marginal protein and probably too wide an energy-to-protein ratio

Stable (2)   Abnormal blood characteristics
              Marginally low protein intake
              Insufficient allowances of zinc, folic acid and salt

General comments

— Oats and poor-quality grass hay frequently constitute over 90% of the diet and their composition is variable and unknown

— Several non-complementary micronutrient supplements are frequently used in the same diet
— Frequently insufficient common salt is consumed in hot weather
— Frequency of feeding is often insufficient
— Rate of ration change and of energy intake frequently inappropriate
— Notion that a rest on Sundays with changes in management and in feeding benefits the horse is misguided, in contrast to human benefits

Appendix C

Chemical composition of feedstuffs used for horses

Composition of feedstuffs: values for feedstuffs (a) assume 880 g of dry matter per kg (dashes imply no value available); values for forages (b) are typical rather than average values.

(a) Feedstuffs

	Crude protein (g/kg)	Oil (g/kg)	Crude fibre (g/kg)	MAD fibre (g/kg)	Ash (g/kg)	Ca (g/kg)	P (g/kg)	K (g/kg)
Oats	96	45	100	170	40	0.7	3.0	5.0
Barley	95	18	50	70	25	0.6	3.3	5.0
Wheat	100	15	22	40	19	0.4	3.2	4.2
Maize	85	38	25	30	15	0.2	3.0	3.1
Sorghum (white)	10.6	25	27	60	18	0.3	2.7	4.1
Rice (rough)	73	17	90	—	52	0.4	2.6	—
Millet	88	13	277	—	81	—	—	—
White fishmeal	660	80	0	0	215	57.0	34.0	8.3
Dried skim milk (spray)	340	6	0	0	80	10.5	9.8	16.0
Linseeds	219	316	76	135	45	2.4	5.2	9.4
Exp. linseed meal	320	60	100	170	60	3.0	7.3	11.0
Extr. soyabean mean	440	10	62	100	60	2.4	6.3	23.5
Extr. sunflower seedmeal	280–450	18	230	—	77	2.9	8.0	14.0
Exp. cottonseed cake	410	37	140	220	65	2.0	10.5	15.0
Maize gluten feed	210	25	70–80	—	70	2.5	7.5	9.7
Field beans	255	90	74	114	29	0.8	4.8	11.0
Peas	229	50	57	82	27	0.7	4.0	11.0
High-protein grassmeal	160	32	220	360	70	6.0	2.3	21.0
Lucerne meal (alfalfa)	170	30	250	400	100	15.0	2.0	22.0
Brewers yeast	450	10	30	—	65	4.4	13.3	18.0
Rapeseed meal extr.	350	24	130	—	70	6.5	11.0	—
Wheatfeed	155	35	85	100	50	1.0	10.0	12.0
Wheat bran	155	30	110	120	70	1.0	12.1	14.0
Extr. rice bran	135	15	120	190	125	1.1	19.0	19.0
Oatfeed	50	22	250	400	70	0.8	1.2	7.0

Lysine (g/kg)	DE (MJ/kg)	α-Toco-pherol (mg/kg)	Free folic acid (mg/kg)	Available biotin (μg/kg)	Thiamin (mg/kg)	Riboflavin (mg/kg)	Pantothenic acid (mg/kg)
3.2	10.9–12.1	9	0.12	50	17.0	1.7	12.0
3.1	12.8	7	0.11	12	5.0	1.8	14.0
2.8	14.1	9	0.12	4	4.0	1.1	11.0
2.6	14.2	9	0.06	65	2.0	1.5	6.0
2.4	13.0	7	0.13	—	3.0	1.1	12.0
2.5	11.1	—	0.20	15	2.5	0.9	—
—	—	—	—	—	—	—	—
48.0	14.1	13	0.22	100	5.0	5.0	8.8
29.0	15.1	10	0.60	330	4.0	20.0	30.0
7.7	18.5	2	—	—	7.0	2.5	—
11.3	13.9	1	—	—	8.0	3.0	12.0
26.0	13.3	2	0.57	280	6.0	3.3	14.0
13.0	9.5	10	—	415	34.0	3.0	—
14.0	12.8	5	0.22	250	7.0	4.0	9.5
6.0	12.8	7	0.20	85	2.0	2.2	14.0
17.0	13.1	—	—	—	—	—	—
15.8	14.1	8	—	—	—	1.4	—
8.0	9.6	25	1.80	300	3.0	15.0	25.0
8.2	9.0	30	3.00	400	3.7	14.0	25.0
29.3	12.2	2	2.31	300	100.0	39.0	105.0
20.0	11.5	—	—	575	1.0	3.6	9.0
6.1	11.0	23	0.67	15	13.0	4.5	16.0
6.0	10.8	21	0.30	15	7.3	5.4	26.0
5.3	10.8	—	—	20	23.0	2.6	23.0
1.9	7.7	—	—	—	—	—	—

	Crude protein (g/kg)	Oil (g/kg)	Crude fibre (g/kg)	MAD fibre (g/kg)	Ash (g/kg)	Ca (g/kg)	P (g/kg)	K (g/kg)
Malt sprouts (culms)	250	20	140	—	65	2.0	7.0	—
Dried brewer's grains	180–250	62	140–170	180	38	2.6	5.1	1.0
Beet pulp	70	10	174	340	50	7.0	0.8	2.0
Molassed beet pulp	90–120	1–6	130	—	60–80	6.0	0.7	16.0
Grass hay	45–80	25	330	380	50	2.9	1.7	17.0
Clover/grass hay	60–100	27	330	380	65	4.0	1.7	19.0
Spring cereal straw	30	19	410	590	70	2.0	0.4	19.0
Vegetable oil	0	100	0	0	0	0	0	0
Molasses (cane)	30	0	0	0	85	7.2	1.0	27.0
NIS*	45	12	340	490	—	4.0	1.0	18.0
Limestone flour	0	0	0	0	100	365	4	0
Dicalcium phosphate	0	0	0	0	100	238	187	0
Steamed bone flour	0	0	0	0	98	323	133	0

* Contains 30 g Na/kg.
Abbreviations: exp., expeller; extr., extracted; MAD, modified acid detergent.

Lysine (g/kg)	DE (MJ/kg)	α-Toco-pherol (mg/kg)	Free folic acid (mg/kg)	Available biotin (μg/kg)	Thiamin (mg/kg)	Riboflavin (mg/kg)	Pantothenic acid (mg/kg)
12.0	10.0	—	—	—	—	—	—
8.3	10.0	10	—	—	0.6	1.0	10.0
2.8	11.0	—	—	—	0.4	0.6	1.5
1.8	7.4	7	—	—	1.5	10.0	—
2.9	7.5	8	—	—	2.0	15.0	—
0	6.0	6	—	—	—	—	—
0	35.0	—	0	0	0	0	0
0	11.4	0	0.08	100	0.8	1.0	35.0
—	—	—	—	—	—	—	—
0	0	0	0	0	0	0	0
0	0	0	0	0	0	0	0
0	0	0	0	0	0	0	0

(b) Forages

	Crude protein (g/kg)	Oil (g/kg)	Crude fibre (g/kg)	MAD fibre (g/kg)	Ash (g/kg)	Ca (g/kg)	P (g/kg)	K (g/kg)	DE (MJ/kg)	pH	NH_3N as % of total N
Alfalfa hay mid-bloom	150	17	270	350	70	11.4	1.9	16	7.6	—	—
Pasture:											
1. First growth	167	38	176	194	97	5.3	3.1	26	9.3	—	—
2. Second growth Pure clover	194	35	195	215	90	5.3	1.9	21	9.6	—	—
Pure grass	220	31	176	229	97	15.8	1.9	18	9.0	—	—
3. Blooming Pure clover	176	54	150	211	85	5.7	1.8	21	10.0	—	—
Pure grass	150	26	211	308	105	14.1	2.0	17	8.8	—	—
4. Winter after close grazing until July,	79	13	264	290	92	3.1	1.8	15	8.4	—	—
and free growth from July to December	136	26	194	—	70	—	—	—	—	—	—
Clamp silage	108	28	299	334	62	5.3	2.6	15	9.2	4.2	12.4
Big-bale silage	100	35	232	273	62	—	—	—	—	5.1	8.9

Appendix D

Estimate of base excess of a diet and of blood plasma

Estimate of (a) the base excess (BE) of a diet from its fixed ion content and of (b) the BE of blood plasma from its bicarbonate [HCO_3] concentration (see Ch. 9 for further details).

(a)

$$\text{(Cations–anions)}_{\text{absorbed}} - \text{(cations–anions)}_{\text{excreted in urine}} - H^+$$
$$\text{endogenous} = \text{BE} \tag{1}$$

Account here is taken only of fixed ions absorbed from the diet and equation (2) shows the principal ones involved.
(Note: fixed ions are those that cannot be degraded by metabolism.)

$$\text{(Cations–anions)}_{\text{absorbed}} = \text{mequiv}$$
$$(0.95\ Na + 0.95\ K + 0.5^*\ Ca + 0.5\ Mg) -\text{mequiv}$$
$$(0.95\ Cl + 0.95\ S + 0.5^*\ P) \tag{2}$$

*Approximate values which will be inversely related to dietary concentration.

In order to avoid a degree of arbitrariness a simplified balance of ions has been proposed and these are shown in equation (3) in relation to their optimum range in the diet of a horse in light work,

$$(Na + K - Cl)_{\text{absorbed}} = 200 \text{ to } 300 \text{ mequiv/kg diet} \tag{3}$$

(Note: $Na^+ + K^+ = 95\%$ of all cations in extracellular fluid and $Cl^- + HCO_3^- = 85\%$ of all anions.)

(b)
Plasma bicarbonate (mequiv/litre) at pH 7.4

$$\cong [HCO_3^-] \text{ measured } -10(7.4\text{-pH measured})$$

BE of plasma at pH 7.4

$$\cong [HCO_3^-] \text{ at pH } 7.4 - 24$$

(Normal bicarbonate of venous blood at pH 7.4 = 24 mequiv/litre.)

If the venous plasma of a horse was found to have a pH of 7.0 and $[HCO_3^-]$ was 30 mequiv/litre then $[HCO_3^-]$ at pH 7.4 would be

$$= 30 - 10\ (7.4 - 7.0)$$
$$= 26\ \text{mequiv/litre}$$

Therefore

$$BE = 26 - 24$$
$$\cong 2\ \text{mequiv/litre}$$

Note: Other organic acid anions could also be included in a base excess calculation.

Glossary

acidaemia An increased hydrogen ion concentration (acidity) and lowered blood bicarbonate, or decreased pH (q.v.) of the blood (see p. 187).

acidosis See **acidaemia**.

acute Applied to a metabolic upset, or a disease, which progresses rapidly to a climax followed by death or rapid recovery. Contrasts with a chronic (q.v.) condition or disease.

ad libitum feeding A system in which feed supply is unrestricted at all times except during exercise. However, usually applies only to growing horses.

adipose tissue The cells of this tissue readily store fat that is drawn on as a source of energy, especially when the blood levels of glucose and volatile fatty acids are low.

adrenal glands A pair of ductless glands, one situated near each kidney and consisting of an internal medulla, which secretes the hormones epinephrine and norepinephrine (adrenaline and noradrenaline), and an external cortex, which secretes corticosterone, cortisol (glucocorticoids) and aldosterone.

aerobic In aerobic respiration energy-yielding nutrients are broken down with the consumption of dissolved oxygen that has reached the tissue cells from the lungs. In anaerobic respiration the breakdown of these nutrients is incomplete, it yields less energy and occurs in the absence of oxygen.

afferent Afferent nerve fibres conduct impulses centripetally, e.g. from sense organs to the central nervous system.

agglutination This is the clumping together of particulate antigen (foreign substance) in the presence of homologous antibody (defence substance).

air dry Under United Kingdom regulations, feed that has been allowed to dry without heating in the air contains 100–140 g moisture per kg.

alfalfa See **lucerne**.

alimentary canal/tract See **gastrointestinal tract**. In addition it includes the buccal cavity and oesophagus.

alkalaemia A decreased hydrogen ion concentration, or elevated pH (q.v.) of the blood, irrespective of changes in blood bicarbonate. Normal arterial blood pH is 7.5. (See p. 187.)

allergy A condition of exaggerated susceptibility, or sensitivity, to a specific substance, usually but not necessarily containing a specific protein. Exposure, especially to large amounts of the allergen through inhalation, ingestion or injection, or even by skin contact, may cause respiratory difficulties, sneezing, a skin rash or diarrhoea of increasing severity through repeated contact.

ALT (also **GPT**) Alanine amino transferase EC 2.6.1.2 (formally termed glutamic pyruvic transminase). The activity of this enzyme in the blood plasma shows a similar reaction to that of AST (q.v.), particularly in respect of exercise and muscle damage.

amino acids These nitrogen-containing compounds are the building-blocks of proteins. There are some twenty-five different kinds in protein, ten of which are known as dietary indispensable (essential) nutrients, the most critical of which is lysine.

α-amylase An important digestive enzyme in the digestion of starch and other polysaccharides containing three or more α-1,4 linked D-glucose units. It hydrolyses the α-1,4 glucan links.

anabolism The process of synthesis of complex organic molecules in the body from simpler precursors (cf. catabolism, q.v.).

anaemia A condition in which there is a reduced number of red cells and/or a reduced haemoglobin content of the blood. The volume of packed red cells is reduced when the equilibrium between blood loss, through bleeding or destruction, and blood production is disturbed.

analgesic A pain-relieving substance.

anodyne A drug used for relieving pain.

anorexia Lack, or loss, of appetite for feed.

anthelmintic A substance used to destroy parasitic worms.

antibiotic A chemical substance produced by and obtained from living cells, especially of lower plants such as moulds, yeasts or bacteria, that is antagonistic to, or destroys, some other form of life. It may be so used to destroy infectious organisms.

antibody A specific substance (immunoglobulin) found in the blood, or in certain secretions, in response to the antigenic stimulus of bacteria, viruses, worm parasites and certain other foreign substances. An antibody has a specific amino acid sequence and can combine specifically with the inducing foreign entity (antigen) helping to inactivate it.

antigen Any substance which is capable under appropriate conditions of inducing the formation of antibodies and of reacting specifically with those antibodies.

antihistamine A drug that counteracts the effects of histamine or certain other amines that cause inflammation.

antitoxin A substance found in blood serum or other body fluid that is antagonistic to a particular toxin (q.v.). For therapeutic, or protective, use it may be injected into horses to neutralize the toxin of a particular disease, but as it does not stimulate the horse to produce its own antitoxin its (passive) effects may last for only a few weeks.

artery A vessel containing smooth muscle through which blood passes away from the heart to the various parts of the body.

ascarids Roundworms. A group of large parasitic intestinal parasites in the phylum Nematoda. They attain 15–20 cm long, are white and infest primarily young horses, as those over 3–4 years of age have usually developed considerable immunity (for life history see Ch. 11). Large numbers of adult worms in the intestines can cause impactions, intestinal perforations and colic.

ash The ash content of feed is determined by ignition of a known weight of it at 500°C until the carbon has been removed. The residue represents approximately the inorganic constituents of the feed – principally calcium, potassium, magnesium, sodium, sulphur, phosphorus and chloride. Some feeds, particularly those contaminated with soil, may contain a significant amount of silica.

AST (also **AAT** and **GOT**) Aspartate amino transferase EC 2.6.1.1, formally called glutamic oxaloacetic transaminase. This enzyme is released into the blood following damage to liver or muscle cells so that the blood level rises sharply. Plasma activity of the enzyme usually increases after hard exercise. The normal maximum blood plasma level in adult horses is 250 iu/litre.

ataxia Failure of muscular co-ordination, or irregularity of muscular action resulting in a staggering gait. It may result particularly from exhaustion or from pathological change in the nerves.

ATP Adenosine triphosphate mediates the transfer of energy, from the breakdown of glucose and fatty acids (exergonic reactions), for the synthetic processes of growth, milk secretion, etc. (endergonic reactions) and for muscular action. ATP is split by the enzyme ATPase, with the liberation of inorganic phosphate.

autogenous Self-generated. It refers particularly in the text to antibodies produced by the dam to blood proteins of the foetus that happen to be circulating in her blood. If colostrum is then taken by the foal within the first 12 hours, there will be an antibody reaction with the foal's blood proteins.

autonomic nervous system This is the self-controlling part of the nervous system – it is not subject to direct influence by the conscious brain. The sympathetic system (thoracolumbar outflow) and the parasympathetic (craniosacral flow) have largely antagonistic actions. The combined system is of importance in regulating the activities of many of the glands, the smooth musculature of the gastrointestinal tract and elsewhere, and the heart and blood vessels.

azotaemia An excess of urea and other nitrogenous compounds in the blood.

azoturia An excess of nitrogenous compounds in the urine. Considered synonymous with exertion myopathy. It frequently occurs within a short interval of beginning exercise after a rest of 2 or 3 days. A reluctance to move and muscular spasms are witnessed and excessive lactic acid accumulates in the muscles. The muscles of the hindquarters become tense and there is a tendency to knuckle-over at the fetlocks. At some stage the horse will pass quantities of urine from light-Burgundy wine to dark-coffee colour. Sometimes the urine is retained and requires relief with a catheter.

beta-carotene See carotene.

big-head A condition seen in horses and ponies given a ration based on cereals, bran and poor hay without adequate mineral supplementation. The bones of the upper jaw and face are particularly enlarged owing to a replacement of their normal mineralized structure by fibrous connective tissue.

bile duct The duct carrying bile, synthesized by the liver, to the duodenum, where the bile facilitates emulsification of fats and contributes to the production of an alkaline reaction of the intestinal contents.

biological values (BV) The amount of nitrogen retained by an animal per unit of nitrogen absorbed from the gastrointestinal tract and from a given feed protein.

biuret A simple organic compound containing three nitrogen atoms, which has been put forward as a source of dietary nonprotein-nitrogen. The structurally related compound, urea, contains only two N atoms per molecule.

bleeders Broken blood vessels. This usually refers to the loss of blood through the nose after hard exercise. Blood is lost from small broken blood vessels in the lungs, or nasal passages, of about half the population of healthy Throughbreds during hard exercise and the condition requires no treatment unless large blood vessels are involved.

blood counts These normally include the numbers of red blood cells (erythrocytes) per unit volume, the packed cell volume (haematocrit), or proportion of cells by volume in the total blood measured after the blood has been centrifuged, and the haemoglobin content of the total blood. Three other characteristics of the red cells (McV, mean corpuscular volume; McH, mean corpuscular haemoglobin; and McHc, mean corpuscular haemoglobin concentration) are calculated from these basic data. The numbers of white blood cells (leucocytes) per unit volume and the differential count (proportion of each type of white cell) may also be measured.

botfly The larvae, or maggots, of *Gastrophilus* spp. cause chronic gastritis and loss of condition in grazing animals. *G. intestinalis* lays eggs of pin-head size on horse's hair in the summer. When the horse licks itself, the eggs are taken into the mouth and hatch there or in the stomach. On rare occasions they cause perforation of the stomach and death (see Ch. 11).

botulism A rapidly fatal motor paralysis caused by the ingestion of the toxin of *Clostridium botulinum*, a spore-forming anaerobic bacterium, which proliferates in decomposing animal tissue and sometimes in plant material. The toxin (a di-chain protein, MW 140 000) is the most neurotoxic substance known. In the text (Ch. 10) botulism is referred to in cases in which horses have consumed silage (q.v.) that has been subject to an abnormal fermentation. The toxin seems to inhibit irreversibly the release of acetylcholine from peripheral nerves, and so impedes neuromuscular transmission. A flaccid descending paralysis develops.

broken-wind, heaves These are outdated expressions applied to longstanding respiratory diseases in which a double expiratory effort is a feature. The causes include bacteria, viruses and, rarely, lung tumours. A common cause is an allergic reaction and that related to the inhalation of spores and particles of the moulds, *Micropolyspora faeni* and *Aspergillus fumigatus*, is discussed in Chapter 11.

buccal cavity The cavity of the mouth between the cheeks bounded at one end by the lips and at the other by the pharynx.

buffers Substances which in aqueous solution increase the amount of acid or alkali that may be added without changing the pH, or degree of acidity or alkalinity. In general, a buffer is made up of two components: (1) a weak acid, e.g. H_2CO_3, and (2) its corresponding base HCO_3^-. Arterial blood is well buffered around a pH of 7.5.

caecum A great cul-de-sac intercalated between the small intestine and the colon. In the adult horse, it is about 1.25 m long, with a capacity of 25−30 litres. For further details see Chapter 1.

calcitonin This hormone is synthesized by the parafollicular cells of the thyroid gland (q.v.). It is secreted when the serum concentration of calcium ions rises, promoting the deposition of Ca in bones, and so counteracting the action of parathyroid hormone (q.v.).

calculi Urinary calculi consist of accumulations of mineral substances in the urinary tract. They form in the bladder, less frequently in the urethra and rarely in the kidneys. They are commonly rough and yellow-brown and are composed of calcium carbonate. These are seen more commonly in horses on high-roughage diets or on pasture. The less-common phosphate calculi are smooth and white and occur with high-cereal rations. A low intake of water may predispose the horse to calculi and the signs include difficulty in urination and incontinence.

calorie A unit of energy, being the amount of heat required to raise 1 g water 1°C. 1000 calories = 1 kilocalorie (kcal). The joule (J) has now been adopted as the unit of energy in nutrition; 4.184 kJ = 1 kcal.

carbohydrates Compounds composed of carbon, hydrogen and oxygen and including the sugars, starches and other storage carbohydrates and the structural (fibre) carbohydrates – cellulose and hemicelluloses also pectins, gums and mucilages. Lignin is included in the structural fibre but is not strictly a carbohydrate.

cardiac output This is the volume of blood expelled by each ventricle per minute. The stroke volume is the volume of blood discharged by each ventricle at each beat (i.e. the cardiac output divided by the heart rate. The cardiac output of the left ventricle after birth is 2–8 per cent higher than that of the right, so the value quoted represents the mean of these two.

carotene Green plants contain a number of yellow carotenoid pigments, the most important of which are α, β and γ-carotenes and hydroxy-β-carotene. The most potent of these, β-carotene, is converted to vitamin A by the intestinal wall.

carpus The 'knee' of the forelimb of the horse between the radius (q.v.) above and metacarpus (q.v.) below.

catabolism The breaking down of tissue nutrients and components to less-complex molecules (cf. anabolism, q.v.).

cathartic A purgative, or medicine, that quickens the evacuation of the bowels (intestines).

cation The positively charged elements of electrolytes, which include all metals and hydrogen ions.

cellulose A structural carbohydrate (fibre) of plant cells.

cellulolysis The breaking down, or digestion, of cellulose. Horses and other mammals depend on certain bacteria in their intestinal tracts to carry this out as they do not secrete enzymes capable of it.

choke Obstruction to the passage of food through the pharynx and oesophagus, either partial or complete. This is frequently caused by a mass of dry impacted feed.

chronic Long-continued, the opposite of acute (q.v.).

CK (or **CPK**) Creatine kinase, EC 2.7.3.2. This enzyme is measured as an indicator of muscle damage. The normal maximum blood-plasma level in adult horses is 105 iu/litre.

coldblooded horse In Europe, two types of horse evolved – the light short-legged Celtic pony and the Great Horse of the Middle Ages (the large powerful but slow heavy horse). Present-day breeds whose major blood lines derive from either of these types are termed 'coldblooded' (cf. 'hotblooded' horse, q.v.).

colic Abdominal pain. Where this originates from part of the gastrointestinal tract it is termed 'true colic', but where it derives from one of the other vital organs or muscles it is defined as 'false colic.'

colon This is made up of the great colon, which originates from the caecocolic orifice and terminates where it joins the small colon, and the latter, which continues to the rectum. In the adult horse, the great colon is 3.0–3.7 m long with an average diameter of 20–25 cm, whereas the small colon has a diameter of 7.5–10.0 cm, and a length of about 3.5 m.

colostrum This is secreted by the mammary gland of the mare shortly before foaling (parturition) and for about the first 24 hours after the birth of the foal. It is rich in gamma globulins (q.v.), which comprise the antibodies that the foal absorbs undigested into its blood during approximately the first 12–18 hours of life providing it with a measure of protection from disease.

compounded feeds Balanced mixtures of ground, or otherwise processed feedstuffs, to which appropriate supplements of vitamins, minerals and trace elements have been added.

concentrates The portion of the horse's ration, or a feedstuff, which is rich in starch, protein, or both, and which contains less than 15–17 per cent crude fibre (q.v.).

contagious Transmissible from one horse to another.

contracted tendons Hyperflexion or flexural deformity of limbs. The condition in foals ranges from uprightness of fore- or hindlegs to knuckling over at the fetlock and/or inability to extend knee joints. Not uncommon in Thoroughbred foals. Correction of slight abnormalities may include the fitting of boots and, in more severe cases by severance of fibres of the flexor muscles and tendons or desmotomy of the superior check ligament.

convulsions A violent involuntary (uncontrollable) contraction, or series of contractions, of the voluntary (skeletal) muscles.

coprophagy The eating by an animal of its faeces. Within 3 weeks of birth, foals will eat their dam's faeces and thereby acquire the species of bacteria and protozoa necessary for the rapid development of a normal microbial population in their gastrointestinal tract so that invasion by harmful microorganisms is prevented.

cornea The transparent structure that forms the anterior part of the eyeball.

coronary band, coronary matrix This runs around the horse's foot just below the hair line and forms part of the sensitive structures from which grows the wall. A permanent defect in the hoof wall usually follows injury to the coronary band.

corticosteroids These comprise the natural glucocorticoids, cortisone and hydrocortisone-hormones secreted by the adrenal cortex (q.v.) and synthetic equivalents, e.g. prednisone, prednisolone and fluoroprednisolone used in the treatment of inflammation, shock, stress and, in other animals, ketosis.

CP Creatine phosphate (phosphocreatine). The fixation of energy in the form of ATP is a transitory phenomenon, and any energy produced in excess of immediate requirements is stored more permanently in compounds such as muscle phosphocreatine. As ATP becomes depleted, more is generated from phosphocreatine by a reverse reaction.

creatine The normal excretory breakdown product of muscle creatine found in horses' urine. As the quantity produced daily is relatively constant for a particular horse, being proportional to muscle mass, its concentration in urine is used in the assessment of the level of other substances excreted in urine (see creatinine clearance tests, Ch. 12).

creatine kinase See **CK**.

creep feed A feed, normally dry pellets, offered to nursing foals behind a barrier, which does not allow access to the mare but permits entry to the foal.

cribbing (aerophagia) An outdated expression for 'wind-sucking', which is a vice of domesticated horses and ponies. This consists of the habitual swallowing of air while the animal bites, or pulls down with its upper incisor teeth on some solid object such as a fencing rail or gate. The neck is slightly arched and gulps of air are swallowed into the stomach with emission of a grunt. The term 'wind-sucking' is also unfortunately used to describe mares that aspirate air, and frequently faecal material, into the vagina. This is rectified by Caslick's operation.

crimping A term used for the pressing of cereal grains between corrugated rollers to rupture the kernels and to increase digestibility slightly.

croup That part of the horse's hindquarters lying immediately behind the loins. The 'point of the croup' is its highest part and corresponds to the internal angles of the ilia.

crude fibre The feed residue identified after subjecting the residual feed from ether extraction to successive treatments with boiling acid and alkali of defined concentrations. The crude fibre contains cellulose, hemicellulose and lignin, but it is not an accurate measure because it underestimates the structural components of vegetative matter (see **NFE**).

cyclic AMP An intracellular hormonal mediator formed from ATP under the influence of the stimulating hormone.

DE Digestible energy. The gross energy (or heat of combustion) of a feed minus the gross energy of the corresponding faeces, expressed as megajoules (MJ) or as kilojoules per kilogram (kJ/kg) total feed. Synonymous with apparent digestible energy.

deamination When amino acids are present in excess of needs for tissue protein synthesis, or when the horse is forced to catabolize tissue to maintain essential functions, amino acids may be degraded to provide energy. This occurs mainly in the liver and to some extent in the kidneys. The first step in the oxidative degradation of amino acids is the removal of the amino group, a process called deamination. This group is then either transferred to a keto acid to produce another amino acid, or is incorporated into urea.

decarboxylation In the text, this term is principally restricted to reactions of amino acids. Bacteria – for example in the intestinal tract and in silage during the early stages of fermentation of the clamp – elaborate enzymes (called carboxylases), which act upon amino acids to yield amines and carbon dioxide. This implies a loss of dietary protein value and the amines, including histamines and tryptamine, may have toxic effects following absorption.

dehydrated Feed from which most of the moisture has been removed. This extends shelf life and greatly retards the rate of, or inhibits, mould spoilage.

diaphragm A thin muscular partition or membrane that separates the thorax (chest cavity) from the abdomen. During its contraction air is drawn into the lungs and when it relaxes air is expelled.

diaphysis The shank of a long bone between the ends, or epiphyses, which are usually wider and articular.

dicoumarol An anticoagulant with similar properties to warfarin except that its action has a slower onset, a longer duration and a less-predictable response. When ribbed melilot or yellow sweet clover (*Melilotus officinalis*) or white melilot or white sweet clover (*M. alba*) plants are damaged by weather, badly harvested, or when as hay they become mouldy, coumarin contained in the sweet clover is broken down to dicoumarol. Both yellow and white sweet clover are found in British pastures, and white sweet clover is grown as a forage crop in North America and the Soviet Union.

digestible energy See **DE**.

dispensable amino acids Amino acids that are synthesized in the tissues of horses, and/or are made available from synthesis by gut microorganisms, in amounts sufficient to meet tissue requirements of horses without a dietary source.

distal Remote, farthest from the centre, or origin, as opposed to proximal (q.v.).

diuresis Increased secretion of urine. A diuretic drug induces diuresis.

duodenum This is the first (proximal) part of the small intestine and is connected to the stomach. In the adult horse it is 1 m long with a diameter of 5–10 cm.

dysphagia Difficulty in swallowing.

dyspnea Difficult or laboured breathing.

dystrophy Faulty nutrition. Dystrophy of muscles causes their atrophy and degeneration.

electrolytes Substances that in water solution break up into particles carrying electrical charges. The principal electrolytes from a nutritional point of view are Na^+, K^+, Cl^-, HCO_3^-, Ca^{2+}, Mg^{2+} and HPO_4^{2+}.

emphysema A swelling, or inflation, of the chest which is caused principally by the presence of air in the intra-alveolar tissue of the lungs following rupture of the alveoli.

endocrine (hormone) system Glands dispersed in various parts of the body, the function of which is to liberate into the blood, lymph or neurosecretory channels specific substances that influence metabolism in other organs and tissues.

endotoxaemia Presence in the blood of endotoxins, which are non-protein, lipopolysaccharide fragments of the cell wall of Gram-negative bacteria. Per mg they are much less toxic than are exotoxins, but appear to play a crucial role in certain diet-related disorders of bowel origin, including forms of colic (q.v.) and laminitis (q.v.).

enema (pl. enemata) A fluid for injection into the rectum, or small colon for cathartic or diagnostic purposes.

enterotoxaemia Presence in the blood of toxins, produced and secreted in the intestines by certain bacteria; e.g. those of *Clostridium perfringens*, which are referred to in the text (Ch. 5). Enterotoxin produced in large quantities by this organism is also specific for cells of the intestinal mucosa and causes enteritis. This anaerobic organism is found in soil and the consumption of relatively small numbers of the spores in herbage is usually without remarkable effect. (See also **botulism**.)

epiphysis A head of a long bone joined to the shaft, or diaphysis, during growth by a cartilaginous growth plate – a metaphysis.

epiphysitis Inflammation of an epiphysis or of the cartilage that separates it from the shaft during growth.

epistaxis A nose bleed. Blood may be present in the nostrils through being coughed up from broken blood vessels in the lung. Small losses can be a normal phenomenon after a race, but persistent losses may occur in chronic bronchitis.

epithelial cells All the body surfaces, including the external surface of the skin, the internal surfaces of the digestive, respiratory and genito-urinary tracts, the inner coats of vessels and ducts of all secreting and excreting glands are covered by one or more layers of cells, called epithelium or epithelial cells.

ergot A fungus that infects and finally replaces the seed of a cereal or grass, especially the sclerotium of *Claviceps purpurea* (ergot of rye). This ergot is small, hard, black and resembles mouse droppings. Ergot contracts the arteriolar and other unstriped muscle fibres. Its toxins are used to arrest haemorrhage after parturition and internal injury. The persistent consumption of ergot of rye sufficiently decreases blood flow to cause gangrene (typically of the ears, tail and legs). Several ergot toxins cause abortions. After eating large amounts of ergotized hay, horses become dull and listless, a cold sweat breaks out on the neck and flanks, the breathing is slow and deep, the body temperature is subnormal and the pulse weak. Death occurs during a deep coma within the first 24 hours. Lesser amounts over a longer period may cause diarrhoea, colic, trembling and loss of condition.

erythrocytes See **red blood cells.**

essential amino acids See **indispensable amino acids.**

ether extract Chemical substances that are soluble in, and extracted by, ether. Whether diethyl ether, or 40:60 petroleum ether, is used should always be specified.

exertion myopathy Azoturia (q.v.) myohaemoglobinuria. An acute condition in which affected muscles, especially of the hindquarters, become hard to the touch, and the hindlimbs rapidly become stiff and weak or staggering. There is a tendency to 'knuckle-over' at the fetlocks. The risk is greatest in horses that have been in continuous work, abruptly rested for a few days on full feed and then returned to work.

extracellular space The fluid space, or volume, within the body that is external to the cells.

fatty acids Fatty acids are composed of a hydrocarbon chain of one to more than twenty units attached to a carboxyl group. In the formation of storage fats fatty acids are neutralized by the trihydric alcohol glycerol. Both neutral fat and fatty acids circulate in the blood (see also **free fatty acids** and **volatile fatty acids**).

feed That which is given to the animal to consume, whether a single feedingstuff or a mixture, but excluding water. Not synonymous with ration (q.v.).

fermentation Decomposition of organic substances by microorganisms. In the horse's gastrointestinal tract this refers especially to bacteria, yeasts and ciliate protozoa; the first are the most significant.

fertilizer Inorganic fertilizers are plant nutrients prepared synthetically or mined as minerals. Organic fertilizers are sources of the same nutrients of animal and vegetable origins bound in organic (q.v.) form.

fetlock The horse's 'ankle' joint in fore- and hindlimbs between the metacarpus or metatarsus (cannon bones) and the first phalanx (long pastern (q.v.) bone).

FFA See **free fatty acids.**

fibre See **crude fibre.**

fibrin The insoluble protein formed from fibrinogen by the proteolytic action of thrombin during normal clotting of blood.

filled legs Oedema or puffiness of the legs. Abnormally large amounts of fluid (exudate) in the intercellular tissue spaces beneath the skin. The most common cause in healthy horses is a period of inactivity in a box following a season of hard exercise, particularly where corn or other concentrated feeds are being given. However, it is not protein poisoning. Oedema can also arise from diseases of the heart, liver or kidneys, or from longstanding malnutrition involving diets impoverished in protein.

flexor tendons Muscles are attached to the long bones of the legs by extensor tendons, which extend or straighten the leg at that point when the muscle contracts, and by flexor tendons, which flex joints by contraction of the appropriate muscles.

founder, laminitis A painful disease of the feet in which there is apparently a transitory inflammation followed by congestion of the laminae of the hooves. It is most frequent in ponies but can be induced readily in both horses and ponies by a sudden increase in the starch or protein content of the diet. Sometimes all four feet are affected, sometimes only the forefeet, and occasionally only the hindfeet, or a single foot. Affected feet feel hot and the body temperature may rise. The stance is unnatural, affected forefeet are thrust forwards and the horse is reluctant to move.

free fatty acids (FFA) During work storage fats are mobilized when lipase enzymes catalyse the production of fatty acids, splitting them from glycerol and leading to a rise in the blood plasma concentration of both components.

frog A wedge-shaped mass of elastic horn projecting downward from the underside of the foot that receives considerable concussion. The frog protects the lower surface of the plantar cushion of which it is an exact mould.

fundus gland region of stomach A large region of the stomach containing glands of two types of cells. The region lies distal to the oesophageal and cardiac regions and proximal to the pyloric gland region.

fungal units Fungi contaminating feed produce a mycelial mat of fine threads and fruiting bodies from which spores are shed. When badly affected feed is disturbed, the mycelial threads may break up into small particles and new growth can be initiated by each of these particles or by germination of spores. The fungal unit is any particle from which new growth can be started.

furlong Forty poles, one-eighth of a mile. Originally the length of the furrow in the common field. The side of a square of 10 statute acres. Equals the Roman stadium, one-eighth of a Roman mile.

β-Galactosidase Neutral, or brush-border lactase. This enzyme is present in the intestinal juice of normal young horses. It is necessary for the cleavage of milk sugar (lactose) to glucose and galactose, which can then be absorbed into the blood.

gamma globulin A protein fraction of the plasma globulin, that has a slow moving electrophoretic mobility. Most antibodies are gamma globulins.

gastrointestinal tract Stomach and intestines. Sometimes used synonymously with alimentary tract, which is the entire tube extending from the lips to the anus.

Gastrophilus See **botfly**.

GGT gamma-glutamyltransferase EC2.3.2.2 is released into the blood following liver damage. The blood activity of this enzyme is elevated in liver cirrhosis, pancreatitis and renal disease. The normal range of plasma activity is 0 to 4liu/l and seneciosis may cause a rise up to 80 to 280iu/l.

glucagon A hormone secreted by the alpha cells of the islets of Langerhans in the pancreas with the function of increasing blood glucose and so opposing the action of insulin, q.v.

glucogenic Giving rise to, or producing, glucose. According to a long-used classification, amino acids are ketogenic if (like leucine) they are converted to acetyl CoA and, when fed to a starved animal, produce ketones in the blood. Glucogenic amino acids such as valine, when fed to a starved animal, promote the synthesis of glucose and glycogen.

glutathione peroxidase (GSH-P$_x$) The activity of this enzyme EC 1.11.1.9 in horse erythrocytes is used as an indication of the horse's selenium status. The activity is determined as μmol NADPH oxidized in 1 min by 1 ml of erythrocytes. In Throughbreds the evidence (Blackmore, Campbell et al. 1982) indicates that the activity should be approximately 25–35 units/ml red cells.

glycolysis The major pathway whereby glucose is metabolised to give energy is a two-stage process; the first stage, called glycolysis, can occur anaerobically and yields pyruvate.

glycosuria The presence of abnormal amounts of glucose in the urine. It arises from failure of renal tubular reabsorption or from an abnormality of hormone status as in diabetes mellitus.

goitrogenic A term applied to substances in certain feeds, derived, for example, from members of the plant genus Brassica (family Cruciferae) which, if consumed persistently in large quantities, cause goitre (an enlargement of the thyroid gland, q.v.). A deficiency of iodine in the diet will cause a similar condition, especially in young horses.

GOT See AST.

GPT See ALT.

grass sickness A disease of horses, ponies and donkeys that seems to be non-contagious. It occurs mainly in the summer months among grazing animals on certain pastures in Europe in a peracute form in which death occurs within 8–16 hours of initiation after some periods of great violence. In a subacute form the horse is dull, listless and salivates. The bowel becomes impacted and distended. Food material may appear in the nostrils. There seems to be some loss of motor control of the gastrointestinal tract. Muscles of the back are hard, the horse is 'tucked-up' and twitching over the shoulders occurs. Recovery is uncommon.

growthplate See metaphyseal plate

haematocrit Packed cell volume (PCV). This is the proportion of blood by volume made up of cells, especially red cells, and expressed as a percentage. It is determined by centrifugation of blood samples containing an anticoagulant.

haemoglobin The oxygen-carrying red pigment of the red cells (erythrocytes) of blood. A conjugated protein consisting of the protein globin combined with an iron-containing prosthetic group (heme).

haemolysis The rupture of red blood cells.

haemolytic icterus of foals Caused by the absorption of colostral antibodies that destroy the foal's red blood cells (see Ch. 7).

haylage Originally registered trade name for a silage containing a high proportion (35–50%) of dry matter made from wilted forage, precision-chopped to $\frac{1}{2}$–1 in ($\sim$ 12–25 mm) nominal length and ensiled in a Harvestore tower silo. However, the name has acquired a more liberal definition to include unchopped vacuum-packed material.

heating feed A concentrated feed, which is readily digestible and fermentable and which leads to a rapid rise in blood metabolites, waste heat production and probably to some increase in metabolic rate.

heaves See **broken-wind.**

hind gut The large intestine consisting of the caecum and colon.

hives Large numbers of small bumps, or raised areas, about 0.5 cm diameter under the skin, which eventually form scabs. They appear suddenly and are caused by an allergic response to specific components of feed, to drugs or to insect bites. The cause (antigen q.v.) can usually be determined by allergy tests and recovery follows removal of the source from the diet, or general environment. The condition sometimes arises when 'rich' feeds are suddenly introduced to the diet.

hock The tarsal joint between the tibia (q.v.) and metatarsus (q.v.) (cannon bone) of the hindlimb.

homeostasis Stability of the normal body states. Refers frequently to the constancy of pH and chemical composition of extracellular fluids.

hormone A discrete chemical substance secreted into the body fluids by an endocrine gland and which influences the action of a tissue, or organ, other than that which produced it.

hotblooded horse Hot and warmblooded horses (cf. coldblooded horses, q.v.) are those derived to a significant extent from breeds originating in Mediterranean countries, which came to be called Arabian, Barb and Turk. Modern breeds of this type include Thoroughbred, Arabian, Standardbred, American Saddle Horse, Morgan, Quarter Horse and Tennessee Walker. The principal North European warmblooded breeds are the Hanovarian, Trakehner, German Holstein, Dutch Warmblooded and Oldenburg breeds.

hyper-, hypo- Prefixes: hyper signifying above normal or excessive and hypo meaning below.

hyper and hypokalaemia Abnormal level of plasma potassium. Hypokalaemia may be caused by a combination of inadequate dietary potassium with excessive losses from the body. The causes of both hypo- and hyperkalaemia can be metabolic derangements, when there are abnormal shifts of K^+ between intracellular and extracellular space, as in acidosis.

hyperlipidaemia An excessively high concentration of blood fats, sometimes seen in starved horses.

hyperparathyroidism Abnormally increased activity of the parathyroids (q.v.), causing loss of calcium from the bones.

hyperplasia The abnormal multiplication, or increase, in the number of normal cells in normal arrangement in a tissue.

hyperpnea Abnormal increase in the depth and rate of the respiratory movements.

hypertension Usually refers to high arterial blood pressure, which may be confined to a specific circulation such as pulmonary or renal.

hypertonic A body fluid with a concentration, or osmotic pressure, above normal (more than isotonic).

hypertrophy An overgrowth of an organ, or part of the body, due to an increase in size of its constituent cells rather than in their number.

hypocalcaemia A reduction of blood calcium below the normal range of concentrations.

hypocupraemia A subnormal concentration of blood copper.

hypoglycaemia Concentration of blood glucose below the normal limit for the breed.

hypoglycaemic shock, insulin shock Occurs when blood glucose falls below the normal range causing nervousness, trembling and sweating.

hypomagnesaemia A reduction of blood magnesium below the normal range of concentrations.

icterus Jaundice. A yellowish discoloration of the visible mucous membranes – eye, mouth, nostrils and genital organs. It can also be detected in blood plasma of stabled animals receiving a diet low in pigments and can be caused by certain infections resulting in destruction of the red blood cells and release of the haem pigment (haemolytic jaundice) or by liver damage.

ileum The distal portion of the small intestine extending from the jejunum to the caecum.

immunity Resistance, to infection by an organism, or to the action of certain poisons. Immunity can be inherited, acquired naturally or acquired artificially.

immunoglobulin Specific proteins, found in blood, colostrum (q.v.) and in most secretions, produced by plasma cells in response to stimulation by specific antigens, which in turn are inactivated. The antigens may be carried by, or released from, bacteria, viruses or even certain parasites.

imprinting An inborn tendency of a neonatal foal to attach itself to a set group of objects, or a single object, such as its mother.

indispensable (essential) amino acids Those amino acids that are not synthesized in the tissues of the horse, or otherwise made available from, for example synthesis by gut microorganisms, in amounts sufficient to meet the requirements of tissues and which must therefore be present in the feed.

infarct An area of tissue necrosis (q.v.) due to local anaemia resulting from obstruction of the blood circulation. In the text this refers to the effects of migrating strongyle (q.v.) larvae on the mesenteric blood vessels, which causes death of a segment of the intestines.

inorganic Not a precise term but can be taken to refer to the ash content of the body remaining after it has been incinerated, which removes hydrogen, carbon and nitrogen as oxides. The minerals (see definition), in an oxidized form, remain, together with a very small proportion of carbonates.

insulin A protein hormone synthesized by the islet cells of Langerhans in the pancreas and secreted into the blood where it regulates glucose metabolism. It is deficient in diabetes mellitus.

international unit (iu) As applied to vitamins it is the internationally agreed unit of potency for a particular vitamin that may have several molecular forms. The unit is now being displaced in favour of gravimetric measurements of each molecular form, or vitaminer.

intracellular space The fluid volume within the cells of the body as distinct from the extracellular fluid space. Movement of ions, water, glucose, etc. from one to the other is under metabolic control.

isoantibody An antibody (q.v.) generated in the body in reaction to an isoantigen (q.v.). An example is one found in the dam's blood and colostrum in response to a foetal protein that has entered her bloodstream.

isoantigen An antigen in the body that will induce the production of an antibody against itself.

isotonic solutions Those solutions that have the same concentration, or more specifically, the same osmotic pressure.

iu See **international unit**.

jaundice See **icterus**.

joint-ill A disease in which organisms enter the body by way of the unclosed navel causing abscesses to form at the umbilicus and in some of the joints.

joule SI (Système Internationale d'Unités) unit of energy; $4.184\,J = 1$ calorie. The joule is defined as the energy expended when 1 kg is moved 1 m by a force of 1 newton (1 N m), $1\,J = 1\,kg\,m^2S^{-2}$, or 1 watt. second (1 Ws). The kilojoule (kJ) $= 10^3$ joules and the megajoule (MJ) $= 10^6$ joules.

keratinization To become horny. Keratin, the principal protein of epidermis, hair and the hoof, is very insoluble and contains a relatively large amount of sulphur. Increased keratinization of epithelia can occur under physiological conditions and under pathological conditions, e.g. vitamin A deficiency.

knuckling-over Usually refers to flexing of the fetlock (q.v.) joint owing to contraction of muscles and tendons or ligaments behind the cannon.

labile Chemically unstable.

lactase An enzyme that hydrolyses the milk sugar lactose to form glucose and galactose. The enzyme accomplishing this in the intestine of the horse is β-galactosidase.

laminitis See **founder**.

Latent period The period or state of seeming inactivity between the time of stimulation and the start of the response e.g. the interval between the injection, or absorption of antigen and the first appearance of antibody.

LDH Lactic dehydrogenase. A tissue enzyme that has several different forms or isoenzymes. It catalyses the transfer of H^+ with the formation of pyruvic acid. The blood activity of LDH is elevated during and following strenuous exercise and following tissue damage.

legumes Plants of the family Leguminosae or Fabaceae, which includes useful forage species (e.g. the clovers and **lucerne**) and seed species (e.g. soya beans, peas and field beans)

level of feeding Weight of complete dry diet eaten daily, not confined to energy. Strictly it should be given as a proportion of metabolic body size ($W^{0.7}$), or more crudely per 100 kg body weight.

ligament A tough fibrous band supporting viscera, or which binds bones together. In different situations they are cord-like or in flat bands, or, in forming the joint capsule, they are in sheets, preventing dislocation.

light horse A loose term for small and large lady's hacks, usually of mixed breeding, which may include Thoroughbred and English Arabian. An alternative definition includes all riding horses of the present day in this classification, which therefore excludes heavy draught horses and ponies.

lipases A class of enzymes, members of which are found in the digestive secretions and in body tissues. Individual lipases have as major functions the hydrolysis of fats to yield fatty acids, monoglycerides, glycerol and cholesterol, and the hydrolysis of phospholipids.

lipids A group of substances found in plant and animal tissues, insoluble in water but soluble in common organic solvents, including petroleum, benzene, ether and chloroform. The crude fat of feed is the material extracted using light petroleum.

lucerne (alfalfa) A legume, *Medicago sativa*. A perennial forage crop with a strong tap root, which grows well on light alkaline soils in warm climates. Weeds must be kept at bay during its establishment.

lymphatics A system of vessels that drains lymph from various body tissues and conveys it to the bloodstream and that conveys neutral fats from the small intestine after they have been absorbed.

lysine An indispensable basic amino acid the concentration, or frequency, of which in most vegetable proteins limits their biological value. L-Lysine hydrochloride is sometimes used as a feed additive to make good any deficit in the dietary protein.

MAD fibre Modified acid detergent fibre. A fibre fraction of feed determined by the MAD fibre procedure isolates principally the lignocellulose complex. This complex is the fraction of plant material that has most influence on energy digestibility of feed. In comparison, during the chemical procedure for crude fibre (q.v.) determination a considerable amount of the lignin may become soluble and hence lost from the residue, so leading to an overestimation of the digestible fraction of feed.

maintenance At the maintenance level of feeding, the requirements of the horse for nutrients for the continuity of vital processes within the body, including the replacement of obligatory losses in faeces and urine and from the skin, are just met so that there is no net gain or loss of nutrients and other tissue substances by the animal.

maize (corn) *Zea Mays*. A member of the grass family, the Gramineae, the seeds of which constitute an excellent high-energy cereal grain for horses both cooked and uncooked. The above-ground vegetative parts, at the milky grain stage, can be made into a good silage for horses.

mandible Lower jawbone. In the adult horse this bone has sockets for three incisors, one canine, three premolars and three molars on each side in the male (in the female the canines are usually absent, or rudimentary). Its grinding movements are controlled by powerful muscles.

ME Metabolizable energy. The digestible energy (DE) of a unit weight of feed less the heats of combustion of the corresponding urine and gaseous products of digestion.

meconium A dark-brown, viscid semi-fluid, or hard, material which accumulates in the intestines of the foal prior to birth. It should be discharged soon after birth. The colostrum (q.v.) has a natural purgative action on it.

metabolism A term embracing the chemical processes of anabolism (q.v.) and catabolism (q.v.) in the body.

metacarpus The part of the forelimb lying between the carpus (wrist) and the digit (cf. metatarsus, q.v.). There are three metacarpal bones: the central, or third, metacarpal ('cannon') is the largest; the other two are rudimentary metacarpals (splint bones).

metaphyseal plate The region of linear growth of the long bones of growing horses lying between the epiphysis, or head, and the diaphysis, or shank.

metatarsus The part of the hindlimb lying between the tarsus, or hock, and the digit. It is similar in layout to the metacarpus (q.v.) with a central large metatarsal bone ('cannon') and rudimentary metatarsals (splint bones).

methaemaglobin A modified form of haemoglobin found in the blood in which the iron has been converted from the ferrous to the ferric state when it can no longer combine with, and transport, oxygen. The lesion can occur following the administration of large doses of certain drugs or after the consumption of nitrites in feed or water.

minerals The essential elements in the diet other than carbon, hydrogen and nitrogen. They include the macrominerals calcium (Ca), phosphorus (P), magnesium (Mg), potassium (K), sodium (Na), chlorine (Cl), and sulphur (S), and the microminerals iron (Fe), zinc (Zn), manganese (Mn), copper (Cu), cobalt (Co), iodine (I), selenium (Se) and fluorine (F). The microminerals are also known as trace elements. Some other elements, such as chromium, nickel and molybdenum, are required in very small amounts. The mineral elements are required for incorporation into compounds different in the main from those in which they may appear in the diet whereas carbon, hydrogen and nitrogen are required as constituents of preformed organic nutrients. A crude approximation to the mineral content of the diet is obtained from the measurement of the ash (q.v.) content of the feed.

mitochondria Minute bodies occurring in the cytoplasm of cells (except bacteria and blue-green algae, or cyanobacteria). They exert important regulatory functions both on catabolic and biosynthetic sequences. They are the seat of the citric acid cycle, the β-oxidation pathway and of oxidative phosphorylation.

MJ 1000 kJ or 10^6 joules.

myoglobin A small oxygen-carrying protein of muscles containing the pigment haem with an atom of iron at its centre. Severe muscle damage leads to its appearance in the urine in azoturia (q.v.). Being a much smaller molecule than haemoglobin, it passes the glomerular filter much more readily causing a dark-brown stain to the urine. Precipitation of myoglobin in the renal tubules, as with haemoglobin, may contribute to terminal uraemia (q.v.).

myopathy Any disease of muscle.

mycotoxins Substances produced under specific conditions by certain fungi or moulds. Their chemical form and their effects on animals are wide ranging. These include hormone-like effects, disruption of intestinal, renal and hepatic function, tumour induction, and antibacterial effects. Most antibiotics are mycotoxins. The best-known include aflatixon produced by *Aspergillus flavus*, T-2 toxin produced by *Fusarium* and *Myrothecium* spp., ergotoxin produced by *Claviceps purpurea*, zearalenone (F-2) produced by *Fusarium* spp, dicoumarin produced on melilots or sweet clover by *Aspergillus* and *Penicillium* spp., and vomitoxin produced by *Fusarium* spp. Several mycotoxins are harmful to the horse in normal husbandry conditions.

N-balance (N-retention) The net gain of nitrogen (N) by the animal. N in feed − (N in faeces + N in urine) − loss of ammonia expired in air per unit time.

NE Net energy, the energy value of animal product formed, or of body substance saved at or below maintenance, per unit weight of feed consumed. NE = metabolizable energy (q.v.) − heat increment.

necrosis Death of a cell, or of a group of cells, which is in contact with living tissue.

nephritis Inflammation of the kidney; a focal or diffuse proliferative, or destructive, process which may involve the glomerulus, tubule or interstitial renal tissue (cf. nephrosis, q.v.).

nephrosis Any disease of the kidney, especially one characterized by degeneration of the renal tubules (cf. nephritis, q.v.).

NFE Nitrogen-free extractives. Measured in g/kg feed this is numerically evaluated as 1000 − (moisture + ash + crude protein + ether extract + crude fibre). It includes some of the feed cellulose, hemicellulose, lignin, sugars, fructans, starch, pectins, organic acids, resins, tannins, pigments and water-soluble vitamins if each of these components was present in significant amounts in the original feed.

nuts, cubes Compounded feed mixtures for horses that have been compressed into solid cylinders 3–10 mm in diameter and two to three times as long by forcing the mixture through the holes of a metal die. The mix may or may not have been previously steamed in a kettle.

obligatory loss The amount of a nutrient that has undergone metabolic change in the body and which is unavoidably excreted in the urine or faeces or in both, irrespective of its dietary presence or absence.

oedema The presence of abnormally large amounts of fluid in the intercellular tissue spaces of the body. Applied in the text especially to an accumulation in the subcutaneous tissue.

oesophagus The gullet, a muscular membranous tube or canal extending from the pharynx to the stomach.

oestrous cycle A cycle in the mare typically lasts 21 days in the breeding season in which there is a pattern of physiological and behavioural events under hormonal control. The cycle, which forms the basis of sexual activity, has two components: oestrus (heat) in which the mare is receptive to the stallion and the egg is shed, and dioestrus, a period of sexual quiescence.

open knees A dished concave appearance to the front of the 'knee', or carpus joint caused by epiphysitis (q.v.) immediately above the knee.

oral Of the mouth.

organic Complex molecules containing at least carbon and hydrogen and synthesized by living tissue.

osmolarity This refers to the total number of dissolved particles, or osmoles, in water solution. Fluids with an osmolarity greater than that of body fluids are hypertonic and those for which it is lower are hypotonic. Osmolarity depends on molar concentration and not upon equivalents per litre (mequiv/litre). For instance, if Mg^{2+} concentration is 30 mequiv/litre, this is 15 mmol/litre, or 15 mosmol/litre.

100 ml of 0.6 M-$NaHCO_3$ provides 60 mmol Na^+ and 60 mmol HCO^-_3, or 120 mosmol, on the assumption that the salt is completely dissociated.

The osmolarity of body fluids is about 285 mosmol/litre.

osteoarthritis Chronic multiple degenerative joint disease characterized by degeneration of the articular cartilage, hypertrophy of the bone at the margins and changes in the synovial membrane.

osteochondritis Inflammation of both bone and cartilage.

osteomalacia Softening of bones in adults from a deficiency of vitamin D or minerals. There is an increased amount of uncalcified bony matrix (osteoid).

oxalate An organic acid anion that combines with calcium and some other positively charged dietary minerals to form a very insoluble precipitate that inhibits digestion and absorption (see tropical grassland, Ch. 10). Oxalates circulating in the blood also react with ionized blood calcium and form a crystalline deposition in the kidneys.

Oxyuris equi Pinworm. A nematode intestinal worm, which is not a serious hazard to horses but it causes intense irritation of the anal region and this encourages tail-rubbing and biting. Piperazine compounds are effective.

pancreas A gland in the abdominal cavity that secretes a juice containing digestive enzymes into the duodenum (see Ch. 1) and secretes the hormones insulin (q.v.) and glucagon (q.v.) into the bloodstream.

Parascaris equorum A nematode intestinal worm that commonly infects foals, yearlings and horses under 3 years old. It causes intestinal problems, colic, coughing and nasal discharge (see Ch. 11).

parathormone Parathyroid hormone synthesized by the parathyroid gland.

parathyroid This endocrine gland is located in the upper neck adjacent to the thyroid gland. When serum concentration of calcium ions falls, parathyroid hormone is secreted into the blood. It induces mobilization of Ca in the bones increases Ca absorption in the intestines, reabsorption of Ca by the renal tubules and the urinary excretion of P. Its effects are counteracted by that of thyrocalcitonin. See also **secondary nutritional hyperparathyroidism**.

paresis (general) A condition short of complete paralysis in which certain muscles are relaxed and weak. If it is more generalized the animal cannot support itself or stumbles. Sometimes there is slight paralysis – that is there has been some injury, or affect, on certain motor nerves and an inability to make purposeful movements.

paresis (parturient). This clinical sign, associated with hypocalcaemia (q.v.) is *not* characteristic of mares. Lactation tetany (q.v.) was commonly observed in draught horse mares. Hypocalcaemia is a consistent characteristic, although hypomagnesaemia (q.v.) has been associated with recent transport. Mares grazing lush pasture and with a heavy milk flow are particularly prone to tetany. For other causes see Chapters 3, 9, 10 and 11.

parturition The act, or process, of foaling.

pastern The region of the leg between the fetlock and the hoof in both fore- and hindlimbs, formed by the long and the short phalanges, which create the pastern joint. The third phalanx is in the hoof.

peristalsis The rhythmic involuntary muscular contractions of the alimentary canal by which it mixes and propels ingesta towards the rectum.

petechial haemorrhage Small blood spot caused by the effusion of blood from a capillary. Frequently numerous spots are present in tissue reacting to toxins, such as endotoxin.

pH The symbol used in expressing the hydrogen ion concentration of water solutions. It signifies the negative logarithm of the hydrogen ion concentration in gram-molecules (mols) per litre (the logarithm of the reciprocal of the hydrogen ion concentration). pH 7 is neutral. Progressively above the value alkalinity increases and below it acidity increases.

photosensitization The development of abnormally heightened reactivity of the skin to sunlight. In the text, reference is made to a skin reaction of horses following the consumption of certain plants. Reports exist of domestic animals reacting to St John's worts (*Hypericum* spp.) or bog asphodel (*Narthecium ossifragum*). White, grey or piebald horses or those with liver damage are most susceptible.

phytates Much of the phosphorus present in cereal grains and other seeds occurs as phytates, which are calcium, magnesium, and other salts of phytic acid, a phosphoric acid derivative, composed of a six-carbon ring structure with a phosphate group bonded indigestibly to each carbon atom. The horse can utilize the phosphate phosphorus only to the extent of about 30 per cent, partly through the activity of bacterial phytase present in the intestines, but probably also as a result of some phytase activity that has been recognized in the wall of the intestines of several species. Phytates reduce the availability of dietary zinc. Wheat bran contains 1 per cent of P present as phytin.

pinworm See *Oxyuris equi.*

pituitary gland (hypophysis) This gland lies at the base of the brain and is connected with the hypothalamus by the pituitary stalk. Physiologically it consists of two parts, the anterior and posterior pituitary. Six hormones are secreted by the anterior portion: growth hormone, adrenocorticotropin, thyroid stimulating hormone, prolactin, follicle-stimulating hormone and luteinizing hormone. The posterior portion secretes antidiuretic hormone (vasopressin) and oxytocin.

pK The symbol used in expressing the dissociation constant of weak acids (or bases) in the form of a negative logarithm. The larger the value the weaker, or less dissociated, is the acid. When equal concentrations of the salt of an acid and the acid are mixed, the pK = pH, (q.v.) and the buffering capacity of the mixture is maximal. Thus, for the primary ionization of carbonic acid to bicarbonate, as in blood, the pK = 6.36, and at pH 6.36 half the molecules of carbonic acid are dissociated forming bicarbonate. At the normal pH of venous blood (7.4) the mixture is an even more effective buffer to acid produced during anaerobic muscular activity with the formation of undissociated carbonic acid.

placenta An organ that develops within the uterus in early pregnancy and that establishes communication between the dam and the developing foetus. It is composed of a maternal portion and a foetal portion attached to the foetus by the umbilical cord. Following parturition, it is passed as the after-birth.

prostaglandins These are grouped in six main series of cyclic compounds derived from unsaturated fatty acids such as arachidonic acid (itself a derivative of the dietary indispensable (essential) linoleic acid) and from fatty acids with one less and one more double bond. Prostaglandins were first recognized in seminal fluid and the prostate gland. They show a variety of biological actions that influence smooth muscle contraction (as in contraction of uterine muscle and in blood-pressure control) and they are mediators in the regulation of the dilatation and permeability of arterioles, capillaries and venules in the inflammatory response. They are involved in immune mechanisms and are used for oestrus synchronization and for abortions in cases of twin foetuses. Prostaglandin $F_{2\alpha}$ ($PGF_{2\alpha}$) which terminates the life of the corpus luteum, has been the one most commonly used.

protease An enzyme that digests proteins by hydrolytically splitting off amino acids. Several kinds are secreted into the alimentary canal (see Ch. 1).

protein True proteins are chains in which the links are amino acids. All amino acids possess at least one N-containing amino-group. The crude protein content of the diet is defined as the N-content times 6.25 as it is assumed that protein contains 16 per cent N. However, this product includes dietary nucleic acids, nitrogenous glycosides, amines, nitrates, etc. and so overestimates the true protein content.

proteolysis The enzymatic digestion of protein, which, if carried out by the horse's own secretions, yields proteoses, peptones and amino acids, but if carried out by intestinal bacteria it includes deamination with a loss of protein value.

prothrombin time The synthesis of prothrombin occurs in the liver and requires vitamin K. It is essential for the clotting of blood and any defect in prothrombin formation, or in the activity of other substances involved in clotting, extends the interval between the initiation of the process and the formation of fibrin from fibrinogen. Fibrin spontaneously coagulates. Dicoumarol (q.v.), which arises from the activity of a mould on coumarin in spoiled sweet clover or melitots, interferes with the metabolism of vitamin K causing an extension of prothrombin time and consequential extensive haemorrhaging (see Ch. 11).

proximal Nearest to the centre or origin, and opposed to distal (q.v.). The duodenum is proximal to the jejunum and the 'knee' is proximal to the fetlock.

purgative Cathartic, a medicine that stimulates peristaltic action and evacuation of the intestines.

pylorus The distal or duodenal aperture of the stomach. It is controlled by a sphincter muscle and through it stomach contents enter the small intestine.

Quarter Horse (American Quarter Horse) A breed devised mainly from dams of Spanish origin, for long bred by American Indians, and from Galloway sires introduced by early settlers of North America.

Quidding The expulsion of partially chewed feed from the mouth. The habit may arise from injuries to the tongue, or cheek, resulting from molar teeth which are too sharp, irregular in height or in alignment, or even from permanent teeth pushing the temporaries out from the gums. Causes also include infections of the mouth, or teeth and paralysis of the throat and consequent inability to swallow.

radius One of the two long bones of the 'fore-arm', between the point of the 'elbow' and the 'knee'. The other long bone is the ulna.

ration This refers to the amount of daily feed rather than to its composition, but it should include all the constituents of the diet apart from water.

rectum The distal portion of the large intestine extending from the small colon to the anus and holding the faeces.

red blood cells (erythrocytes) The most numerous cells in the blood (6.8–12.9 × 10^{12} per litre), there being only 5.4–14.3 × 10^9 per litre white blood cells (leucocytes). The cellular portion of blood makes up 32–53 per cent of the total, the remainder being the plasma. About 35 per cent of each red cell is the protein haemoglobin, which transports oxygen from the lungs to the various body tissues.

renal clearance This the ratio of the concentration of a substance in the urine to that in the blood times the rate of urine formation. As the latter is normally unknown, the creatinine clearance ratio is measured (see Ch. 12). To calculate the glomerular filtration rate, when the rate of urine formation is known, creatinine (q.v.) or insulin may be used; these substances are readily filtered by the glomerulus but not secreted or absorbed by the renal tubules.

renal tubules These run from the glomerulus through the cortex of the kidney, as convoluted tubules, then through the medulla as collecting tubules, and they open into the pelvis of the kidney at the apices of the renal pyramids. Their main function is the reabsorption of water and various solutes – glucose, chloride, calcium, phosphorus, etc. required by the animal.

reproductive cycle The time from the conception of one foal to the conception of the next.

requirement (nutrient) The requirement for any given nutrient is the amount of that nutrient that must be supplied in the diet to meet the net requirement of a normal healthy animal given a completely adequate diet in an environment compatible with good health. The net requirement is the quantity of that nutrient that should be absorbed to meet the needs of maintenance (q.v.), including the replacement of obligatory losses, and of any work, growth, production or reproduction taking place.

rhinopneumonitis (equine) A mild viral disease of the upper respiratory tract of horses, which also commonly causes abortion.

rickets Defective calcification of the epiphyseal cartilage of growing horses owing to inadequate dietary vitamin D, calcium and phosphorus or an incorrect proportion of calcium to phosphorus in the diet.

Ringer's solution An isotonic solution, devised by Sydney Ringer, containing sodium chloride, potassium chloride and calcium chloride. However, it is but little more physiological than physiological saline as its chloride concentration is even higher and therefore it can cause metabolic acidosis of the same magnitude (see Ch. 9).

roughage There are several types of roughage, which fall broadly into the following categories: (1) long and dry, e.g. hay and straw; (2) ground and pelleted hay, straw and oatfeed; (3) ensiled long grass and comparable succulent forages; and (4) chopped succulent ensiled material. Within each category only feeds analysed to contain more than 20 per cent crude fibre on an air-dry basis should be included. Roughages tend to reduce the intake of dry matter in horses fed *ad libitum* and they decrease net energy intake in these animals in comparison with those also receiving concentrates (q.v.). The fibre is useful in maintaining the microbial populations of the large intestine in a steady state.

ruminant Herbivorous species that possess an enlarged forestomach (rumen) and that chew the cud by regurgitation of ingesta from the forestomach to the mouth.

sclerosis An induration or hardening, especially resulting from persistent inflammation.

scours Diarrhoea.

secondary nutritional hyperparathyroidism The increased secretion of parathyroid hormone (q.v.) as a compensatory mechanism directed against a disturbance in mineral homeostasis induced by nutritional imbalances. A loss of calcium from the bones is induced, resulting in a condition marked by pain spontaneous fractures, muscular weakness and osteofibrosis (see Ch. 3 and 11).

septicaemia A serious condition in which bacteria circulate in the blood stream and become widely distributed throughout practically every organ. The horse becomes distressed in severe septicaemia, respiration rate and heart action are accelerated and body temperature is elevated.

serum (blood) The clear liquid that separates from the clot and the corpuscles in the clotting of blood.

set-fast see **tying-up**.

shank (of long bone) See **diaphysis**.

silage (ensilage) Succulent feed preserved either by adding acid or allowing natural fermentation to occur under anaerobic conditions in compacted material. The pH achieved is approximately 4.0–4.2. Too low a pH limits intake and too high a pH provokes protein breakdown. Materials ensiled include fresh grass, forage crops, the above-ground growth of young cereal crops and a variety of byproducts from beet pulp to fish waste. The dry matter content is 30–50 per cent.

spleen A glandlike, but ductless organ in the anterior part of the abdominal cavity on the left side. Its functions are at least threefold. First it disintegrates red cells setting free the haemoglobin, which the liver converts to bilirubin, conserving the iron. Second, it acts as a storehouse of red cells, which it releases into the blood during times of higher oxygen demand. Third, evidence (mainly from other species) indicates a role for the spleen in immunological responses. The spleen is an antibody-forming tissue and macrophages constitute a predominant cell form in it.

splints Bony enlargements that occur on the cannon bones or in connection with the small metacarpals or metatarsals (splint bones) as the result of localized inflammation of the bone or periosteum (periostitis or osteitis).

stifle The joint corresponding to the human knee at the top of the hindlimb.

stocking up Swelling, or oedema (q.v.), of the legs, owing to the accumulation of fluid beneath the skin frequently caused by a period of inactivity on rich feed immediately after an extended period of activity. Exercise, purging and a reduction in energy intake normally bring rapid relief. The condition is also occasionally seen in horses with damaged livers, especially where the diet is of low quality and deficient in protein.

strangles An acute contagious fever of horses, donkeys and mules caused by the bacterium *Streptococcus equi*, characterized by catarrhal inflammation of the mucous membranes of the nasal passages and pharynx, and frequently accompanied by abscess formation in the submaxillary or pharyngeal lymphatic glands noticeable under the jaw.

stroke volume See **cardiac output**.

strongyles A group of strongyloid nematodes or roundworms widely distributed in the intestinal contents of mammals. In horses, strongyles are commonly called redworm (see Ch. 11).

sucrase (invertase). A digestive enzyme secreted by the small intestine, which hydrolyses sucrose (cane and beet sugar), forming glucose and fructose, both of which are readily absorbed.

sweetfeed An American term for a concentrate mix containing molasses.

synchronous diaphragmatic flutter Contraction of the diaphragm in synchrony with the heart beat. It is observed in fatigued horses following severe exercise in hot weather when excessive sweating may precipitate a large decrease in the plasma concentrations of ionized calcium, chloride and potassium.

tachycardia Excessive rapidity of heart action and pulse.

tapeworm A parasitic intestinal cestode composed of numerous flattened segments and attached to the gut wall by a head. Not infrequently occurring in horses, but generally causes no major problem. Occasionally it may block the ileocaecal sphincter and cause severe colic. This can be accomplished by *Anoplocephala perfoliata* the only species to have been observed infect horses in the United Kingdom.

tarsus Bones of the hock. There are usually six bones of the hock, which join the tibia above to the two metatarsal bones below in the hindleg.

tetany A condition in which there are localized spasmodic contractions, or twitching, of muscles (see Ch. 11 for stress tetany).

Thoroughbred A hotblooded (q.v.) breed of about 16 hands, which originated in the United Kingdom and which has been used to improve many other breeds.

thrush A degenerative condition of the horn in the central cleft of the frog (q.v.) of the horse's foot caused by a bacterial infection and often resulting from their standing in dirty wet boxes with insufficient clean dry bedding.

thumps See **Synchronous diaphragmatic flutter**.

thyroid gland Situated in the neck in connection with the upper extremity of the trachea. The gland secretes two hormones: an iodine containing thyroxine and calcitonin (thyrocalcitonin) (see Ch. 3).

tibia The major long bone between the stifle (q.v.) and the hock (q.v.) of the hind limb.

tillering The process of forming side shoots from the base of the stem of graminaceous plants (cereals and grasses). Grazing, or cutting, in the vegetative phase of growth encourages this process so thickening the 'bottom' or base of young pasture swards and increasing their suitability for grazing and exercising horses.

α-tocopherol The principal tocopherol with vitamin E potency (see Ch. 4).

toxin Any poisonous substance of microbial, vegetable or animal origin.

trace elements See **minerals.**

tricarboxylic acid (TCA) cycle The series of metabolic reactions by which acetyl coenzyme A (coA) is oxidized to carbon dioxide and water. The energy released is stored as ATP (q.v.) and the process occurs only in the mitochondria of cells (see Ch. 9). Acetyl CoA is generated by the catabolism (q.v.) of fatty acids, carbohydrates and amino acids.

trypsin This enzyme acts on peptide linkages that involve the carboxyl groups of lysine and arginine and ruptures protein chains at these points. It is one of the proteolytic enzymes secreted by the pancreas, but in the inactive form of trypsinogen. This is activated by the enzyme enterokinase liberated from the duodenal mucosa.

tying-up (set-fast) A condition in racehorses in which stiffness, blowing and sweating occurs after a period of hard extended exercise said to be caused by a depletion of muscle glycogen (see Ch. 9 and 11).

tympany Distension of the stomach, or intestines, by gas usually caused by rapid microbial fermentation of ingesta, leading to a drum-like condition of the abdomen and colic.

ulna A bone behind the radius in the foreleg that forms the point of the elbow. The shaft of the ulna is vestigial and the tapered end is fused to the radius.

umbilical cord The nourishment of the foetus mainly passes to it through the cord from the placenta. After birth, the cord should not be interfered with as early severance deprives the newborn foal of 1000–1500 ml of placental foetal blood, whereas under 'natural' conditions the amount concerned is under 200 ml.

uraemia The presence of urinary constituents in the blood and the toxic condition produced thereby. Normal blood urea level is greatly exceeded, indicating a failure of normal renal function.

urea The chief nitrogenous wasteproduct synthesized in the liver, discharged from the body in the urine and also secreted into the small intestine. It is highly soluble in water and the amount produced by the healthy horse receiving regular meals is proportional to the crude protein content of the diet.

urticaria See **hives.**

uterus A hollow muscular organ lying in the abdominal cavity below the rectum. In the mare it has a large body and small horns. It carries the foetal foal and nourishes it during pregnancy through the placenta attached to its wall.

vagus (pneumogastric) nerve. A major parasympathetic nerve (tenth cranial) of the autonomic nervous system (q.v.) possessing both efferent and afferent fibres distributed to the larynx, lungs, heart, oesophagus, stomach, liver, intestines and in fact most of the abdominal viscera. It therefore plays a considerable part in digestion and physical exercise.

vein Thinner walled vessels than arteries (q.v.), in which deoxygenated blood under lower pressure is carried back to the heart. Veins possess a system of valves that control the direction of blood flow. Muscular contraction and relaxation of the limbs therefore provides the main force by which blood is lifted back up the legs against the force of gravity.

vertebrae A chain of bones running from the base of the skull to the tip of the tail and which carries in the spinal canal the spinal cord – the posterior part of the central nervous system.

VFA See **volatile fatty acids.**

villus (pl. villi) Small vascular processes covering the mucous epithelium of the small intestine. They greatly enlarge the surface area for the absorption of nutrients into branches of the portal blood system and into the lymphatic system.

viscera In this text the term refers to the abdominal viscera – the organs of the abdominal cavity.

vitamins These are a group of unrelated organic substances that occur in many foods in small amounts and that are necessary for normal metabolism of the body. They have been arbitrarily divided into a group of four major fat-soluble vitamins and at least ten water-soluble ones (see Ch. 4).

volatile fatty acids These are short-chain steam volatile acids, principally acetic, propionic, butyric and smaller quantities of higher acids, which are the microbial wasteproducts of fermentation of dietary polysaccharides and protein within the alimentary canal. They are absorbed into the bloodstream and constitute a major energy source to the horse.

warmblooded See **hotblooded.**

wasting disease A state of chronic emaciation.

windsucking (crib-biting) In the English literature these terms refer to the act of swallowing gulps of air into the stomach. They are habitual vices. The crib-biter effects the action by grasping the edge of the manger, fence, etc. with the incisor teeth and by the coincidence of the raising of the floor of the mouth, opening the soft palate and a swallowing action, a gulp of air passes into the stomach (Pl. 11.1). The wind-sucker achieves the same end without a resting place for the teeth. Young idle animals are said to acquire the habit from individuals in their company with a confirmed habit, which can initiate repeated bouts of mild colic. Excessive wear of the incisor teeth may compromise the individual's grazing powers. Sometimes a cribbing strap is fitted snugly around the upper neck of persistent offenders. The American literature restricts the term 'windsucker' to mares that aspirate air and faecal material into the vagina. This is corrected by Caslick's operation. The term 'cribbing' is then reserved for the aspiration of air into the stomach.

withers The ridge on the back of the horse over the dorsal processes of the thoracic vertebrae and the shoulder blades and directly in front of the saddle.

wobbler The name given to a horse showing a slight swaying action of the hindquarters, or stumbling, occurring mainly between 1 and 3 years of age. The signs can become progressively worse over 6–9 months when the horse is unable to trot without rolling from side to side and falling. The condition apparently results from damage to the spinal cord in the neck through injury and/or nutritionally induced abnormalities of the vertebrae caused by imbalances or inadequacies of calcium and phosphorus.

wood chewing A habit developed by many horses probably as a result of boredom. The horse normally does not swallow the wood and the habit is unlikely to have any dietary implications.

zone of thermal neutrality This is the range of environmental temperatures over which heat production by the animal is minimized. Below this range the animal must increase heat production by shivering and other means in order to maintain a normal body temperature. Above the range normal cooling mechanisms prove inadequate, body temperature rises and with it metabolic rate.

Bibliography

Abrams J T (1979) The effect of dietary vitamin A supplements on the clinical condition and track performance of racehorses. *Biblthca Nutr Dieta* **27**, 113–20

Adam K M G (1951) The quantity and distribution of the ciliate protozoa in the large intestine of the horse. *Parasitology* **41**, 301–11

Agricultural Research Council (1980) *The Nutrient requirements of ruminant livestock*, p. 293. Commonwealth Agricultural Bureaux: London

Agricultural Research Council (1981) *The nutrient requirements of pigs*, p. 146, Commonwealth Agricultural Bureaux: London

Aitken M M, Anderson M G, Mackenzie G & Sanford J (1974) Correlations between physiological and biochemical parameters used to assess fitness in the horse. *JS Afr vet Ass* **45**, 361–70

Aldred T, Fontenot J P & Webb K E Jr (1978) Availability of phosphorus from three sources for ponies. *Va Polytech Inst State Univ Res Div Rep* **174**, 152–7

Alexander F (1963) Digestion in the horse, in *Progress in nutrition and allied sciences*, ed. Cuthbertson D P, pp. 259–68. Oliver & Boyd: Edinburgh

Alexander F (1972) Symposium (1) Certain aspects of the physiology and pharmacology of the horse's digestive tract. *Equine vet J* **4**, 166–9

Alexander F (1978) The effect of some anti-diarrhoeal drugs on intestinal transit and faecal excretion of water and electrolytes in the horse. *Equine vet J* **10**, 229–34

Alexander F & Benzie D (1951) A radiological study of the digestive tract of the foal. *Q J exp Physiol* **36**, 213–17

Alexander F & Hickson J C D (1969) The salivary and pancreatic secretions of the horse, in *Physiology of digestion and metabolism in the ruminant* (*Proc 3rd Int Symp*, August 1969, *Cambridge, England*), ed. A T Phillipson, pp. 375–89. Oriel Press: Stocksfield, Northumberland

Alexander F, Macpherson J D & Oxford A E (1952) Fermentative activities of some members of the normal coccal flora of the horse's large intestine. *J comp Path* **62**, 252–9

Aluja A S, de Gross D R, McCosker P J & Svendsen J (1968) Effect of altitude on horses. *Vet Rec* **82**, 368–72

Anderson C E, Potter G D, Kreider J L & Courtney C C (1981) Digestible energy requirements for exercising horses. *J Anim Sci* **53** (suppl. 1) 42, abstr. 101

Anderson C E, Potter G D, Kreider J L & Courtney C C (1983) Digestible energy requirements for exercising horses. *J Anim Sci* **56**, 91–5

Anderson M G (1975a) The effect of exercise on blood metabolite levels in the horse. *Equine vet J* **7**, 27–33

Anderson M G (1975b) The influence of exercise on serum enzyme levels in the horse. *Equine vet J* 7, 160–5

Anderson M G (1976) The effect of exercise on the lactic dehydrogenase and creatine kinase isoenzyme composition of horse serum. *Res vet Sci* 20, 191–6

Anderson P H, Patterson D S P & Berrett S (1978) Selenium deficiency. *Vet Rec* 103, 145–6

Anderson R A, Bryden N A, Polansky M M & Patterson K Y (1983) Strenuous exercise: effects on selected clinical values and chromium, copper and zinc concentrations in urine and serum of male runners. *Fedn Proc* 42, 804, abstr. 2998

Angsubhakorn S, Poomvises P, Romruen K & Newberne P M (1981) Aflatoxicosis in horses. *J Am vet med Ass* 178, 274–8

Anon (1976a) Drugs and doping in horses. *Vet Rec* 98, 453

Anon (1976b) Salmonellosis in horses. *Vet Rec* 99, 19–20

Answer M S, Chapman T E & Gronwell R (1976) Glucose utilization and recycling in ponies. *Am J Physiol* 230, 138–42.

Answer M S, Gronwall R, Chapman T E & Klentz R D (1975) Glucose utilization and contribution to milk components in lactating ponies. *J Anim Sci* 41, 568–71

Archer M (1973) Variations in potash levels in pastures grazed by horses: a preliminary communication. *Equine vet J* 5, 45–6

Archer M (1978a) Studies on producing and maintaining balanced pastures for studs. *Equine vet J* 10, 54–9

Archer M (1978b) Further studies on palatability of grasses to horses. *J Br Grassld Soc* 33, 239–43

Argenzio R A (1975) Functions of the equine large intestine and their interrelationship in disease. *Cornell Vet* 65, 303–29

Argenzio R A & Hintz H F (1970) Glucose tolerance and effect of volatile fatty acid on plasma glucose concentration in ponies. *J Anim Sci* 30, 514–18

Argenzio R A & Hintz H F (1972) Effect of diet on glucose entry and oxidation rates in ponies. *J Nutr* 102, 879–92

Argenzio R A, Lowe J E, Hintz H F & Schryver H F (1974) Calcium and phosphorus homeostasis in horses. *J Nutr* 104, 18–27

Argenzio R A, Southworth M, Lowe J E & Stevens C E (1977) Interrelationship of Na, HCO_3, and volatile fatty acid transport by equine large intestine. *Am J Physiol* 233, E469–78

Argiroudis S A, Kent J E & Blackmore D J (1982) Observations on the isoenzymes of creatine kinase in equine serum and tissues. *Equine vet J* 14, 317–21

Asmundsson T, Gunnarsson E & Johannesson T (1983) 'Haysickness' in Icelandic horses: precipitin tests and other studies. *Equine vet J* 15, 229–32

Austic R E (1980) Acid–base interrelationships in nutrition. *Proc Cornell Nutr Conf Feed Manuf* 12–17. Cornell University: Ithaca, NY

Baker J P (1971) Horse nutritive requirements. *Feed Management* (Sept), 10–15

Baker J P, Lieb S, Crawford B H Jr & Potter G D (1972) Utilization of energy sources by the equine. *Proc 27th Distillers Feed Conf* 27, 28–33. Distillers Co. Ltd

Baker J P, Sutton H H, Crawford B H Jr & Lieb S (1969) Multiple fistulation of the equine large intestine. *J Anim Sci* 29, 916–20

Baker J R (1970) Salmonellosis in the horse. *Br vet J* 126, 100–5

Baker J R & Leyland A (1973) Diarrhoea in the horse associated with stress and tetracycline therapy. *Vet Rec* **93**, 583–4

Baker J P & Quinn P J (1978) A report on clinical aspects and histopathology of sweet itch. *Equine vet J* **10**, 243–8

Balls D (1976) Notes on equine toxicology. *Vet Pract* **8**, 5–6

Banach M A & Evans J W (1981a) The effects of energy intake during gestation and lactation on reproductive performance in mares. *Proc West Sect Am Soc Anim Sci* **32**, 264–7

Banach M A & Evans J W (1981b) The effects of energy intake during gestation and lactation on reproductive performance in mares. *J Anim Sci* **53** (suppl. 1) 500, abstr. 94

Baranova D (1977) Vitamins in the feeding of weaned foals. *Konevodstvoi Konnyi Sport* no. 10, 29–30

Barratt M E J, Strachan P J & Porter P (1979) Immunologically mediated nutritional disturbances associated with soya-protein antigens. *Proc Nutr Soc* **38**, 143–50

Bartel D L, Schryver H F, Lowe J E & Parker R A (1978). Locomotion in the horse: a procedure for computing the internal forces in the digit. *Am J vet Res* **39**, 1721–7

Barth K M, Williams J W & Brown D G (1977) Digestible energy requirements of working and non-working ponies. *J Anim Sci* **44**, 585–9

Bartley E E, Avery T B, Nagaraja T G, Watt B R, Davidovich A, Galitzer S & Lassman B (1981) Ammonia toxicity in cattle v. ammonia concentration of lymph and portal, carotid and jugular blood after the ingestion of urea. *J Anim Sci* **53**, 494–8

Basler S E & Holtan D W (1981) Factors affecting blood selenium levels in Oregon horses and association of blood selenium level with disease incidence. *Proc West Sect Am Soc Anim Sci* **32**, 399–400

Bauer J E (1983) Plasma lipids and lipoproteins of fasted ponies. *Am J vet Res* **44**, 379–84

Belko A Z & Roe D A (1983) Exercise effects on riboflavin status. *Fedn Proc* **42**, 804, abstr. 2995

Bendroth M (1981) A survey of reasons for some trotters being non-starters as 2-, 3-, and 4-year olds, *Proc 32nd A Meet Eur Assn Anim Prod* IIa–1

Bentley O E, Burns S J, McDonald D R, Drudge J H, Lyons E T, Kruckenberg S M & Vaughn J T (1978) Safety evaluation of pyrantel pamoate administered with Trichlorfon as a broad-spectrum anthelmintic in horses. *Vet Med Small Anim Clin* **73**, 70–3

Bergsten G, Holmbäck R & Lindberg P (1970) Blood selenium in naturally fed horses and the effect of selenium administration. *Acta vet Scand* **11**, 571–6

Berliner V R (1942) Seasonal influences on the reproductive performance of mares and jennets in Mississippi. *J Anim Sci*, 63–4

Beuchat L R (1978) Microbial alternations of grains, legumes and oil seeds. *Food Tech* (May), 193–6

Blackmore D J & Brobst D (1981) *Biochemical values in equine medicine*. Animal Health Trust: Newmarket, Suffolk

Blackmore D J & Elton D (1975) Enzyme activity in the serum of Thoroughbred horses in the United Kingdom. *Equine vet J* **7**, 34–9

Blackmore D J, Henley M I & Mapp B J (1983) Colorimetric measurement of albumin in horse sera. *Equine vet J* **15**, 373–4

Blackmore D J, Willett K & Agness D (1979) Selenium and gamma-glutamyl transferase activity in the serum of thoroughbreds. *Res vet Sci* **26**, 76–80

Blackmore D J, Campbell C, Dant C, Holden J E & Kent J E (1982) Selenium status of thoroughbreds in the United Kingdom. *Equine vet J* **14**, 139–43

Blaney B J, Gartner R J W & McKenzie R A (1981a) The effect of oxalate in some tropical grasses on the availablility to horses of calcium, phosphorus and magnesium. *J agric Sci Camb* **97**, 507–14

Blaney B J, Gartner R J W & McKenzie R A (1981b) The inability of horses to absorb calcium from calcium oxalate. *J agric Sci Camb* **97**, 639–41

Blaxter K L (1962) *The energy metabolism of ruminants.* Hutchinson: London

Bolton J R, Merritt A M, Cimprich R E, Ramberg C F & Streett W (1976) Normal and abnormal xylose absorption in the horse. *Cornell Vet* **66**, 183–90

Bouwman H (1978) Digestibility trials with extruded feeds and rolled oats in ponies. *Landbouwkundig Tijdschr* **90**, 2–6

Bouwman H & Schee W van der (1978) Composition and production of milk from Dutch warmblooded saddle horse mares. *Z Tiephysiol Tiernähr Futtermittelk* **40**, 39–53

Bowland J P & Newell J A (1974) Fatty acid composition of shoulder fat and perinephric fat from pasture-fed horses. *Can J Anim Sci* **54**, 373–6

Bowman V A, Meacham T N, Dana G R & Fontenot J P (1978) Pelleted complete rations containing different roughage bases for horses. *Va Polytech Inst State Univ Res Div Rep* **174**, 179–82

Brady P S, Ku P K & Ullrey D E (1978) Lack of effect of selenium supplementation on the response of the equine erythrocyte glutathione system and plasma enzymes to exercise. *J Anim Sci* **47**, 492–6

Brady P S, Shelle J E & Ullrey D E (1977) Rapid changes in equine erythrocyte glutathione reductase with exercise. *Am J vet Res* **38**, 1045–7

Breuer L H (1970) Horse nutrition and feeding. *Feedstuffs* **42**, 44–5

Breukink H J (1974) Oral mono- and disaccharide tolerance tests in ponies. *Am J Vet Res* **35**, 1523–7

British Equine Veterinary Association [*c.* 1978] Working party report, *Veterinary guidelines on equine endurance competitions.* BEVA: London

British Horse Society (n.d.) *Grassland management for horse and pony owners.* BHS: Kenilworth, Warwickshire

Brobst D F & Bayly W M (1982) Response of horses to a water deprivation test. *J Equine vet Sci,* **2** 51–6

Brook D & Schmidt G R (1979) Pre-renal azotaemia in a pony with an oesophageal obstruction. *Equine vet J,* **11** 53–5

Brophy P O (1981) Assessment of the immunological status of the newborn foal. *Proc 32nd A Meet Eur Assn Anim Prod* IIIa–5

Brown R H (1978) Horses can digest high fat levels, Georgians told. *Feedstuffs* **50**(10), 15

Brown R F, Houpt K & Schryver H F (1976) Stimulation of food intake in horses by diazepam and promazine. *Pharmac Biochem Behav* **5**, 495–7

Brownlow M A & Hutchins D R (1982) The concept of osmolality; its use in the evaluation of 'dehydration' in the horse. *Equine vet J* **14**, 106–10

Buntain B J & Coffman J R (1981) Polyuria polydypsia in a horse induced by psychogenic salt consumption. *Equine vet J* **13**, 266–8

Burke D J & Albert W W (1978) Methods for measuring physical condition and energy expenditure in horses. *J Anim Sci* **46**, 1666–72

Burridge J C, Reith J W S & Berrow M L (1983) Soil factors and treatments affecting trace elements in crops and herbage, in *Trace elements in animal production and veterinary practice* eds NF Suttle, R G Gunn, W M Allen, K A Linklater & G Weiner, pp. 77–85. British Society of Animal Production (Occasional Publication no. 7): Edinburgh

Burrows G E (1981) Endotoxaemia in the horse. *Equine vet J* **13**, 89–94

Butler K D Jr & Hintz H F (1977) Effect of level of feed intake and gelatin supplementation on growth and quality of hoofs of ponies. *J Anim Sci* **44**, 257–63

Butler P & Blackmore D J (1982) Retinol values in the plasma of stabled thoroughbred horses in training. *Vet Rec* **111**, 37–8

Butler P & Blackmore D J (1983) Vitamin E values in the plasma of stabled thoroughbred horses in training. *Vet Rec* **112**, 60

Callear J F F & Neave R M S (1971) The clinical use of the anthelmintic membendazole. *Br vet J* **127**, xli–xliii

Cameron I R & Hall R J C (1975) The effect of dietary K⁺ depletion and subsequent repletion on intracellular K⁺ concentration and pH of cardiac and skeletal muscle in rabbits. *J Physiol* **251**, 70–71P

Campbell J R (1977) Bone growth in foals and epiphyseal compression. *Equine vet J* **9**, 116–21

Campbell J R & Lee R (1981) Radiological estimation of differential growth rates of the long bones of foals. *Equine vet J* **13**, 247–50

Candau M & Bueno L (1977) Motricité caecale et transit chez le poney: influence de l'état de réplétion du caecum et des fermentations microbiennes. *Annls Biol anim Biochim Biophys* **17**, 503–8

Cape L & Hintz H F (1982) Influence of month, color, age, corticosteroids, and dietary molybdenum on mineral concentration of equine hair. *Am J vet Res* **43**, 1132–6

Caple I W, Edwards S J A, Forsyth W M, Whiteley P, Selth R H & Fulton L J (1978) Blood glutathione peroxidase activity in horses in relation to muscular dystrophy and selenium nutrition. *Aust vet J* **54**, 57–60

Cardinet G H, Fowler M E & Tyler W S (1963) Heart rates, respiratory rates for evaluating performance in horses during endurance trial ride competition. *J Am vet med Ass* **143**, 1303–9

Cardinet G H, Littrell J F & Freedland R A (1967) Comparative investigations of serum creatine phosphokinase and glutamic-oxaloacetic transaminase activities in equine paralytic myoglobinuria. *Res vet Sci* **8**, 219–26

Carlson G P (1975) Hematological alterations in endurance-trained horses. *Proc 1st Int Symp Equine Hematol* 444–9

Carlson G P (1983a) Response to saline solution of normally fed horses and horses dehydrated by fasting. *Am J vet Res* **44**, 964–8

Carlson G P (1983b) Thermoregulation and fluid balance in the exercising horse in *Equine exercise physiology* (*Proc 1st Int Conf, Oxford 1982*) eds Snow D H, Persson S G B & Rose R J, pp. 291–309. Granta Editions: Cambridge

Carlson G P & Mansmann R A (1974) Serum electrolyte and plasma protein alterations in horses used in endurance rides. *J Am vet med Ass* **165**, 262–4

Carlson G P & Ocen P O (1979) Composition of equine sweat following exercise in high environmental temperatures and in response to intravenous epinephrine administration. *J Equine med Surg* **3**, 27–31

Carlson G P, Ocen P O & Harrold D (1976) Clinicopathological alterations in normal and exhausted endurance horses. *Theriogenology* **6**, 93–104

Carlson L A, Fröberg S & Persson S (1965) Concentration and turnover of the free fatty acids of plasma and concentration of blood glucose during exercise in horses. *Acta physiol Scand* 63, 434–41

Carroll C L and Huntingdon P J (1988). Body condition scoring and weight estimation of horses. *Equine Vet J*, 20, 41–45.

Carroll F D, Goss H & Howell C E (1949) The synthesis of B vitamins in the horse. *J Amin Sci* 8, 290–6

Carson Katherine & Wood-Guch D G M (1983) Behaviour of Thoroughbred foals during nursing. *Equine vet J* 15, 257–62

Chachula J & Chachulowa J (1969) The use of the concentrates in feeding arden and fjord stallions. *Roczn Nauk Roln* 91B–4, 635–56

Chachula J & Chachulowa J (1970) Further investigations on the use of mixed feeds in feeding of various groups of breeding horses. *Roczn Nauk Roln* 92B–3, 351–75

Chachula J & Chrzanowski S (1972) Investigations upon several factors affecting the results of breeding of the fjording horse at the Nowielice state stud. *Biul Inst Genetyki i Hodowli Zwierzat Warszawa* no. 26, 71–85

Chapman D I, Haywood P E & Lloyd P (1981) Occurrence of glycosuria in horses after strenuous exercise. *Equine vet J* 13, 259–60

Chrichlow E C, Yoshida K & Wallace K (1980) Dust levels in a riding stable. *Equine vet J* 12, 185–8

Clark I (1969) Metabolic interrelations of calcium, magnesium and phosphorus. *Am J Physiol* 217, 871–8

Clater F (1786) *Every man his own farrier, or, the whole art of farriery laid open.* J Tomlinson for Baldwin and Bladon: Newark

Codazza D, Maffeo G & Redaelli G (1974) Serum enzyme changes and haematochemical levels in Thoroughbreds after transport and exercise. *J S Afr vet Assn* 45, 331–4

Coffman J (1979a) Blood glucose 1–Factors affecting blood levels and test results. *Vet Med Small Anim Clin* 74, 719–23

Coffman J (1979b) Blood glucose 2–Clinical application of blood glucose determination *per se*. *Vet Med Small Anim Clin* 74, 855–8

Coffman J (1979c) Plasma lactate determinations. *Vet Med Small Anim Clin* 74, 997–1002

Coffman J (1979d) The plasma proteins. *Vet Med Small Anim Clin* 74, 1168–70

Coffman J (1980a) Calcium and phosphorus physiology and pathophysiology. *Vet Med Small Anim Clin* 75, 93–6

Coffman J (1980b) Adrenocortical pathophysiology and consideration of sodium, potassium and chloride. *Vet Med Small Anim Clin* 75, 271–5

Coffman J (1980c) Acid:base balance. *Vet Med Small Anim Clin* 75, 489–98

Coffman J (1980d) Percent creatinine clearance ratios. *Vet Med Small Anim Clin* 75, 671–6

Coffman J (1980e) Urology–2. Testing for renal disease. *Vet Med Small Anim Clin* 75, 1039–44

Coffman J (1980f) Hemostasis and bleeding disorders. *Vet Med Small Anim Clin* 75 1157–64

Coffman J (1980g) A data base for abdominal pain–1. *Vet Med Small Anim Clin* 75 1583–8

Coffman J R, Hammond L S, Garner H E, Thawley D G & Selby L A (1980) Haematology as an aid to prognosis of chronic laminitis. *Equine vet J* 12 30–1

Colles C M (1979) A preliminary report on the use of Warfarin in the treatment of navicular disease. *Equine vet J* **11**, 187–90

Comben N, Clark R J, Sutherland D J B (1983) Improving the integrity of hoof horn in equines by high-level dietary supplementation with biotin. *Proc A Cong Br Equine Vet Assn* 1–17

Comben N, Clark R J & Sutherland D J B (1984) Clinical observations on the response of equine hoof defects to dietary supplementation with biotin. *Vet Rec* **115**, 642–5

Comerford P M, Edwards R L, Hudson L W & Wardlaw F B (1979) Supplemental lysine and methionine for the equine. *Tech Bull S Carolina Agric exp Stn* No. 1073

Comline R S, Hall L W, Hickson J C D, Murillo A & Walker R G (1969) Pancreatic secretion in the horse. *Proc physiol Soc* **204**, 10–11P

Comline R S, Hickson J C D & Message M A (1963) Nervous tissue in the pancreas of different species. *Proc physiol Soc* **170**, 47–48P

Cook W R (1973) Diarrhoea in the horse associated with stress and tetracycline therapy. *Vet Rec* **93**, 15–17

Cooper J P, Green J D & Haggar R (1981) *The management of horse paddocks. A booklet of instructions*. Horserace Betting Levy Board: London

Cornell C N, Garner G B, Yates S G & Bell S (1982) Comparative fescue foot potential of fescue varieties. *J Anim Sci* **55**, 180–4

Cornwell R L & Jones R M (1968) Critical tests in the horse with the anthelmintic pyrantel tartrate. *Vet Rec* **82**, 483–4

Cunha T J (1969) Horse feeding and nutrition. *Feedstuffs* **41**(28), 19–24

Cunha T J (1971) The mineral needs of the horse. *Feedstuffs* **43**(46), 34–8

Cygax A & Gerber H (1973) Normal values of and the effect of age on haematocrit, total bilirubin, calcium, inorganic phosphates and alkaline phosphatase in the serum of horses. *Schweiz Arch Tierheilkunde* **115**, 321–31

Cymbaluk N F, Fretz P B & Loew F M (1978) Amprolium-induced thiamine deficiency in horses: clinical features. *Am J Vet Res* **39**, 255–61

Cymbaluk N F, Schryver H F & Hintz H F (1981) Copper metabolism and requirement in mature ponies. *J Nutr* **111**, 87–95

Cymbaluk N F, Schryver H F, Hintz H F, Smith D F & Lowe J E (1981) Influence of dietary molybdenum on copper metabolism in ponies. *J Nutr* **iii**, 96–106

Cysewski S J, Pier A C, Baetz A L & Cheville N F (1982) Experimental equine aflatoxicosis. *Toxicol Appl Pharmac* **65**, 354–65

Darlington F G & Chassels J B (1960) The final inclusive report on a 5-year-study on the effect of administering alpha-tocopherol to Thoroughbreds. *The Summary* **12**, 52

Datt S C & Usenik E A (1975) Intestinal obstruction in the horse. Physical signs and blood chemistry. *Cornell Vet* **65**, 152–72

Davies J V, Gerring E L, Goodburn R & Manderville P (1984) Experimental ischaemia of the ileum and concentrations of the intestinal isoenzyme of alkaline phosphatase in plasma and peritoneal fluid. *Equine vet J* **16**, 215–7

Davies M E (1968) Role of colon liquor in the cultivation of cellullolytic bacteria from the large intestine of the horse. *J appl Bact* **31**, 286–9

Davies M E (1971) The production of vitamin B_{12} in the horse. *Br vet J* **127**, 34–6

Davies W (1952) *The grass crop*. E. & F.N. Spon: London

Dawson W M, Phillips R W & Speelman S R (1945) Growth of horses under Western range conditions. *J Anim Sci* **4**, 47–51

De Gray T (1639) *The compleat horseman and expert farrier. In two bookes*. Thomas Harper: London

Demarquilly C (1970) Feeding value of green forages as influenced by nitrogen fertilization . *Annls Zootech* **19**, 423–37

Denman A M (1979) Nature and diagnosis of food allergy. *Proc Nutr Soc* **38**, 391–402

Dixon P M & Brown R (1977) Effects of storage on the methaemoglobin content of equine blood. *Res vet Sci* **23**, 241–3

Dixon P M, McPherson E A & Muir A (1977) Familial methaemoglobinaemia and haemolytic anaemia in the horse associated with decreased erythrocytic glutathione reductase and glutathione. *Equine vet J* **9**, 198–201

Dollahite J W, Younger R L, Crookshank H R, Jones L P, Peterson H D (1978) Chronic lead poisoning in horses. *Am J vet Res* **39**, 961–4

Donoghue S, Kronfield D S, Berkowitz S J & Copp R L (1981) Vitamin A nutrition of the equine: growth serum biochemistry and hematology. *J Nutr* **111**, 365–74

Doreau M (1978) Comportement alimentaire du cheval à l'écurie *Annls Zootech* **27**, 291–302

Dorn C R, Garner H E, Coffman J R, Hahn A W & Tritschler L G (1975) Castration and other factors affecting the risk of equine laminitis *Cornell Vet* **65**, 57–64

Drew B, Barber W P & Williams D G (1975) The effect of excess dietary iodine on pregnant mares and foals. *Vet Rec* **97**, 93–5

Drudge J H & Lyons E T (1972) Critical tests of a resin-pellet formulation of dichlorvos against internal parasites of the horse. *Am J vet Res* **33**, 1365–75

Drummond R O (1981) Biology and control of insect pests of horses. *Pony of the Americas* **26** (August), 30–2

Duncan J L & Reid J F S (1978) An evaluation of the efficacy of oxfendazole against the common nematode parasites of the horse. *Vet Rec* **103**, 332–4

Duncan J L, McBeath D G & Preston N K (1980) Studies on the efficacy of fenbendazole used in a divided dosage regime against strongyle infections in ponies. *Equine vet J* **12**, 78–80

Duncan J L, McBeath D G, Best J M J & Preston N K (1977) The efficacy of fenbendazole in the control of immature strongyle infections in ponies. *Equine vet J* **9**, 146–9

Dunsmore J D (1985) Integrated control of *Strongylus vulgaris* infection in horses using ivermectin. *Equine vet J* **17**, 191–5

Dybdal N O, Gribble D, Madigan J E & Stabenfeldt G H (1980) Alterations in plasma corticosteroids, insulin and selected metabolites in horses used in endurance rides. *Equine vet J* **12**, 137–40

Dyce K M, Hartman W & Aalfs R H G (1976) A cinefluoroscopic study of the caecal base of the horse. *Res vet Sci* **20**, 40–6

Egan D A & Murrin M P (1973) Copper concentration and distribution in the livers of equine fetuses, neonates and foals. *Res vet Sci* **15**, 147–8

El Shorafa W M (1978) Effect of vitamin D and sunlight on growth and bone development of young ponies. *Diss Abstr Int B* **30**, 1556–7, no. 7817436

El Shorafa W M, Feaster J P, Ott E A & Asquith R L (1979) Effect of vitamin D and sunlight on growth and bone development of young ponies. *J Anim Sci* **48**, 882–6

Ellis R N W & Lawrence T L J (1978a) Energy under-nutrition in the weanling filly foal. I. Effects on subsequent live-weight gains and onset of oestrus. *Br vet J* **134**, 205–11

Ellis R N W & Lawrence T L J (1978b) Energy under-nutrition in the weanling filly foal. II. Effects on body conformation and epiphyseal plate closure in the fore-limb. *Br vet J* **134**, 322–32

Ellis R N W & Lawrence T L J (1978c) Energy under-nutrition in the weanling filly foal. III. Effects on heart rate and subsequent voluntary food intake. *Br vet J* **134**, 333–41

Ellis R N W & Lawrence T L J (1979) Energy and protein under-nutrition in the weanling filly foal. *Br vet J* **135**, 331–7

Ellis R N W & Lawrence T L J (1980) The energy and protein requirements of the light horse. *Br vet J* **136**, 116–21

Elsden S R, Hitchcock M W S, Marshall R A & Phillipson A T (1946) Volatile acid in the digesta of ruminants and other animals. *J exp Biol* **22** 191–202

Epstein V (1984) Relationship between potassium administration, hyperkalaemia and the electrocardiogram: an experimental study. *Equine vet J* **16**, 453–6

Essén B, Lindholm A & Thornton J (1980) Histochemical properties of muscle fibre types and enzyme activities in skeletal muscles of Standardbred trotters of different ages. *Equine vet J* **12**, 175–80

Essén-Gustavsson B, Karlström K & Lindholm A (1984) Fibre types, enzyme activities and substrate utilisation in skeletal muscles of horses competing in endurance rides. *Equine vet J* **16**, 197–202

Eyre P (1972) Equine pulmonary emphysema: a bronchopulmonary mould allergy. *Vet Rec* **91**, 134–40

Fagan T W (1928) Factors that influence the chemical composition of hay. *Welsh J Agric* **4**, 92

Fonnesbeck P V (1981) Estimating digestible energy and TDN for horses with chemical analysis of feeds. *J. Anim Sci* **53** (suppl. 1) 241, abstr. 290

Fonnesbeck P V & Symons L D (1967) Utilization of the carotene of hay by horses. *J Anim Sci* **26**, 1030–8

Forbes T J, Dibb C, Green J O, Hopkins A & Peel S (1980) *Factors affecting the productivity of permanent grassland. A national farm study.* The GRI–ADAS Joint Permanent Pasture Group, pp. 141. The Grassland Research Institute: Hurley, Maidenhead

Ford C W, Morrison I M & Wilson J R (1979) Temperature effects on lignin hemicellulose and cellulose in tropical and temperate grasses. *Aust J agric Res* **30**, 621–33

Ford E J H & Evans J (1982) Glucose utilization in the horse. *Br J Nutr* **48**, 111–17

Ford J & Lokai M D (1979) Complications of sand impaction colic. *Vet Med Small Anim Clin* **74**, 573–8

Franceso L L, Saurin L L & Dibana G F (1981) Mechanism of negative potassium balance in the magnesium deficient rat. *Proc Soc exp Biol Med* **168**, 382–8

Francis-Smith K & Wood-Gush D G M (1977) Coprophagia as seen in Thoroughbred foals. *Equine vet J* **9**, 155–7

Frank C J (1970) Equine colic – a routine modern approach. *Vet Rec* **87**, 497–8

Frape D L (1975) Recent research into the nutrition of the horse. *Equine vet J* **7**, 120–30

Frape D L (1980) Facts about feeding horses. *In Practice* **2**, 14–21

Frape D L (1981) Digestibility studies in horses and ponies. *Proc 32nd A Meet Eur Assn Anim Prod* IVb

Frape D L (1983) Nutrition of the horse, in *Pharmacological basis of large animal medicine*, eds Bogan J A, Lees P & Yoxall A T. Blackwell Scientific: Oxford

Frape D L (1984a) Straw in the diet of other ruminants and non-ruminant herbivores, in *Straw and other fibrous byproducts for food* (*Developments in Animal and Veterinary Sciences* (14) eds Owen E & Sundstl F, pp. 487–532. Elsevier Scientific: Amsterdam

Frape D L (1984b) The relevance of red cell potassium in diagnosis. *Equine vet J* **16**, 401–2

Frape D L & Boxall R C (1974) Some nutritional problems of the horse and their possible relationship to those of other herbivores. *Equine vet J* **6**, 59–68

Frape D L & Pringle J D (1984) Toxic manifestations in a dairy herd consuming haylage contaminated by lead. *Vet Rec* **114**, 615–6

Frape D L, Cash R S G & Ricketts S W (1983) Panda food allergy. *Lancet* i, 870–1

Frape D L, Peace C K & Ellis Pat M (1979) Some physiological changes in a fit and an unfit horse associated with a long distance ride. *Proc 30th A Meet Eur Ass Anim Prod* H-VI-5

Frape D L, Tuck M G, Sutcliffe N H & Jones D B (1982) The use of inert markers in the measurement of the digestibility of cubed concentrates and of hay given in several proportions to the pony, horse and white rhinoceros (*Diceros simus*). *Comp Biochem Physiol* **72A**, 77–83

Fricker Ch, Riek W & Hugelshofer J (1982) Occlusion of the digital arteries–a model for pathogenesis of navicular disease. *Equine vet J* **14**, 203–7

Garner H E, Coffman J R, Hahn A W, Hutcheson D P & Tumbleson M E (1975) Equine laminitis of alimentary origin: an experimental model. *Am J vet Res* **36**, 441–4

Garner H E, Hutcheson D P, Coffman J R, Hahn A W & Salem C (1977) Lactic acidosis: a factor associated with equine laminitis. *J Anim Sci* **45**, 1037–41

Garner H E, Moore J N, Johnson J H, Clark L, Amend J F, Tritschler L G, Coffman J R, Sprouse R F, Hutcheson D P & Salem C A (1978) Changes in the caecal flora associated with the onset of laminitis. *Equine vet J* **10**, 249–52

Gartner R J W, Blaney B J & McKenzie R A (1981) Supplements to correct oxalate induced negative calcium and phosphorus balances in horses fed tropical grass hays. *J agric Sci Camb* **97**, 581–9

Gibbs P G, Potter G D, Blake R W & McMullan W C (1982) Milk production of Quarterhorse mares during 150 days of lactation. *J Anim Sci* **54**, 496–9

Gibson W (1726) *The true method of dieting horses*. Osborn and Longman: London

Giddings R F, Argenzio R A & Stevens C E (1974) Sodium and chloride transport across the equine cecal mucosa. *Am J vet Res* **35**, 1511–14

Giles C J (1983) Outbreak of ragwort (*Senecio jacobea*) poisoning in horses. *Equine vet J* **15**, 248–50

Gillespie J R, Kauffman A, Steere J & White L (1975) Arterial blood gases and pH during long distance running in the horse. *Proc 1st Int Symp Equine Hematol* 450–68

Gilmour J (1973a) Grass sickness: the paths of research. *Equine vet J* **5**, 102–4

Gilmour J S (1973b) Observations on neuronal changes in grass sickness of horses. *Res vet Sci* **15**, 197–200

Gilmour J S & Jolly G M (1974) Some aspects of the epidemiology of equine grass sickness. *Vet Rec* **95**, 77–81

Gilmour J S, Brown R & Johnson P (1981) A negative serological relationship between cases of grass sickness in Scotland and *Clostridium perfringens* type A enterotoxin. *Equine vet J* **13**, 56–8

Glade M J (1983a) Nutrition and performance of racing Thoroughbreds. *Equine vet J* **15**, 31–6

Glade M J (1983b) Nitrogen partitioning along the equine digestive tract. *J Anim Sci* **57**, 943–53

Glade M J (1984) The influence of dietary fiber digestibility on the nitrogen requirements of mature horses. *J. Anim Sci* **58**, 638–46

Glade M J & Bell P I (1981) Nitrogen partitioning along the equine digestive tract. *J Anim Sci* **53** (suppl. 1), 243, abstr. 294

Glade M J, Gupta S & Reimers T J (1984) Hormonal responses to high and low planes of nutrition in weanling Thoroughbreds. *J Anim Sci* **59**, 658–65

Glade M J, Krook L, Schryver H F & Hintz H F (1982) Calcium metabolism in glucocorticoid-treated pony foals. *J Nutr* **112**, 77–86

Glendinning S A (1974) A system of rearing foals on an automatic calf feeding machine. *Equine vet J* **6**, 12–16

Glinsky M J, Smith R M, Spires H R & Davis C L (1976) Measurement of volatile fatty acid production rates in the cecum of the pony. *J Anim Sci* **42**, 1465–71

Goater L E, Meacham T H, Gwazdauskas F C & Fontenot J P (1981) Effect of dietary energy level in mares during gestation. *J Anim Sci* **53** (suppl 1), 243, abstr. 295

Goater L E, Snyder J L, Huff A N & Meacham T N (1982) A review of recurrent problems in feeding horses. *J Equine vet Sci* **2**, 58–61

Godbee R G & Slade L M (1979) Range blocks with urea for broodmares increase the nutritional value of pasture feeding. *Feedstuffs* **51**(16), 34–5

Godbee R G & Slade L M (1981) The effect of urea or soybean meal on the growth and protein status of young horses. *J Anim Sci* **53**, 670–6

Godbee R G, Slade L M & Lawrence L M (1979) Use of protein blocks containing urea for minimally managed broodmares. *J Anim Sci* **48**, 459–63

Goodman H M, Noot G W van der, Trout J R & Squibb R L (1973) Determination of energy source utilized by the light horse. *J Anim Sci* **37**, 56–62

Green D A (1961) A review of studies on the growth rate of the horse. *Br vet J* **117**, 181

Green D A (1969) A study of growth rate in Thoroughbred foals. *Br vet J* **125**, 539–46

Green D A (1976) Growth rate in Thoroughbred yearlings and two year olds. *Equine vet J* **8**, 133–4

Green J O (1982) *A sample survey of grassland in England and Wales 1970–1972*. The Grassland Research Institute: Hurley, Maidenhead

Greene H J & Oehme F W (1976) A possible case of equine aflatoxicosis. *Clin Toxicol* **9**, 251–4

Greet T R C (1982) Observations on the potential role of oesophageal radiography in the horse. *Equine vet J* **14**, 73–9

Gronwall R (1975) Effects of fasting on hepatic function in ponies. *Am J vet Res* **36**, 145–8

Gronwall R, Engelking L R & Noonan N (1980) Direct measurement of biliary bilirubin excretion in ponies during fasting. *Am J vet Res* **41**, 125–6

Gronwall R, Engelking L R, Answer M S, Erichsen D F & Klentz R D (1975) Bile secretion in ponies with biliary fistulas. *Am J vet Res* **36**, 653–8

Güllner H, Gill J R & Bartler F C (1981) Correction of hypokalemia by magnesium repletion in familial hypokalemic alkalosis with tubulopathy. *Am J Med* **71**, 578–82

Guy P S & Snow D H (1977) The effect of training and detraining on muscle composition in the horse. *J Physiol* **269**, 33–51

Guy P S & Snow D H (1977) The effect of training and detraining on lactate dehydrogenase isoenzymes in the horse. *Biochem biophys Res Commun* **75**, 863–9

Haenlein G F W (1969) Nutritive value of a pelleted horse ration *Feedstuffs* **41**(26), 19–20

Hails M R & Crane T D (1982) *Plant poisoning in animals: a bibliography from the world literature, 1960–1979.* Commonwealth Agricultural Bureaux: Slough, Bucks

Hall G M, Adrian T E, Bloom S R & Lucke J N (1982) Changes in circulating gut hormones in the horse during long distance exercise. *Equine vet J* **14**, 209–12

Hanna C J, Eyre P, Wells P W & McBeath D G (1982) Equine immunology 2: immunopharmacology-biochemical basis of hypersensitivity. *Equine Vet J* **14**, 16–24

Harbers L H, McNally L K & Smith W H (1981) Digestibility of three grass hays by the horse and scanning electron microscopy of undigested leaf remnants. *J Anim Sci* **53**, 1671–7

Harmon D L, Britton R A & Prior R L (1983) Rates of utilization of L(+) and D(−) lactate in bovine tissues. *Fedn Proc* **42**, 815, abstr. 3065

Harper O F & Noot G W V (1974) Protein requirement of mature maintenance horses. *J Anim Sci* **39**, 183, abstr. 181

Harrington D D (1974) Pathologic features of magnesium deficiency in young horses fed purified rations. *Am J vet Res* **35**, 503–13

Harrington D D (1975) Influence of magnesium deficiency on horse foal tissue concentrations of Mg, calcium and phosphorus. *Br J Nutr* **34**, 45–57

Harrington D D (1982) Acute vitamin D_2 (ergocalciferol) toxicosis in horses: case report and experimental studies. *J Am vet med Ass* **180**, 867–73

Harrington D D & Walsh J J (1980) Equine magnesium supplements: evaluation of magnesium oxide, magnesium sulphate and magnesium carbonate in foals fed purified diets. *Equine vet J* **12**, 32–3

Harrison R J (1974) Vitamin B_{12} content in erythrocytes in horses and sheep. *Res vet Sci* **17**, 259–60

Hart G H, Goss H & Guilbert H R (1943) Vitamin A deficiency not the cause of joint lesions in horses. *Am J vet Res* **4**, 162–8

Harthoorn A M & Young E (1974) A relationship between acid–base balance and capture myopathy in zebra (*Equus burchelli*) and an apparent therapy. *Vet Rec* **95**, 337–42

Hatak J (1977) Effect of the degree of training and the physiological condition of horses on the dynamics of some urine metabolites. *Veterinarstvi* **27**, 301–2

Hatfull R S, Milner I & Stanway V (1980) Determination of theobromine in animal feeding stuffs. *J Ass Publ Analysts* **18**, 19–22

Hathaway R L, Oldfield J E & Buettner M R (1981) Effect of selenium in a mineral salt mixture on heifers grazing tall fescue and quack grass pastures. *Proc West Sect Am Soc Anim Sci* **32**, 32–3

Hawkes J, Hedges M, Daniluk P, Hintz H F & Schryver H F (1985) Feed preferences of ponies. *Equine vet J* **17**, 20–2.

Hemken R W, Jackson J A, Jr & Boling J A (1984) Toxic factors in tall fescue. *J Anim Sci* **58**, 1011–16

Hemken R W, Boling J A, Bull L S, Hatton R H, Buckner R C & Bush L P (1981) Interaction of environmental temperature and anti-quality factors on the severity of summer fescue toxicosis. *J Anim Sci* **52**, 710–4

Heneke D R, Potter G D & Kreider J L (1981) Rebreeding efficiency in mares fed different levels of energy during late gestation. *Proc Equine Nutr Phys Soc*, 101

Heusner G L, Albert W W & Norton H W (1976) Energy and systems of feeding for weanlings. *J Anim Sci* **43**, 253, abstr. 170

Hill K R (1960) The world-wide distribution of seneciosis in man and animals. *Proc R Soc Med* **53**, 281–3

Hinton M (1978) On the watering of horses: a review. *Equine vet J* **10**, 27–31

Hintz H F (1980a) Growth in the horse, in *Stud manager's handbook*, vol. 16, pp. 59–66. The International Stockman's School, Tucson Arizona, Agriservices Foundation: Clovis, California

Hintz H F (1980b). Diagnosis of nutritional status, in *Stud manager's handbook*, vol. 16, pp. 185–7. Agriservices Foundation: Clovis, California

Hintz H F (1983) Nutritional requirements of the exercising horse – a review, in *Equine exercise physiology* (*Proc 1st Int Conf. Oxford 1982*), eds Snow D H, Persson S G B & Rose R J, pp. 275–90. Granta Editions: Cambridge

Hintz H F & Kallfelz F A (1981) Some nutritonal problems of horses. *Equine vet J* **13**, 183–6

Hintz H F & Meakim D W (1981) A comparison of the 1978 National Research Council's recommendations of nutrient requirements of horses with recent studies. *Equine vet J* **13**, 187–91

Hintz H F & Schryver H F (1972) Magnesium metabolism in the horse. *J Anim Sci* **35**, 755–9

Hintz H F & Schryver H F (1973) Magnesium, calcium and phosphorus metabolism in ponies fed varying levels of magnesium. *J Anim Sci* **37**, 927–30

Hintz H F & Schryver H F (1975a) The evolution of commercial horse feeds. *Feedstuffs* **47**(36), 26–8

Hintz H F & Schryver H F (1975b) Recent developments in equine nutrition. *Proc Cornell Nutr Conf Feed Manuf* 95–8. Cornell University: Ithaca, NY

Hintz H F & Schryver H F (1976a) Corrugated paper boxes can be ingredients in complete pelleted diets for horses. *Feedstuffs* **48**(44) 51–3

Hintz H F & Schryver H F (1976b) Current status of mineral nutrition in the equine. *Proc Univ Maryland Nutr Conf Feed Manuf* 42–4. Univ. Maryland Agriculture Experimental Station

Hintz H F &. Schryver H F (1976c) Nutrition and bone development in horses. *J Am vet med Ass* **168**, 39–44

Hintz H F & Schryver H F (1976d) Potassium metabolism in ponies. *J Anim Sci* **42**, 637–43

Hintz H F, Hintz R L & Van Vleck L D (1979) Growth rate of Thoroughbreds. Effect of age of dam, year and month of birth and sex of foal. *J Anim Sci* **48**, 480–7

Hintz H F, Lowe J E & Schryver H F (1969) Protein sources for horses. *Proc Cornell Nutr Conf Feed Manuf* 65–8. Cornell University: Ithaca, NY

Hintz H F, Schryver H F & Cymbaluk N F (1979) Feeding dehydrated forages to horses, in *Proc 2nd Int Green Crop Drying Cong*, ed. Howarth R E, pp. 314–8. University of Saskatchewan

Hintz H F, Schryver H F & Lowe J E (1973) Digestion in the horse. *Feedstuffs* **45**(27), 25–31

Hintz H F, Schryver H F & Lowe J E (1976) Delayed growth response and limb conformation in young horses. *Proc Cornell Nutr Conf Feed Manuf* 94–6. Cornell University, Ithaca, NY

Hintz H F, Sedgewick C J & Schryver H F (1976) Some observations on digestion of a pelleted diet by ruminants and non-ruminants. *Int Zoo Yearbk* **16**, 54–7

Hintz H F, Ross M W, Lesser F R, Leids P F, White K K, Lowe J E, Short C E & Schryver H F (1978a) The value of dietary fat for working horses I. Biochemical and hematological evaluations. *J equine Med Surg* **2**, 483–8

Hintz H F, Ross M W, Lesser F R, Leids P F, White K K, Lowe J E, Short C E & Schryver H F (1978b) Value of supplemental fat in horse rations. *Feedstuffs* **50**(12), 27–8

Hintz H F, Schryver H F, Doty J, Lakin C & Zimmerman R A (1984) Oxalic acid content of alfalfa hays and its influence on the availability of calcium, phosphorus and magnesium. *J Anim Sci* **58**, 939–42

Hintz H F, Williams A J, Rogoff J & Schryver H F (1973) Availability of phosphorus in wheat bran when fed to ponies. *J Anim Sci* **36**, 522–5

Hintz R L, Hintz H F & Van Vleck L D (1978) Estimation of heritabilities for weight, height and front cannon bone circumference of Thoroughbreds. *J Anim Sci* **47**, 1243–5

Hodge S L, Kreider J L, Potter G D & Harms P G (1981) Influence of photoperiod on the pregnant and postpartum mare. *J. Anim Sci* (suppl 1) 330, abstr. 505

Holmes J R (1982) A superb transport system. The Circulation. *Equine vet J* **14**, 267–76

Holt P E & Pearson H (1984) Urolithiasis in the horse – a review of 13 cases. *Equine vet J* **16**, 31–4

Holton D W, Garrett B J & Cheeke P R (1983) Effects of dietary supplementation with BHA, cysteine and B vitamins on tansy ragwort toxicity in horses. *Proc West Sect Am Soc Anim Sci* **34**, 183–6

Houpt K A (1983) Taste preferences in horses. *Equine Pract* **5**, 22–5

Houpt T R & Houpt K A (1971) Nitrogen conservation by ponies fed a low protein ration. *Am J vet Res* **32**, 579–88

Houpt K A, Hintz H F & Pagan J D (1981) The response of mares and their foals to brief separation. *J Anim Sci* **53** (suppl 1) 129, abstr. 6

Houpt K A, Parsons M S & Hintz H F (1982) Learning ability of orphan foals, of normal foals and of their mothers. *J Anim Sci* **55**, 1027–32

Householder D D, Potter G D, Lichtenwainer R E & Hesby J H (1976) Growth and digestion in horses fed sorghum or oats. *J Anim Sci* **43**, 254

Huchton J D, Potter G D, Sorensen Jr A M & Orts F A (1976) Foal development related to prepartum nutrition. *J. Anim Sci* **43**, 253, abstr. 171

Jacobs K A & Bolton J R (1982) Effect of diet on the oral glucose tolerance test in the horse. *J Am vet med Ass* **180**, 884–6

Jacobs K A, Norman P, Hodgson D R G & Cymbaluk N (1982) Effect of diet on oral D-xylose absorption test in the horse. *Am J vet Res* **43**, 1856–8

Jaeschke G & Keller H (1978) The ascorbic acid status in horses. 2. Clinical aspects and deficiency symptoms. *Berl Münch Tierärztl Wschr* **91**, 375–9

Jeffcott L B (1974a) Some practical aspects of the transfer of passive immunity to newborn foals. *Equine vet J* **6**, 109–15

Jeffcott L B (1974b) Studies on passive immunity in the foal. 1. γ-Globulin and antibody variations associated with the maternal transfer of immunity and the onset of active immunity. *J. comp Path* **84**, 93–101

Jeffcott L B (1974c) Studies on passive immunity in the foal. II. The absorption of ^{125}I-labelled PVP (polyvinyl pyrrolidone) by the neonatal intestine. *J comp Path* **84**, 279–89

Jeffcott L B (1974d) Studies on passive immunity in the foal. III. The characterization and significance of neonatal proteinuria. *J comp Path* **84**, 455–65

Jeffcott L B & Kold S E (1982) Stifle lameness in the horse: a survey of 86 referred cases. *Equine vet J* **14**, 31–9

Jeffcott L B, Rossdale P D & Leadon D P (1982) Haematological changes in the neonatal period of normal and induced premature foals. *J Reprod Fert Suppl* **32**, 537–44

Jeffcott L B, Dalin G, Drevemo S, Fredricson I, Björne K & Bergquist A (1982) Effect of induced back pain on gait and performance of trotting horses. *Equine vet J* **14**, 129–33

Jeffcott L B, Rossdale P D, Freestone J, Frank C J & Towers-Clark P F (1982) An assessment of wastage in Thoroughbred racing from conception to 4 years of age. *Equine vet J* **14**, 185–98

Johnson R J & Hart J W (1974a) Influence of feeding and fasting on plasma free amino acids in the equine. *J Anim Sci* **38**, 790–8

Johnson R J & Hart J W (1974b) Utilization of nitrogen from soybean meal, biuret and urea by equine. *Nutr Rep Int* **9**, 209–15

Johnson R J & Hughes I M (1974) Alfalfa cubes for horses. *Feedstuffs* **46**(43), 31

Jones D G C, Greatorex J C, Stockman M J R & Harris C P J (1972) Gastric impaction in a pony: relief via laparotomy. *Equine vet J* **4**, 98–9

Jones S & Blackmore D J (1982) Observations on the isoenzymes of aspartate aminotransferase in equine tissues and serum. *Equine vet J* **14**, 311–16

Jordan R M (1979) Effect of corn silage and turkey litter on the performance of gestating pony mares and weanlings. *J Anim Sci* **49**, 651–3

Jordan R M (1982) Effect of weight loss of gestating mares on subsequent production. *J Anim Sci* **55** (suppl. 1), 208

Jordan R M & Marten G C (1975) Effect of three pasture grasses on yearling pony weight gains and pasture carrying capacity *J Anim Sci* **40**, 86–9

Jordan R M, Myers V S, Yoho B & Spurell F A (1975) Effect of calcium and phosphorus levels on growth, reproduction and bone development of ponies. *J Anim Sci* **40**, 78–84

Judson G J & Mooney G J (1983) Body water and water turnover rate in Thoroughbred horses in training, in *Equine exercise physiology* (*Proc 1st Int Conf. Oxford 1982*), eds Snow D H, Persson S G B & Rose R J, pp. 354–61. Granta Editions: Cambridge

Kane E, Baker J P & Bull L S (1979) Utilization of a corn oil supplemented diet by the pony. *J Anim Sci* **48**, 1379–84

Keenan D M (1978) Changes of plasma uric acid levels in horses after galloping. *Res vet Sci* **25**, 127–8

Kellock E M (1982) The origins of the Thoroughbred. *Equi* no. 13 (Nov/Dec), 4–5

Kelly W R & Lambert M B (1978) The use of cocoa-bean meal in the diets of horses: pharmacology and pharmacokinetics of theobromine. *Br vet J* **134**, 171–80

Kennedy L G & Hershberger T V (1974) Protein quality for the non-ruminant herbivore. *J Anim Sci* **39**, 506–11

Kilshaw P J & Sissons J W (1979) Gastrointestinal allergy to soyabean protein in preruminant calves. Allergenic constituents of soyabean products. *Res vet Sci* **27**, 366–71

Kim H L, Herrig B W, Anderson A C, Jones L P & Calhoun, M C (1983) Elimination of adverse effects of ethoxyquin (EQ) by methionine hydroxy analog

(MHA). Protective effects of EQ and MHA for bitterweed poisoning in sheep . *Toxicol Lett* **16**, 23–9

Knaus E (1981) Diseases of the foal in the first days of life. *Proc A 32nd Meet Eur Assn Anim Prod* IIIa

Knight D A & Tyznik W J (1985) The effect of artificial rearing on the growth of foals. *J Anim Sci* **60**, 1–5

Kohn C W, Muir W W & Sams R (1978) Plasma volume and extracellular fluid volume in horses at rest and following exercise. *Am J vet Res* **39**, 871–4

Körber H-D (1971) Zur Kolikstatistic des Pferdes. *Berl Münch Tierärztl Wschr* **84**, 75–7

Kosiniak K (1981) Some properties of semen from young stallions and its value for preservation in liquid nitrogen. *Proc A 32nd Meet Eur Assn Anim Prod* I–10

Kossila V & Ljung G (1976) Value of whole oat plant pellets in horse feeding. *Annl Agric Fenn* **15**, 316–21

Kossila V, Tanhuanpaa E, Virtanen E & Luoma E (1972) Hb value, blood glucose, cholesterol, minerals and trace elements in saddle horses. 1. Differences due to age and maintenance. *J Scient Agric Soc Finland* **44**, 249–57

Kownacki M (1983) The development of type in the Polish Konik. *Rocznik Nauk Rolniczych* **B82-1**, 71–104

Kozak A & Bickel H (1981) Die Verdaulichkeit von Gras dreier Weidetypen Ostafrikas. *Proc 32nd A Meet Eur Assn Anim Prod* I–11

Kronauer M & Bickel H (1981) Estimation of the energy value of East African pasture grass. *Proc 32nd A Meet Eur Assn Anim Prod* I–12

Krook L (1968) Dietary calcium-phosphorus and lameness in the horse. *Cornell Vet* **58**, 59–73

Krook L, Bélanger L F, Henrikson P, Lutwak L & Sheffy B E (1970) Bone flow. *Rev Can Biol* **29**, 157–67

Kruczynska H, Ponikiewska T & Berthold St (1981) Einfluss der mit Gülle gedüngten Weide auf die Milchleistung und Mineralbilanz bei Kühen. *Proc 32nd A Meet Eur Assn Anim Prod* IV–14

Krzywanek H (1974) Lactic acid concentrations and pH values in trotters after racing. *J S Afr Vet Ass* **45**, 355–60

Langlands J P & Cohen R D H (1978) The nutrition of ruminants grazing native and improved pastures. III. Mineral composition of bones and selected organs from grazing cattle. *Aust J Agric Res* **29**, 1301–11

Lawrence L M, Slade L M, Nockels C F, Shideler R K (1978) Physiologic effects of vitamin E supplementation on exercised horses. *Proc West Sect Am Soc Anim Sci* **29**, 173–7

Lawson G H K, McPherson E A, Murphy J R, Nicholson J M, Wooding P, Breeze R G & Pirie H M (1979) The presence of precipitating antibodies in the sera of horses with chronic obstructive pulmonary disease (COPD). *Equine vet J* **11**, 172–6

Ledgard S F, Steele K W & Saunders W H M (1982) Effects of cow urine and its major constituents on pasture properties. *N Z J agric Res* **25**, 61–8

Leonard T M, Baker J P & Willard J G (1974) Effect of dehydrated alfalfa on equine digestion. *J Anim Sci* **39**, 184, abstr. 188

Leonard T M, Baker J P & Willard J G (1975) Influence of distillers feeds on digestion in the equine. *J Anim Sci* **40**, 1086–90

Levine S B, Myhre G D, Smith G L, Burns J G & Erb H (1982) Effect of a nutritional supplement containing *N, N*-dimethylglycine (DMG) on the racing Standardbred. *Equine Practice* **4**, 17–20

Lindholm A (1974) Glycogen depletion pattern and the biochemical response to varying exercise intensities in Standardbred trotters. *J S Afr vet Ass* **45**, 341–3

Lindholm A & Piehl K (1974) Fibre composition, enzyme activity and concentrations of metabolites and electrolytes in muscles of Standardbred horses. *Acta vet Scand* **15**, 287–309

Lindholm A & Saltin B (1974) The physiological and biochemical response of Standardbred horses to exercise of varying speed and duration. *Acta vet Scand* **15**, 310–24

Lindholm A, Johansson H-E & Kjaersgaard P (1974) Acute rhabdomyolysis ('Tying-up') in Standardbred horses. A morphological and biochemical study. *Acta vet Scand* **15**, 325–39

Littlejohn A, Kruger J M & Bowles F (1977) Exercise studies in horses: 2. The cardiac response to exercise in normal horses and in horses with chronic obstructive pulmonary disease. *Equine vet J* **9**, 75–83

Loew F M (1973) Thiamin and equine laryngeal hemiplegia. *Vet Rec* **92**, 372–3

Loew F M & Bettany J M (1973) Thiamine concentrations in the blood of Standardbred horses. *Am J vet Res* **34**, 1207–8

Lolas G M & Markakis P (1975) Phytic acid and other phosphorus compounds of beans (*Phaseolus vulgaris*). *J agric Fd Chem* **23**, 13–15

Lord K A & Lacey J (1978) Chemicals to prevent the moulding of hay and other crops. *J Sci Fd Agric* **29**, 574–5

Löscher W, Jaeschke G & Keller H (1984) Pharmacokinetics of ascorbic acid in horses. *Equine Vet J* **16**, 59–65

Lucke J N & Hall G M (1978) Biochemical changes in horses during a 50-mile endurance ride. *Vet Rec* **102**, 356–8

Lucke J N & Hall G M (1980) Further studies on the metabolic effects of long distance riding: Golden Horseshoe Ride 1979. *Equine Vet J* **12**, 189–92

Lunn P G & Austin S (1983) Dietary manipulation of plasma albumin concentration. *J Nutr* **113**, 1791–802

Lyons E T, Drudge J H & Tolliver S C (1980) Antiparasitic acitivity of parbendazole in critical tests in horses. *Am J vet Res* **41**, 123–4

McCall C A, Potter G D, Friend T H & Ingram R S (1981) Learning abilities in yearling horses using the Hebb–Williams closed field maze. *J Anim Sci* **53**, 928–33

MacCallum F J, Brown M P & Goyal H O (1978) An assessment of ossification and radiological interpretation in limbs of growing horses. *Br vet J* **134**, 366–73

MacCarthy D D, Spillane T A & O'Moore L B (1976) Type and quality of oats used for bloodstock feeding in Ireland. *Ir J agric Res* **15**, 47–54

McDonald P, Edwards R A & Greenhalgh J F D (1981) *Animal Nutrition*. Longman: London & New York

McGavin M D & Knake R (1977) Hepatic midzonal necrosis in a pig fed aflatoxin and a horse fed moldy hay. *Vet Pathol* **14**, 182–7

McKenzie R A, Blaney B J, Gartner R J W, Dillon R D & Standfast N F (1979) A technique for the conduct of nutritional balance experiments in horses. *Equine vet J* **11**, 232–4

McLaughlin C L (1982) Role of peptides from gastrointestinal cells in food intake regulation. *J Anim Sci* **55**, 1515–27

McMiken D F (1983) An energetic basis of equine performance. *Equine vet J* **15**, 123–33

McPherson E A & Thomson J R (1983) Chronic obstructive pulmonary disease in the horse: nature of the disease. *Equine vet J* **15**, 203–6

McPherson E A, Lawson G H K, Murphy J R, Nicholson J M, Breeze R G & Pirie H M (1979a) Chronic obstructive pulmonary disease (COPD) in horses: aetiological studies: responses to intradermal and inhalation antigenic challenge. *Equine vet J* **11**, 159–66

McPherson E A, Lawson G H K, Murphy J R, Nicholson J M, Breeze R G & Pirie H M (1979b) Chronic obstructive pulmonary disease (COPD): factors influencing the occurrence. *Equine vet J* **11**, 167–71

McPherson R (1978) Selenium deficiency. *Vet Rec.* **103**, 60

Madigan J E & Evans J W (1973) Insulin turnover and irreversible loss rate in horses. *J Anim Sci* **36**, 730–3

Malone J C (1969) Hazards to domestic pet animals from common toxic agents. *Vet Rec* **84**, 161–5

Maloiy G M O (1970) Water economy of the Somali donkey. *Am J Physiol* **219**, 1522–7

Markham G (1636) *Markhams Maister-Peece. In two bookes.* Nicholas and John Okes: London

Marquardt R R, McKirdy J A, Ward T & Campbell L D (1975) Amino acid, hemagglutinin and trypsin inhibitor levels, and proximate analyses of faba beans (*Vicia faba*) and faba bean fractions. *Can J Anim Sci* **55**, 421–9

Martin B, Robinson S & Robertshaw D (1978) Influence of diet on leg uptake of glucose during heavy exercise. *Am J Clin Nutr* **31**, 62–7

Martin-Rosset W, Boccard R & Robelin J(1979) Relative growth of different organs, tissues and body regions in the foal from birth to 30 months. *Proc 30th A Meet Eur Ass Anim Prod* H6.1

Martin-Rosset W, Doreau M & Cloix J (1978) Grazing behaviour of a herd of heavy brood mares and their foals, *Annals Zootech* **27**, 33–45

Mason D K & Kwok H W (1977) Some haematological and biochemical parameters in racehorses in Hong Kong. *Equine Vet J* **9**, 96–9

Matthews H & Thornton I (1982) Seasonal and species variation in the content of cadmium and associated metals in pasture plants at Shipham. *Pl Soil* **66**, 181–93

Meadows D G (1979) Utilization of dietary protein or non-protein nitrogen by lactating mares fed soybean meal or urea. *Diss Abstr Int B* **40**, 999

Meakin D W, Ott E A, Asquith R L & Feaster J P (1981) Estimation of mineral content of the equine third metacarpal by radiographic photometry. *J Anim Sci* **53**, 1019–26

Mehring J S & Tyznik W J (1970) Equine glucose tolerance. *J Anim Sci* **30**, 764–6

Merritt A M (1975) Treatment of diarrhoea in the horse. *J S Afr vet Ass* **46**, 89–90

Merkt H & Günzel A–R (1979) A survey of early pregnancy losses in West German Thoroughbred mares. *Equine vet J* **11**, 256–8

Meyer H (1982) *Contributions to digestive physiology of the horse.* Paul Parey: Hamburg and Berlin

Meyer H & Ahlswede L (1976) Intrauterine growth and the body composition of foals, and the nutrient requirements of pregnant mares. *Ubersichten zur Tierernährung* **4**, 263–92

Meyer H & Ahlswede L (1977) Studies on Mg metabolism in the horse. *Zentlb*

Veterinärmed **24A** (2), 128–39

Meyer H, Ahlswede L & Reinhardt H J (1975) Duration of feeding, frequency of chewing and physical form of the feed for horses. *Dt Tierärztl Wschr* **82**, 54–8

Meyer H, Lindemann G & Schmidt M (1982) Einfluss unterschiedlicher Mischfuttergaben pro Mahlzeit auf praecaecale – und postileale Verdauungsvorgänge beim Pferde, in *Advances in animal physiology and animal nutrition*, ed. Meyer H *(Contributions to digestive physiology of the horse* no. 13), pp. 32–90 Paul Parey: Hamburg & Berlin

Meyer H, Pferdekamp M & Huskamp B (1979) Studies on the digestibility and tolerance of different feeds by typhlektomized ponies. *Dt Tierärztl Wschr* **86**, 384–90

Meyer H, Schmidt M & Güldenhaupt V (1981) Untersuchungen über Mischfutter für Pferde. *Dt Tierärztl Wschr* **88**, 2–5

Meyer H, Schmidt M, Lindemann G & Muuss H (1982) Praecaecale und postileale Verdaulichkeit von Mengen – (Ca, P, Mg) und Spurenelementen (Cu, Zn, Mn) beim Pferd, in *Advances in animal physiology and animal nutrition*, ed. Meyer H *(Contributions to digestive physiology of the horse*, no. 13), pp. 61–9. Paul Parey: Hamburg & Berlin

Meyer H, Winkel C, Ahlswede L & Weidenhaupt C (1978) Untersuchungen über Schweissmenge und Schweisszusammensetzung beim Pferd. *Tierärztl Umsch* **33**, 330–6

Milić B Lj (1972) Lucerne tannins. I. Content and composition during growth. *J Sci Fd Agric* **23**, 1151–6

Milić B Lj & Stojanović S (1972) Lucerne tannins. III. Metabolic fate of lucerne tannins in mice. *J Sci Fd Agric* **23**, 1163–7

Milić B Lj, Stojanovićs & Vučurević N (1972) Lucerne tannins. II. Isolation of tannins from lucerne, their nature and influence on the digestive enzymes in vitro. *J Sci Fd Agric* **23**, 1157–62

Millward D J, Davies C T M, Halliday D, Wolman S L, Matthews D & Rennie M (1982) Effect of exercise on protein metabolism in humans as explored with stable isotope. *Fedn Proc* **41**, 2686–91

Milne D W (1974) Blood gases, acid-base balance and electrolyte and enzyme changes in exercising horses. *J S Afr vet Ass* **45**, 345–54

Milne D W, Skarda R J, Gabel A A, Smith L G & Ault K (1976) Effects of training on biochemical values in Standardbred horses. *Am J vet Res* **37**, 285–90

Milner J & Hewitt D (1969) Weight of horses: improved estimate based on girth and length. *Can vet J* **10**, 314–17

Ministère de l'agriculture (1980) *Aménagement et équipement des centres équestres. Section technique des équipements hippiques.* Service des Haras et de l'équitation: Le lion d'angers

Ministry of Agriculture, Fisheries and Food (1977) *Drainage of grassland* (ADAS leaflet no. 7), pp. 8. HMSO: London

Ministry of Agriculture, Fisheries and Food (1978) *Drainage maintenance* (ADAS leaflet no. 7), pp.9. HMSO: London

Ministry of Agriculture, Fisheries and Food (1979) *Lime and fertiliser recommendations.* 1. *Arable crops and grassland* (ADAS booklet no. 2191), pp. 47 HMSO: London

Moise L L & Wysocki A A (1981) The effect of cottonseed meal on growth of young horses. *J Anim Sci* **53**, 409–13

Moise L L & Wysocki A A (1983) The effect of cottonseed meal on growth of young horses. *J Anim Sci* **53**, 409–13

Moore J N, Garner H E & Coffman J R (1981) Haematological changes during development of acute laminitis hypertension. *Equine vet J* **13**, 240–2

Moore J N, Owen R ap R & Lumsden J H (1976) Clinical evaluation of blood lactate levels in equine colic. *Equine vet J* **8**, 49–54

Moore J N, Garner H E, Berg J N & Sprouse R F (1979) Intracecal endotoxin and lactate during the onset of equine laminitis: a preliminary report. *Am J vet Res* **40**, 722–3

Moore J N, Garner H E, Shapland J E & Hatfield D G (1980) Lactic acidosis and arterial hypoxemia during sublethal endotoxemia in conscious ponies. *Am J vet Res* **41**, 1696–8

Moore J N, Garner H E, Shapland J E & Hatfield D G (1981) Prevention of endotoxin-induced arterial hypoxaemia and lactic acidosis with flunixin meglumine in the conscious pony. *Equine vet J* **13**, 95–8

Moraillon R, De Faucompret P & Cloche D (1978) Results of the long-term administration of a flaked or granulated complete feed into saddle horses. *Recl Med Vet* **154**, 999–1007

Morse E V, Duncan Margo A, Page E A & Fessler J F (1976) Salmonellosis in Equidae: a study of 23 cases. *Cornell Vet* **66**, 198–213

Moss M S (1975) Recent advances in the field of doping detection. *Equine vet J* **7**, 173–4

Moss M S & Clarke E G C (1977) A review of drug 'clearance times' in racehorses. *Equine Vet J* **9**, 53–6

Moss M S & Haywood P E (1984) Survey of positive results from racecourse antidoping samples received at Racecourse Security Services' Laboratories. *Equine vet J* **16**, 39–42

Mullen P A (1970) Variations in the albumin content of blood serum in Thoroughbred horses. *Equine vet J* **2**, 118–20

Mullen P A, Hopes R & Sewell J (1979) The biochemistry; haematology, nutrition and racing performance of two-year-old Thoroughbreds throughout their training and racing season. *Vet Rec* **104**, 90–5

Mundt H C (1978) Untersuchungen über die Verdaulichkeit von aufgeschlossenem Stroh beim Pferd. Published thesis; Tierarztliche Hochschule Hannover, pp. 109

Murphy J R, McPherson E A & Dixon P M (1980) Chronic obstructive pulmonary disease (COPD): effects of bronchodilator drugs on normal and affected horses. *Equine vet J* **12**, 10–14

Mussman H C & Rubiano A (1970) Serum protein electrophoregram in the Thoroughbred in Bogota, Colombia. *Br vet J* **126**, 574–8

Muuss H, Meyer H & Schmidt M (1982) Entleerung und Zusammensetzung des Ileumchymus beim Pferd, in *Advances in animal physiology and nutrition*, ed Meyer H (*Contributions to the digestive physiology of the horse* no. 13), pp. 13–23. Paul Parey: Hamburg & Berlin

Muylle E & Hende C van den (1983) The concept of osmolality. *Equine vet J* **15**, 80–1

Muylle E, Nuytten J, Hende C van den, Deprez P, Vlaminck K & Oyaert W (1984) Determination of red cell potassium content in horses with diarrhoea. A practical approach for therapy. *Equine vet J* **16**, 450–2

Muylle E, Oyaert W, Roose P de, Hende C van den (1973) [Hypocalcaemia in the horse.] *Vlaams Diergeneeskundig Tijdschr* **42**, 44–51 (in Dutch)

Muylle E, Hende C. van den, Nuytten J, Deprez P, Vlaminck K & Oyaert W (1984) Potassium concentration in equine red blood cells: normal values and correlation with potassium levels in plasma. *Equine vet J* **16**, 447–9

Muylle E, Hende C van den, Nuytten J, Oyaert W & Vlaminck K (1983) Preliminary studies on the relationship of red blood cell potassium concentration and performance, in *Equine exercise physiology (Proc 1st Int Conf, Oxford, 1982)* eds Snow D H, Persson S G B & Rose R J pp. 366–70. Granta Editions: Cambridge

Muylle E, Hende C van den, Oyaert W, Thoonen H & Vlaminck K (1981) Delayed monensin sodium toxicity in horses. *Equine vet J* **13**, 107–8

Nagata Y, Takagi S & Kubo K (1972a) Studies on gas metabolism in light horses fed a complete pelleted ration. I. Gas metabolism at rest (the effects of diets and season on gas metabolism at rest). *Expl Rep Equine Hlth Lab* No. 9, 84–9

Nagata Y, Takagi S & Kubo K (1972b) Studies on gas metabolism in light horses fed a complete pelleted ration. II. Gas metabolism at excitement (the effect of epinephrine infusion on gas metabolism) *Expl Rep Equine Hlth Lab* No. 9, 90–5

Nahani F & Atiabt N (1977) Electrophoretic analysis of blood serum protein of normal horses. *J vet Fac Univ Tehran* **33**, 75–9

Nahapetian A & Bassiri A (1975) Changes in concentrations and interrelationships of phytate, phosphorus, magnesium, calcium and zinc in wheat during maturation. *J agric Fd Chem* **23**, 1179–83

National Institute of Agricultural Botany (1983–4a) *Grasses and legumes for conservation* (Technical leaflet no. 2), pp. 11. NIAB: Cambridge

National Institute of Agricultural Botany (1983–4b) *Recommended varieties of herbage legumes* (Farmers leaflet no. 4), pp. 15. NIAB: Cambridge

National Institute of Agricultural Botany (1983–4c) *Recommended varieties of grasses* (Farmers leaflet no. 16), pp. 23 NIAB: Cambridge

National Research Council (1978) *Nutrient requirements of domestic animals*, no. 6. *Nutrient requirements of horses*. National Academy of Sciences: Washington DC

Naylor J M, Kronfield D S & Acland H (1980) Hyperlipemia in horses: effects of undernutrition and diseases. *Am J vet Res* **41**, 899–905

Neave R M S & Callear J F F (1973) Further clinical studies on the uses of mebendazole (R17635) as an anthelmintic in horses. *Br vet J* **129**, 79–82

Nimmo M A, Snow D H & Munro C D (1982) Effects of nandrolone phenylpropionate in the horse: (3) skeletal muscle composition in the exercising animal. *Equine vet J* **14**, 229–33

Noot G W V, Symons L D, Lydman R K & Fonnesbeck P V (1967) Rate of passage of various feedstuffs through the digestive tract of horses. *J Anim Sci* **26**, 1309–11

Ödberg F O & Francis Smith K (1976) A study on eliminative and grazing behaviour – the use of the field by captive horses. *Equine vet J* **8**, 147–9

Oftedal O T, Hintz H F & Schryver H F (1983) Lactation in the horse: milk composition and intake by foals. *J Nutr* **113**, 2096–106

O'Moore L B (1972) Nutritional factors in the rearing of the young Thoroughbred horse. *Equine vet J* **4**, 9–16

Ordidge R M, Schubert F K & Stoker J W (1979) Death of horses after accidental feeding of monensin. *Vet Rec* **104**, 375

Orr J A, Bisgard G E, Forster H V, Rawlings C A, Buss D D & Will J A (1975) Cardiopulmonary measurements in nonanesthetized, resting normal ponies. *Am J vet Res* **36**, 1667–70

Orskov E R & Hovell F D De B (1981) Principles and appropriate technology for improving the nutritive value of tropical feeds. *Proc 32nd A Meet Eur Assn Anim Prod* I–6

Orton R G (1978) Biochemical changes in horses during endurance rides. *Vet Rec* **102**, 469

Orton R K, Hume I D & Leng R A (1985) The effects of level of dietary protein and exercise on growth rates of horses. *Equine vet J* **17**, 381–5

Osborne M (1981) Rearing the orphan foal. *Proc 32nd A Meet Eur Assn Anim Prod* IIIa–1

Østblom L C, Lund C & Melsen F (1982) Histological study of navicular bone disease. *Equine vet J* **14**, 199–02

Østblom L C, Lund C & Melsen F (1984) Navicular bone disease: results of treatment using egg-bar shoeing technique. *Equine vet J* **16**, 203–6.

Osweiler G D, Gelder G D van & Buck G A (1978) Epidemiology of lead poisoning in animals, in *Toxicity of heavy metals in the environment*, ed. Oehme F W, pt 1, pp 143–71. Marcel Dekker: New York

Ott E A (1981) Influence of level of feeding on digestive efficiency of the horse. *Proc Equine Nutr Physiol Symp* 37–43

Ott E A, Asquith R L & Feaster J P (1981) Lysine supplementation of diets for yearling horses. *J Anim Sci* **53**, 1496–503

Ott E A, Feaster J P & Lieb S (1979) Acceptability and digestibility of dried citrus pulp by horses. *J Anim Sci* **49**, 983–7

Ott E A, Asquith R L, Feaster J P & Martin F G (1979) Influence of protein level and quality on the growth and development of yearling foals. *J. Anim Sci* **49**, 620–8

Owen J M (1975) Abnormal flexion of the corono-pedal joint or 'contracted tendons' in unweaned foals. *Equine vet J* **7**, 40–5

Owen J M (1977) Liver fluke infection in horses and ponies. *Equine vet J* **9**, 31

Owen J M, McCullagh K G, Crook D H & Hinton M (1978) Seasonal variations in the nutrition of horses at grass. *Equine vet J* **10**, 260–6

Pagan J D, Hintz H F & Rounsaville T R (1984) The digestible energy requirements of lactating pony mares. *J Anim Sci* **58**, 1382–7

Palmer E & Driancourt M A (1981) Consequences of foaling at different seasons and under different photoperiods. *Proc 32nd A Meet Eur Assn Anim Prod* I–1

Parry B W (1983) Survey of 79 referral colic cases. *Equine vet J* **15**, 345–8

Parry B W, Anderson G A & Gay C C (1983) Prognosis in equine colic: a study of individual variables used in case assessment. *Equine vet J* **15**, 337–44

Pashen R L & Allen W R (1976) Genuine anoestrus in mares. *Vet Rec* **99**, 362–3

Pedersen E J N & Møller E (1976) Perennial ryegrass and clover in pure stand and in mixture. The influence of mixture, nitrogen fertilization and number of cuts on yield and quality. *Beretning fra Faellesudvalget for Statens Planteavls-og Husdyrbrugsforsøg, København* no. 6, 1–27

Persson S G B (1983) Evaluation of exercise tolerance and fitness in the performance horse, in *Equine exercise physiology*, eds Snow D H, Persson S G B & Rose R J, pp. 441–57. Granta Editions: Cambridge

Peterson A J, Bass J J & Byford M J (1978) Decreased plasma testosterone concentrations in rams affected by ryegrass staggers. *Res vet Sci* **25**, 266–8

Platt H (1978) Growth and maturity in the equine foetus. *J R Soc Med* **71**, 658–61
Platt H (1982) Sudden and unexpected deaths in horses: a review of 69 cases. *Br vet J* **138**, 417–29
Prinz K (1978) Effect of vitamin A–E emulsion on stallion semen. *Tierartz Umschau* **33**, 27–30
Prior R L, Hintz H F, Lowe J E & Visek W J (1974) Urea recycling and metabolism of ponies. *J Anim Sci* **38**, 565–71
Pusztai A, Clarke E M W & King T P (1979) The nutritional toxicity of *Phaseolus vulgaris* lectins. *Proc Nutr Soc* **38**, 115–20
Putnam M (1973) Micronization – a new feed processing technique. *Flour Anim Feed Milling* (June), 40–1

Quinn P J, Baker K P & Morrow A N (1983) Sweet itch: responses of clinically normal and affected horses to intradermal challenge with extracts of biting insects. *Equine vet J* **15**, 266–72

Ralston S L (1984) Controls of feeding in horses. *J Anim Sci* **59**, 1354–61.
Ralston S L & Baile C A (1981) Feeding behaviour of ponies after intragastric nutrient and intravenous glucose infusions. *J Anim Sci* **53**, (Suppl. 1), 131, abstr. 12
Ralston S L & Baile C A (1982a) Plasma glucose and insulin concentrations and feeding behavior in ponics. *J Anim Sci* **54**, 1132–7
Ralston S L & Baile C A (1982b) Gastrointestinal stimuli in the control of feed intake in ponies. *J. Anim Sci* **55**, 243–53
Ralston S L & Baile C A (1983) Effects of intragastric loads of xylose, sodium chloride and corn oil on feeding behavior of ponies. *J. Anim Sci* **56**, 302–8
Ralston S L, Freeman D E & Baile C A (1983) Volatile fatty acids and the role of the large intestine in the control of feed intake in ponies. *J Anim Sci* **57**, 815–25
Ralston S L, Van den Broek G & Baile C A (1979) Feed intake patterns and associated blood glucose, free fatty acid and insulin changes in ponies. *J Anim Sci* **49**, 838–45
Randall R P & Pulse R E (1974) Taste reactions in the immature horse. *J Anim Sci* **38**, 1330, abstr. 45
Randall R P, Schurg W A & Church D C (1978) Response of horses to sweet, salty, sour and bitter solutions. *J Anim Sci* **47**, 51–5
Rasmussen R A, Cole C L & Miller M S (1944) Carotene, vitamin A and ascorbic acid in mare's milk. *J Anim Sci* **3**, 346–52
Reid R L & Horvath D J (1980) Soil chemistry and mineral problems in farm livestock. A review. *Anim Feed Sci Technol* **5**, 95–167
Reid J T & Tyrrell H F (1964) Effect of level of intake on energetic efficiency of animals *Proc Cornell Nutr Conf Feed Manuf* 25–38. Cornell University: Ithaca, NY
Reitnour C M (1978) Response to dietary nitrogen in ponies. *Equine vet J* **10**, 65–8
Reitnour C M (1979) Effect of cecal administration of corn starch on nitrogen metabolism in ponies. *J Anim Sci* **49**, 988–91
Reitnour C M (1982) Protein utilization in response to caecal corn starch in ponies. *Equine vet J* **14**, 149–52
Reitnour C M & Salsbury R L (1972) Digestion and utilization of cecally infused protein by the equine. *J Anim Sci* **35**, 1190–3
Reitnour C M & Salsbury R L (1975) Effect of oral or caecal adminstration of

protein supplements on equine plasma amino acids. *Br vet J* **131**, 466–71

Reitnour C M & Salsbury R L (1976) Utilization of proteins by the equine species. *Am J vet Res* **37**, 1065–7

Reitnour C M, Baker J P, Mitchell G E Jr, Little C O & Kratzer D D (1970) Amino acids in equine cecal contents, cecal bacteria and serum. *J Nutr* **100**, 349–54

Rerat A (1978) Digestion and absorption of carbohydrates and nitrogenous matters in the hindgut of the omnivorous nonruminant animal. *J Anim Sci* **46**, 1808–37

Revington M (1983a) Haematology of the racing Thoroughbred in Australia. 1. Reference values and the effect of excitement. *Equine vet J* **15**, 141–4

Revington M (1983b) Haematology of the racing Thoroughbred in Australia. 2. Haematological values compared to performance. *Equine vet J* **15**, 145–8

Ribeiro J M C R, MacRae J C & Webster A J F (1981) An attempt to explain differences in the nutritive value of spring and autumn harvested dried grass. *Proc Nutr soc* **40**, 12A

Ricketts S W, Greet T R C, Glyn P J, Ginnett C D R, McAllister E P, McCaig J, Skinner P H, Webbon P M, Frape D L, Smith G R & Murray L G (1984) Thirteen cases of botulism in horses fed big bale silage. *Equine vet J* **16**, 515–18

Roberts M C (1974a) The D(+) xylose absorption test in the horse. *Equine vet J* **6**, 28–30

Roberts M C (1974b) Total serum cholesterol levels in the horse. *Br vet J* **130**, xvi–xviii

Roberts M C (1974c) The development and distribution of alkaline phosphatase activity in the small intestine of the horse. *Res vet Sci* **16**, 110–1

Roberts M C (1974d) Amylase activity in the small intestine of the horse. *Res vet Sci* **17**, 400–1

Roberts M C (1975) Carbohydrate digestion and absorption studies in the horse. *Res vet Sci* **18**, 64–9

Roberts M C (1983) Serum and red cell folate and serum vitamin B_{12} levels in horses. *Aust Vet J* **60**, 101–5

Roberts M C & Hill F W G (1973) The oral glucose tolerance test in the horse. *Equine vet J* **5**, 171–3

Roberts M C & Norman P (1979) A re-evaluation of the D(+) xylose absorption test in the horse. *Equine vet J* **11**, 239–43

Roberts M C & Seawright A A (1983) Experimental studies of drug-induced impaction colic in the horse. *Equine vet J* **15**, 222–8

Roberts M C, Hill F W G & Kidder D E (1974) The development and distribution of small intestinal disaccharidases in the horse. *Res vet Sci* **17**, 42–8

Roberts M C, Kidder D E & Hill F W G (1973) Small intestinal beta-galactosidase activity in the horse. *Gut* **14**, 535–40

Robinson D W & Slade L M (1974) The current status of knowledge on the nutrition of equines. *J Anim Sci* **39**, 1045–66

Rodd J G (1979) Exercise: a factor to be considered in the determination of the thiamin requirement. MSc thesis, Cornell University, Ithaca, NY

Romić S (1974) Changes in some blood properties of horses during fattening. *Poljoprivredna Znanstvena Smotra* **33**, 17–24

Romić S (1978) Dietary and protective value of lucerne in the feeding of horses. *Poljoprivredna Znanstvena Smotra* **45**, 5–17

Ronéus B O, Hakkarainen R V J, Lindholm C A & Työppönen J T (1985) Vitamin

E requirement of adult Standardbred horses evaluated by tissue depletion and repletion. *Equine vet J* In press

Rooney J R (1968) Biomechanics of equine lameness. *Cornell Vet* **58**, 49–58

Rose R J (1979) Studies on some aspects of intravenous fluid infusion in the dog. PhD thesis, University of Sydney

Rose R J (1981) A physiological approach to fluid and electrolyte therapy in the horse. *Equine vet J* **13**, 7–14

Rose R J & Hodgson D R (1982) Haematological and plasma biochemical parameters in endurance horses during training. *Equine vet J* **14**, 144–8

Rose R J & Sampson D (1982) Changes in certain metabolic parameters in horses associated with food deprivation and endurance exercise. *Res vet Sci* **32**, 198–202

Rose R J, Ilkiw J E & Martin I C A (1979) Blood-gas, acid–base and haematological values in horses during an endurance ride. *Equine vet J* **11**, 56–9

Rose R J, Purdue R A & Hensley W (1977) Plasma biochemistry alterations in horses during an endurance ride. *Equine vet J* **9**, 122–6

Rose R J, Allen J R, Hodgson D R & Kohnke J R (1983) Studies on isoxsuprine hydrochloride for the treatment of navicular disease. *Equine vet J* **15**, 238–43

Rose R J, Arnold K S, Church S & Paris R (1980) Plasma and sweat electrolyte concentrations in the horse during long distance exercise. *Equine vet J* **12**, 19–22

Rose R J, Ilkiw J E, Arnold K S, Backhouse J W & Sampson D (1980) Plasma biochemistry in the horse during 3-day event competition. *Equine vet J* **12**, 132–6

Rose R J, Ilkiw J E, Sampson D & Backhouse J W (1980) Changes in blood gas, acid–base and metabolic parameters in horses during three-day event competition. *Res vet Sci* **28**, 393–5

Ross M W, Lowe J E, Cooper B J, Reimers T J & Froscher B A (1983) Hypoglycemic seizures in a Shetland pony. *Cornell Vet* **73**, 151–69

Rossdale P D (1971) Experiences in the use of corticosteroids in horse practice, in *The application of corticosteroids in veterinary medicine* (Symp Royal Soc Medicine, London), pp. 29–31. Glaxo Laboratories Ltd: Greenford, Middlesex

Rossdale P D (1972) Modern concepts of neonatal disease in foals. *Equine vet J* **4**, 117–28

Rossdale P D & Ricketts S W (1980) *Equine stud farm medicine*. Baillière Tindall: London

Rossdale P D, Burguez P N & Cash R S G (1982) Changes in blood neutrophil/lymphocyte ratio related to adrenocortical function in the horse. *Equine vet J* **14**, 293–8

Roughan P G & Slack C R (1973) Simple methods for routine screening and quantitative estimation of oxalate content of tropical grasses. *J Sci Fd Agric* **24**, 803–11

Round M C (1968a) The diagnosis of helminthiasis in horses. *Vet Rec* **82**, 39–43

Round M C (1968b) Experiences with thiabendazole as an anthelmintic for horses. *Br vet J* **124**, 248–58

Roždestvenskaja G A (1961) Growth of the bony tissue in skeleton of extremities in foals from birth to one year age. *Trudy vses Inst Konevodstva* **23**, 321–30

Rudra M M (1946) Vitamin A in the horse. *Biochem J* **40**, 500

Saba N, Symons, A M & Drane H M (1974) The effects of feeding white clover pellets and red clover hay on teat length, plasma gonadotrophins and pituitary function in wethers. *J agric Sci Camb* **82**, 357–61

Sainsbury D W B (1981) Ventilation and environment in relation to equine respiratory disease. *Equine vet J* **13**, 167–70

St John, Sir Paulet (1780) *Every man his own farrier.* J Wilkes for S Crowder: Winton

Salimem K (1975) Cobalt metabolism in horse serum levels and biosynthesis of vitamin B_{12}. *Acta vet Scand* **16**, 84–94

Sandersleben J von & Schlotke B (1977) Muscular dystrophy (white muscle disease) in foals, apparently a disease of increasing prevalence. *Dt tierärztl Wschr* **84**, 105–7

Sandford J & Aitken M M (1975) Effects of some drugs on the physiological changes during exercise in the horse. *Equine vet J* **7**, 198–202

Santidrian S (1981) Intestinal absorption of D-glucose, D-galactose and L-leucine in male growing rats fed raw field bean (*Vicia faba*) diet. *J Anim Sci* **53**, 414–9

Sato T, Oda K & Kubo M (1978) Hematological and biochemical values of Thoroughbred foals in the first six months of life. *Cornell Vet* **69**, 3–19

Schmidt M, Lindemann G & Meyer H (1982) Intestinaler N-umsatz beim Pferd, in *Advances in animal physiology and nutrition* (*Contributions to the digestive physiology of the horse* no. 13), ed. Meyer H, pp. 40–51. Paul Parey: Hamburg & Berlin

Schmidt S P, Hoveland C S, Clark E M, Davis N D, Smith L A, Grimes H W & Holliman J L (1983) Association of an endophytic fungus with fescue toxicity in steers fed Kentucky 31 tall fescue seed or hay. *J Anim Sci* **55**, 1259–63

Schoental R (1960) The chemical aspects of seneciosis. *Proc R Soc Med* **53**, 284–8

Schryver H F (1975) Intestinal absorption of calcium and phosphorus by horses. *J S Afr vet Ass* **46**, 39–45

Schryver H F & Hintz H F (1972) Calcium and phosphorus requirements of the horse: a review. *Feedstuffs* **44** (28), 35–8

Schryver H F, Craig P H & Hintz H F (1970) Calcium metabolism in ponies fed varying levels of calcium. *J Nutr* **100**, 955–64

Schryver H F, Hintz H F & Craig P H (1971) Calcium metabolism in ponies fed a high phosphorus diet. *J Nutr* **101**, 259–64

Schryver H F, Hintz H F & Lowe J E (1971) Calcium and phosphorus interrelationships in horse nutrition. *Equine vet J* **3**, 102–9

Schryver H F, Hintz H F & Lowe J E (1974) Calcium and phosphorus in the nutrition of the horse. *Cornell Vet* **64**, 493–515

Schryver H F, Hintz H F & Lowe J E (1975) The effect of exercise on calcium metabolism in horses. *Proc Cornell Nutr Conf Feed Manuf* 99–101. Cornell University: Ithaca, NY

Schryver H F, Hintz H F & Lowe J E (1978) Calcium metabolism, body composition and sweat losses of exercised horses. *Am J vet Res* **39**, 245–8

Schryver H F, Bartel D L, Langrana N & Lowe J E (1978) Locomotion in the horse: kinematics and external and internal forces in the normal equine digit in the walk and trot. *Am J vet Res* **39**, 1728–33

Schryver H F, Craig P H, Hintz H F, Hogue D E & Lowe J E (1970) The site of calcium absorption in the horse. *J Nutr* **100**, 1127–31

Schryver H F, Hintz H F, Lowe J E, Hintz R L, Harper R B & Reid J T (1974) Mineral composition of the whole body, liver and bone of young horses. *J Nutr* **104**, 126–32

Schryver H F, Van Wie S, Daniluk P & Hintz H F (1978) the voluntary intake of calcium by horses and ponies fed a calcium deficient diet. *J Equine med Surg* **2**, 337–40

Schulz E & Peterson U (1978) Evaluation of horse beans (*Vicia faba* L. *minor*), sweet lupins (*Lupinus lutens* L.) and solvent-extracted rapeseed oil meal. *Landw Forsch* **31**, 218–32

Schurg W A & Pulse R E (1974) Grass straw: an alternative roughage for horses. *J Anim Sci* **38**, 1330, abstr. 46

Schurg W A, Frei D L, Cheeke P R & Holtan D W (1977) Utilization of whole corn plant pellets by horses and rabbits. *J Anim Sci* **45**, 1317–21

Seawright A A, Groenendyk S & Silva K I N G (1970) An outbreak of oxalate poisoning in cattle grazing *Setaria sphacelata*. *Aust vet J* **46**, 293–6

Shupe J L, Eanes E D & Leone N C (1981) Effect of excessive exposure to sodium fluoride on composition and crystallinity of equine bone tumors. *Am J vet Res* **42**, 1040–2

Skarda R T, Muir W W, Milne D W, Gabel A A (1976) Effects of training on resting and postexercise ECG in Standardbred horses, using a standardized exercise test. *Am J vet Res* **37**, 1485–8

Sklan D & Donoghue S (1982) Serum and intracellular retinol transport in the equine. *Br J Nutr* **47**, 273–80

Slade L M, Robinson D W & Casey K E (1970) Nitrogen metabolism in non-ruminant herbivores. I. The influence of nonprotein nitrogen and protein quality on the nitrogen retention of adult mares. *J Anim Sci* **30**, 753–60

Slade L M, Bishop R, Morris J G & Robinson D W (1971) Digestion and the absorption of ^{15}N-labelled microbial protein in the large intestine of the horse. *Br vet J* **127**, xi–xiii

Slade L M, Lewis L D, Quinn C R & Chandler M L (1975) Nutritional adaptations of horses for endurance performance. *Proc Equine Nutr Physiol Soc* 114–28

Slagsvold P, Hintz H F & Schryver H F (1979) Digestibility by ponies of oat straw treated with anhydrous ammonia. *Anim Prod* **28**, 347–52

Smith A, Allcock P J, Cooper E M & Forbes T J (1982) *Permanent grassland studies. 4. An investigation into the influence of sward composition and environment on stocking rate, using census and survey data.* The GRI–ADAS Joint Permanent Pasture Group, pp. 51. The Grassland Research Institute: Hurley, Maidenhead

Smith B L & O'Hara P J (1978) Bovine photosensitization in New Zealand. *NZ vet J* **26**, 2–5

Smith J D, Jordan R M & Nelson M L (1975) Tolerance of ponies to high levels of dietary copper. *J Anim Sci* **41**, 1645–9

Smith J E, Moore K, Cipriano J E & Morris P G (1984). Serum ferritin as a measure of stored iron in horses. *J Nutr* **114**, 677–81

Smith J F, Jagusch K T, Brumswick L F C & Kelly R W (1979) Coumestans in lucerne and ovulation in ewes. *NZ J agric Res* **22**, 411–16

Smith T K, James L J & Carson M S (1980) Nutritional implications of *Fusarium* mycotoxins. *Proc Cornell Nutr Confr Feed Manuf* 35–42. Cornell University: Ithaca, NY

Snow D H (1977) Identification of the receptor involved in adrenaline mediated sweating in the horse. *Res vet Sci* **23**, 246–7

Snow D H & Guy P S (1980) Muscle fibre type composition of a number of limb muscles in different types of horse. *Res vet Sci* **28**, 137–44

Snow D H & Mackenzie G (1977a) Some metabolic effects of maximal exercise in the horse and adaptations with training. *Equine vet J* **9**, 134–40

Snow D H & MacKenzie G (1977b) Effect of training on some metabolic changes associated with submaximal endurance exercise in the horse. *Equine vet J* **9**, 226–30

Snow D H & Rose R J (1981) Hormonal changes associated with long-distance exercise. *Equine vet J* **13**, 195–7

Snow D H & Summers R J (1977) The actions of the β-adrenoceptor blocking agents propranolol and metoprolol in the maximally exercised horse. *J Physiol* **271**, 39–40P

Snow D H, Baxter P B & Rose R J (1981) Muscle fibre composition and glycogen depletion in horses competing in an endurance ride. *Vet Rec* **108**, 374–8

Snow D H, Ricketts S W & Mason D K (1983) Haematological response to racing and training exercise in Thoroughbred horses, with particular reference to the leucocyte response. *Equine vet J* **15**, 149–54

Snow D H, Kerr M G, Nimmo M A & Abbott E M (1982) Alterations in blood, sweat, urine and muscle composition during prolonged exercise in the horse. *Vet Rec* **110**, 377–84

Sobel A E (1955) Composition of bones and teeth in relation to blood and diet. *Voeding* **16**, 567–75

Soldevila M & Irizarry R (1977) Complete diets for horses. *J agric Univ Puerto Rico* **61**, 413–5

Solleysel S de (1711) *The compleat horseman: or, perfect farrier. In two parts.* R. Bonwicke and others: London

Sommer H & Felbinger U (1983) The influence of racing on selected serum enzymes, electrolytes and other constituents in Thoroughbred horses, in *Equine exercise physiology*, eds Snow D H, Persson S G B & Rose R J, pp. 362–5. Granta Editions: Cambridge

Spais A G, Papasteriadis A, Roubiës N, Agiannidis A, Yantzis N & Argyroudis S (1977) Studies on iron, manganese, zinc, copper and selenium retention and interaction in horses. *Proc 3rd Int Symp Trace Element Metabolism in Man and Animals* 501–5

Srivastava V K & Hill D C (1976) Effect of mild heat treatment on the nutritive value of low glucosinolate–low erucic acid rapeseed meals. *J Sci Fd Agric* **27**, 953–8

Staley T E, Jones E W, Corley L D & Anderson I L (1970) Intestinal permeability to *Escherichia coli* in the foal. *Am J vet Res* **31**, 1481–3

Stanek Ch (1981) Conservative therapy for correction of poor limb conformation in foals. *Proc 32nd A Meet Eur Assn Anim Prod* IIIa–4

Stick J A, Robinson N E & Krehbiel J D (1981) Acid–base and electrolyte alterations associated with salivary loss in the pony. *Am J vet Res* **42**, 733–7

Stoker J W (1975) Monensin sodium in horses. *Vet Rec* **97**, 137–8

Stowe H D (1968) Effects of age and impending parturition upon serum copper of thoroughbred mares. *J Nutr* **95**, 179–84

Stowe H D (1982) Vitamin A profiles of equine serum and milk. *J Anim Sci* **54**, 76–81

Stubley D, Campbell C, Dant C & Blackmore D J (1983) Copper and zinc levels in the blood of Thoroughbreds in training in the United Kingdom. *Equine vet J* **15**, 253–6

Suttle N F (1983) The nutritional basis for trace element deficiencies in ruminant livestock, in *Trace elements in animal production and veterinary practice*, eds Suttle

N F, Gunn R G, Allen W M, Linklater K A & Wiener G, pp 19–25. British Society of Animal Production (Occasional Paper no.7): Edinburgh

Suttle N F, Gunn R G, Allen W M, Linklater K A & Weiner G (eds) (1983) *Trace elements in animal production and veterinary practice* (Proc Symp British Society of Animal Production and British Veterinary Association). BSAP (Occasional Paper no. 7): Edinburgh

Sutton E I, Bowland J P & McCarthy J F (1977) Studies with horses comparing 4 N-HCl insoluble ash as an index material with total fecal collection in the determination of apparent digestibilities. *Can J Anim Sci* 57, 543–9

Sutton E I, Bowland J P & Ratcliff W D (1977) Influence of level of energy and nutrient intake by mares on reproductive performance and on blood serum composition of the mares and foals. *Can J Anim Sci* 57, 551–8

Swartzman J A, Hintz H F & Schryver H F (1978) Inhibition of calcium absorption in ponies fed diets containing oxalic acid. *Am J vet Res* 39, 1621–3

Sweeting M P, Houpt C E & Houpt K A (1985) Social facilitation of feeding and time budgets in stabled ponies. *J Anim Sci* 60, 369–74

Talukdar A H, Calhoun M L & Stinson A W (1970) Sensory end organs in the upper lip of the horse. *Am J vet Res* 31, 1751–4

Tasker J B (1967) Fluid and electrolyte studies in the horse. III: Intake and output of water, sodium and potassium in normal horses. *Cornell Vet* 57, 649–57.

Taylor M C, Loch W E, Heimann E D & Morris J S (1981) Effect of nitrogen fertilization and selenium supplementation on the hair selenium concentration in pregnant pony mares grazing fescue. *J Anim Sci* 53 (suppl. 1), 266, abstr. 350

Teeter S M, Stillions M C & Nelson W E (1967) Maintenance levels of calcium and phosphorus in horses. *J Am vet Med Ass* 151, 1625–8

Thijssen H H W, Van Der Bogaard A E J M, Wetzel J M, Maes J H J & Muller A P (1983) Warfarin pharmacokinetics in the horse. *Am J vet Res* 44, 1192–6

Thomas B, Thompson A, Oyenuga V A & Armstrong R H (1952) The ash constituents of some herbage plants at different stages of maturity *Emp J exp Agric* 20, 10–13

Thomas P T (1963) Breeding herbage plants for animal production and well-being. *Proc Br Vet Ass A Congr* 1–4

Thomson J R & McPherson E A (1981) Prophylactic effects of sodium cromoglycate on chronic obstructive pulmonary disease in the horse. *Equine vet J* 13, 243–6

Thomson J R & McPherson E A (1983) Chronic obstructive pulmonary disease in the horse: therapy. *Equine vet J* 15, 207–10

Thomson J R & McPherson E A (1984) Effects of environmental control on pulmonary function of horses affected with chronic obstructive pulmonary disease. *Equine vet J* 16, 35–8

Thornton I (1983) Soil–plant–animal interactions in relation to the incidence of trace element disorders in grazing livestock, in *Trace elements in animal production and veterinary practice*, eds Suttle N F, Gunn R G, Allen W M, Linklater K A & Wiener G, pp. 39–49. British Society of Animal Production (Occasional Publication no.7): Edinburgh

Thornton J R & Lohni M D (1979) Tissue and plasma activity of lactic dehydrogenase and creatine kinase in the horse. *Equine vet J* 11, 235–8

Tisserand J L, Candau M, Houiste A & Masson C (1977) Evolution de quelques paramètres physico-chimiques due contenu caecal d'un poney au cours du nycthémère. *Annls Zootech* 26, 429–34

Tisserand J L, Masson C, Ottin-Pecchio M & Creusot A (1977) Mesure du pH et la concentration en AGV dans le caecum et le colon du poney. *Ann Biol anim Biochim Biophys* **17**, 553–7

Tobin T & Combie J (1984) Some reflections on positive results from medication control tests in the USA. *Equine vet J* **16**, 43–6

Topliff D R, Potter G D, Kreider J L & Cregan C R (1981) Thiamin supplementation for exercising horses. *Proc equine Nutr Physiol Symp* 167–72

Torún B, Scrimshaw N S & Young V R (1977) Effect of isometric exercises on body potassium and dietary protein requirements of young men. *Am J clin Nutr* **30**, 1983–93

Townley P, Baker K P & Quinn P J (1984) Preferential landing and engorging sites of *Culicoides* species landing on a horse in Ireland. *Equine vet J* **16**, 117–20

Tyznik W K (1968) Nutrition, in *Care and training of the trotter and pacer*, ed. Harrison J C. USTA: Columbus, Ohio

Tyznik W K (1975) Recent advances in horse nutrition. *Proc 35th semi-annual meeting AFMA Nutrition Council* 32–8. American Feed Manufacturers Association: Arlington, Va

UKASTA – United Kingdom Agricultural Supply Trade Association Ltd (1984) *Code of practice for cross-contamination in animal feedingstuffs manufacture. Amended code – June 1984*. London

Ullrey D E, Ely W T, Covert R L (1974) Iron, zinc and copper in mare's milk. *J Anim Sci* **38**, 1276–7

Ullrey D E, Struthers R D, Hendricks D G & Brent B E (1966) Composition of mare's milk. *J Anim Sci* **25**, 217–22

Urch D L & Allen W R (1980) Studies on fenbendazole for treating lung and intestinal parasites in horses and donkeys. *Equine vet J* **12**, 74–7

Van Dam B (1978) Vitamins and sport. *Br J Sports Med* **12**, 74–9

Van der Merwe J A (1975) Dietary value of cubes in equine nutrition *J S Afr Vet Ass* **46**, 29–37

Verberne L R M & Mirck M H (1976) A practical health programme for prevention of parasitic and infectious diseases in horses and ponies. *Equine vet J* **8**, 123–5

Vogel C (1984) Navicular disease and equine insurance. *Vet Rec* **115**, 89

Waite R & Sastry K N S (1949) The composition of timothy (*Phleum pratensi*) and some other grasses during seasonal growth. *Emp J exp Agric* **17**, 179–82

Walker D & Knight D (1972) The anthelmintic activity of mebendazole: a field trial in horses. *Vet Rec* **90**, 58–65

Wallace W M & Hastings A B (1942) The distribution of the bicarbonate ion in mammalian muscle. *J biol Chem* **144**, 637–49

Waterman A (1977) A review of the diagnosis and treatment of fluid and electrolyte disorders in the horse. *Equine vet J* **9**, 43–8

Webb J S, Lowenstein P L, Howarth R J, Nichol I & Foster R (1973) *Provisional geochemical atlas of Northern Ireland* (Appl Geochem Res Grp Tech Commun no.60). Imperial College: London

Webb J S, Thornton I, Howarth R J, Thompson M & Lowenstein P L (1978) *The Wolfson geochemical atlas of England and Wales*. Oxford University Press

Webster A J F (1980) The energetic efficiency of growth. *Livestock Prod Sci* **7**, 243–52

Weiss T (1982) Equine Nutrition in New Mexico. *Vet Med Small Anim Clin* **77**, 817–9

Welch K J, Perry T W, Adams S B & Battaglia R A (1981) Effect of partial typhlectomy on nutrient utilization in ponies. *J Anim Sci* **53** (suppl 1), 92, abstr. 58

Wells P W, McBeath D G, Eyre P & Hanna C J (1981) Equine immunology: an introductory review. *Equine vet J* **13**, 218–22

Whang R, Morasi H T & Rogers D (1967) The influence of sustained magnesium deficiency on muscle potassium repletion. *J Lab clin Med* **70**, 895–902

White G (1789) *Natural history and antiquities of Selbourne.* London

White J (1823). *A compendium of the veterinary art*, vol. 3. London

White K K, Short C E, Hintz H F, Ross M W, Lesser F R, Leids P F, Lowe J E & Schryver H F (1978) The value of dietary fat for working horses. II. Physical evaluation. *J. Equine med Surg* **2**, 525–30

White N A, Moore J N & Douglas M (1983) SEM study of *Strongylus vulgaris* larva-induced arteritis in the pony. *Equine vet J* **15** 349–53

Whitehead C C & Bannister D W (1980) Biotin status, blood pyruvate carboxylase (EC 6.4.1.1) activity and performance in broilers under different conditions of bird husbandry and diet processing. *Br J Nutr* **43**, 541–9

Willard J G (1976) Feeding behavior in the equine fed concentrate versus roughage diets. *Diss Abstr Int* **36**, 4772–B–3–B

Willard J G, Bull L S & Baker J P (1978) Digestible energy requirements of the light horse at two levels of work. *Proc 70th A Meet Am Soc Anim Sci* 324

Willard J G, Willard J C, Wolfram S A & Baker J P (1977) Effect of diet on cecal pH and feeding behavior of horses. *J Anim Sci* **45**, 87–93

Williams, M (1974) The effect of artificial rearing on the social behaviour of foals. *Equine vet J* **6**, 17–18

Williamson H M (1974) Normal and abnormal electrolyte levels in the racing horse and their effect on performance. *J S Afr Vet Ass* **45**, 335–40

Willoughby R A, MacDonald E & McSherry B J (1972) The interaction of toxic amounts of lead and zinc fed to young growing horses. *Vet Rec* **91**, 382–3

Willoughby R A, MacDonald E, McSherry B J & Brown G (1972) Lead and zinc poisoning and the interaction between Pb and Zn poisoning in the foal. *Can J Comp Med* **36**, 348–52

Wilsdorf G, Berschneider F & Mill J (1976) Oriented determination of enzyme activities in racehorses and studies of selenium levels in their feed. *Mschr Veterinaermed* **31**, 741–6

Wilson T M, Morrison H A, Palmer N C, Finley G G & van Dreumel A A (1976) Myodegeneration and suspected selenium/vitamin E deficiency in horses. *J Am vet Med Ass* **169**, 213–17.

Winter L (1980) A survey of feeding practices at 2 Thoroughbred race tracks. MSc thesis, Cornell University, Ithaca, NY

Wirth B L, Potter G D & Broderick G A (1976) Cottonseed meal and lysine for weanling foals. *J Anim Sci* **43**, 261, abstr. 200

Wiseman A, Dawson C O, Pirie H M, Breeze R G & Selman I E (1973) The incidence of precipitins to *Micropolyspora faeni* in cattle fed hay treated with an additive to suppress bacterial and mould growth. *J agric Sci Camb* **81**, 61–4

Witherspoon D M (1971) The oestrous cycle of the mare. *Equine vet J* **3**, 114–17

Wolfram S A, Willard J C, Willard J G, Bull L S & Baker J P (1976) Determining the energy requirements of horses. *J Anim Sci* **43**, 261, abstr. 201

Wolter R & Wehrle P (1977) Appreciation of the principal mineral complements destined for horses. *Revue Méd Vét* **128**, 467–8, 473–83

Wolter R, Durix A & Letourneau J-C (1974) Influence du mode de présentation du fourrage sur la vitesse du transit digestif chez le poney. *Annls Zootech* **23**, 293–300

Wolter R, Durix A & Letourneau J-C (1975) Influence du mode de présentation du fourrage sur la digestibilité chez le poney. *Annls Zootech* **24**, 237–42

Wolter R, Moraillon R & Taulat B (1971) Aliments complets pour chevaux: nouveaux essais. *Recl Méd vét* **147**, 565–76

Wolter R, Gouy D, Durix A, Letourneau J-C, Carcelen M & Landreau J (1978) Digestibilité et activité biochimique intracaecale chez le poney recevant un même aliment complet présenté sous forme granulée, expansée ou semi-expansée. *Annls Zootech* **27**, 47–60

Wolter R, Meunier B, Faucompret R de, Durix A & Landreau J (1977) A complete feed, pelleted or expanded compared to traditional feeding for riding horses. *Revue Méd Vét* **128**, 71–81

Wooden G R, Knox K L & Wild C L (1970) Energy metabolism in light horses. *J Anim Sci* **30**, 544–8

Wootten J F & Argenzio R A (1975) Nitrogen utilization within equine large intestine. *Am J Physiol* **229**, 1062–7

Worden A N, Sellers K C & Tribe D E (1963) *Animal health, production and pasture.* Longmans Green: London

Yamamoto M, Tanaka Y & Sugano M (1978) Serum and liver lipid composition and lecithin:cholesterol acyltransferase in horses, *Equus caballus. Comp Biochem Physiol* B **62**, 185–93

Yoakam S C, Kirkham W W & Beeson W M (1978) Effect of protein level on growth in young ponies. *J Anim Sci* **46**, 983–991

Index